A Modern Approach to Comprehensive Chemistry

D1438187

Also by Dr. G. N. Gilmore
A COMPLETE 'O' LEVEL CHEMISTRY
for 'O' Level J.M.B. and A.E.B. syllabuses

A Modern Approach to Comprehensive Chemistry

Second Edition

G. N. Gilmore

Ph.D. C.Chem M.R.I.C.

Lecturer in Chemistry,
Bury College of Further Education

Stanley Thornes (Publishers) Ltd

First published 1975 by Stanley Thornes (Publishers) Ltd.,
EDUCA House, Liddington Estate, Leckhampton Road,
CHELTENHAM GL53 0D4
Reprinted with minor corrections 1977
Second Reprint 1978
Second Edition entirely reset 1979
Reprinted 1980
Reprinted with minor corrections 1982

ISBN 0 85950 458 1

The Publishers and the Author gratefully acknowledge the
permission of those bodies named to reproduce copyright
material used in this book as follows:

The University of London Entrance and School Examinations Council for
permission to reproduce examination questions. The Joint Matriculation Board
for authorising the reproduction of examination questions and for their
permission to amend the wording in certain of these questions by the deletion of
the words 'molecular weight' and the substitution of the words 'relative
molecular mass'. Mr J. E. White et al., compilers of TABLES FOR
STATISTICIANS, 2nd Edition (Stanley Thornes (Publishers) Ltd., 1979) for
permission to reproduce certain tables from that compilation. The Nuffield
Foundation Science Teaching Project, and the Longman Group for their
permission to reproduce selected physical properties from their PHYSICAL
SCIENCES BOOK OF DATA; also for their permission to reproduce overhead
projection original diagrams Nos. 4, 5, 6, 7, 8, 14, 19, 21, 23, 24 from the
same source.

Typeset in 10/13 Times by
Malvern Typesetting Services Ltd, Malvern, Worcs.,
and printed and bound in Great Britain at
Ebenezer Baylis & Son Ltd, Worcester, and London

Preface to the First Edition

The object of this book is to present, in a concise manner, a comprehensive text to meet the requirements of the G.C.E. 'A' level chemistry syllabuses of the Joint Matriculation Board and the University of London. Obviously, it will also cover a large part of the syllabuses of similar examining bodies. The treatment is in line with current trends. Emphasis is therefore placed on general unifying principles rather than on a mass of unrelated facts.

The physical chemistry section opens with a consideration of atomic structure followed by states of matter, energetics, structure and bonding, equilibria, and kinetics. SI units are used along with those exceptions permitted by the above Examining Boards.

The physical and chemical properties of the elements and their compounds are largely determined by the electronic configuration of the constituent atoms. In the inorganic section, therefore, the book endeavours to show how the relationship arises and how it is applied in correlating the chemistry of the elements.

The organic section is treated as the chemistry of the functional groups and mechanisms are given wherever practicable. The book is not intended to present an exhaustive account of preparations and properties.

Nomenclature throughout is that recommended by the 1972 report of the Association for Science Education. However, the oxidation number of the elements has been omitted if it is the only one normally encountered. In the case of oxoacids, the oxygen content has been included (e.g. tetraoxosulphuric(VI) acid) but these names may be abbreviated if desired. Trivial names are given (in brackets) in order to aid the transition to systematic naming. For convenience, both systematic and trivial names are given in the index.

Physical data in the text has been extracted, wherever possible, from the Nuffield Book of Data and I am grateful to the Nuffield Foundation for permission to do this. I am further indebted to the same body for permission to reproduce 11 of their diagrams on periodicity. I also wish to thank the Joint Matriculation Board (J.M.B.) and the University of London for permission to use questions from their Advanced level examinations and to include in these I.U.P.A.C. nomenclature where necessary.

G.N.G.

Preface to the Second Edition

The basic structure of the first edition has been maintained. However, a number of minor modifications have been made as well as including, wherever possible, the states of substances in equations. Additional material includes the van der Waals equation, liquefaction of gases, radius ratios, calculation of the Avogadro constant from unit cell measurements, dipole moments, steam distillation calculations, and methods for deriving half-reactions for oxidations and reductions. The section on mass spectroscopy has been extended and its use in structure determination of organic molecules has been indicated. An extra chapter on selected reactions of some cations and anions has been included. Finally, for the sake of simplicity, the oxygen content of oxoacids and their salts has been omitted.

In conclusion, I would like to take this opportunity of thanking all those who have written in support of the book and for a number of helpful suggestions they have made.

G.N.G.

Contents

		Page
Chapter 1	Atomic Structure	1
Chapter 2	States of Matter	21
Chapter 3	Electronic Theory and Chemical Bonding	56
Chapter 4	Energetics	69
Chapter 5	Phase Equilibria	85
Chapter 6	Chemical Equilibria	111
Chapter 7	Ionic Equilibria	120
Chapter 8	Chemical Kinetics	149
Chapter 9	Periodicity	163
Chapter 10	A Study of Some of the Elements and Groups in the Periodic Table	181
Chapter 11	Some Elements of Industrial Importance	227
Chapter 12	The First Row Transition Elements	241
Chapter 13	Selected Chemistry of Some Cations and Anions	262
Chapter 14	Fundamentals of Organic Chemistry	270
Chapter 15	Molecular Geometry	287
Chapter 16	The Chemistry of the Functional Groups	296
Chapter 17	Carbohydrates	351
Chapter 18	Aminoacids, Polypeptides, and Proteins	356
Chapter 19	Synthetic Macromolecules	359
Chapter 20	Acidity and Basicity	365
	Answers to Numerical Questions	371
	Units, Conversion Factors, and Physical Constants	373
	Relative Atomic Masses (Atomic Weights)	375
	Logarithms	376
	Index	380

1 Atomic Structure

ATOMS, MOLECULES, AND IONS

Atoms are the smallest particles of elements that can take part in chemical changes. It is known that they consist of a number of fundamental particles, the most important of which are *electrons, neutrons,* and *protons.* The neutrons and protons make up the nucleus and this is surrounded some distance away by the electrons. The properties of these particles are summarised in Table 1.1.

Table 1.1. Properties of the fundamental particles

Name of Particle		Mass	Charge
Proton ⎱ Sometimes referred		1 unit	Positive
Neutron ⎰ to as nucleons		1 unit	Neutral
Electron		Negligible compared with neutrons and protons, i.e. 1/1836 unit	Negative

Since all atoms are electrically neutral, they must contain the same number of electrons and protons. There is, however, no simple relationship between the number of protons and neutrons in an atom.

In between the nucleus and electrons is space, in fact the bulk of the volume of an atom is space, the radius of the nucleus being about $1/10\,000$ of the radius of the atom as a whole.

The number of protons in the nucleus is known as the *atomic number, Z,* of the element. The atomic number is important since it determines the identity of the element and also its properties.

A number of elements have atoms which are not stable under normal conditions and, in many of these cases, the atoms link up to form *molecules* which are stable. A molecule is the smallest particle of a substance that can exist by itself. Atoms of sodium, magnesium, iron, helium, etc. are stable and so atoms and molecules of these elements are the same; the molecules are said to be *monatomic.* Molecules of oxygen, chlorine, nitrogen, etc. contain two atoms and are said to be *diatomic.* Higher atomicities also occur, for example, P_4 and S_8.

Ions are produced when atoms lose or gain electrons, for example

$$Na - e^- \rightarrow Na^+, \qquad Mg - 2e^- \rightarrow Mg^{2+}$$
$$Cl + e^- \rightarrow Cl^-, \qquad S + 2e^- \rightarrow S^{2-}$$

Isotopes

Isotopes are atoms of the same element that differ only in their mass; that is, they have different numbers of neutrons in their nuclei. For example, there are two naturally occurring isotopes of chlorine: chlorine atoms containing 17 protons, 17 electrons, and 18 neutrons and chlorine atoms with 17 protons, 17 electrons, and 20 neutrons. These two isotopes are represented as $^{35}_{17}Cl$ and $^{37}_{17}Cl$ respectively, the subscript being the atomic number, Z, and the superscript the *mass number,* A, which is the mass of the isotope to the nearest whole number.

In general, it is found that elements with even atomic numbers have many more isotopes than those with odd numbers. In fact, if the atomic number is odd, it is rare to find an element with more than two stable isotopes.

It quite often happens that isotopes of different elements have the same mass numbers and such isotopes are termed *isobars*; $^{40}_{18}Ar$, $^{40}_{19}K$, and $^{40}_{20}Ca$ are a set of isobars.

Relative atomic mass, A_r

The mass of a hydrogen atom is $1.673 \times 10^{-27}\,kg$ and, although some elements have atoms over two hundred times as heavy as this, the figures involved are far too small to have any general practical use. For this reason, the masses of atoms relative to one particular atom are determined, the standard being the $^{12}_{6}C$ isotope. *Relative atomic mass* may be defined as the number of times an atom of the element is heavier than 1/12 of the mass of the $^{12}_{6}C$ isotope. Most of the values so obtained are not integers because elements obtained from natural sources consist of mixtures of their isotopes. A further reason is that the electrons have a small mass, and neutrons and protons do not have precisely the same mass (see under radioactivity).

EVIDENCE FOR SUB-ATOMIC PARTICLES

a. Electrons

The discovery of electrons was due to J. J. Thomson at the end of the nineteenth century. He found that, if an electric discharge is passed through a gas at low pressure (10^{-6} atmospheres), a beam of rays is emitted perpendicularly from the cathode. These rays can be deflected by applying a magnetic field across their path (Figure 1.1); this indicates that the rays are charged.

If an electric field is applied to them, the rays are deflected towards the positive pole indicating that they are negatively charged. Thomson was able to determine the ratio of charge to mass, e/m, of the rays by measuring the angles through which they were deflected by known magnetic and electric fields. This ratio was found to be constant regardless of the gas used and so the uniformity of the particles, i.e. electrons, was apparent.

The charge on an electron was determined a few years later by Millikan using an apparatus similar to that in Figure 1.2.

The method involves letting very small drops of oil fall between the two metal

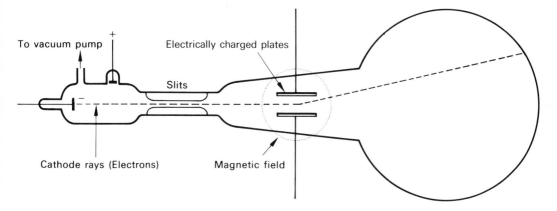

Figure 1.1   Deflection of cathode rays (electrons) by electric and magnetic fields

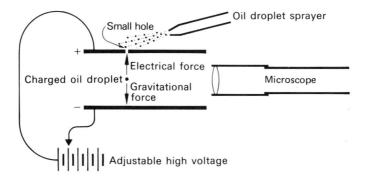

Figure 1.2   Apparatus for the Millikan oil drop experiment

plates by means of the small hole. Electrons are then produced by passing a beam of X-rays through the air above the upper plate and some of them are accepted by the drops so giving them a negative charge. The rate of fall of a drop is determined by observing it through a microscope fitted with a transparent scale, and from this the mass of the drop can be calculated. The upper and lower plates are then connected to the positive and negative terminals respectively of an adjustable high-voltage source. By adjusting the strength of the electrical field, it is possible to balance the gravitational force on the negatively charged drop against the electrical force tending to make it rise, and so the drop remains stationary.

The electrical force on the drop can be calculated in terms of its charge and the strength of the electrical field between the plates. When the charged drop is held stationary, the electrical force equals the weight of the drop and so the charge on the drop can be calculated.

No matter how many times the experiment is repeated, it is always found that the charge on the drop is some integer times 1.602×10^{-19} coulomb. Since the smallest charge Millikan found on a drop was 1.602×10^{-19} coulomb, this was taken to be the charge on a single electron.

Once Millikan had determined the charge and Thomson the charge/mass ratio, it was possible to calculate the mass of an electron. The electron mass is 9.109×10^{-31} kg.

b. Protons

As well as giving a stream of electrons from the cathode, the passage of an electric discharge through a gas at low pressure also gives a stream of positive rays from the anode. The charge on these particles is illustrated by their behaviour in magnetic and electric fields. The mass of these positive ions can be determined very accurately by means of a mass spectrograph, the basis of which is shown in Figure 1.3.

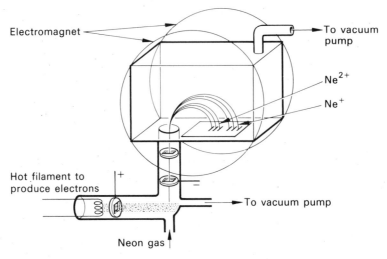

Figure 1.3   A mass spectograph

In this apparatus the positive ions are produced by bombarding the gas with high-speed electrons. Thus the beam of electrons from the filament collides with some of the neon atoms (or atoms of any other gas or vapour being tested) and dislodges electrons from them so producing positive ions.

These ions, for example, Ne^+ and Ne^{2+}, are accelerated to the negative electrode by its high charge. The beam emerging through the slit in the negative electrode passes through another slit into a chamber which is situated between the poles of a powerful electromagnet. This causes the positive ions to move in a circular path, those with high charges moving along an arc of small radius whilst those with high masses move along an arc of large radius. Each positive ion therefore has its own characteristic path determined by its charge and mass. The positive ions then impinge on a photographic plate which, after development, shows a series of lines. The mass of an ion can then be calculated from a knowledge of the electron charge, the radius of the circular path followed, the number of charges on the ion, the strength of the magnetic field, and the voltage between the positive and negative electrodes (this determines the velocity of the ion).

The mass spectrograph for neon, as shown in Figure 1.3, gives a series of three lines for Ne^+ and for Ne^{2+} ions. This indicates that neon is a mixture of three isotopes, i.e. neon has atoms having three different masses depending upon the number of neutrons present.

Mass spectrographs have been replaced by mass spectrometers, the essentials of which are shown in Figure 1.4. The sample, if not gaseous, is vaporised and then it passes at very low pressure into the main chamber. The vapour is bombarded

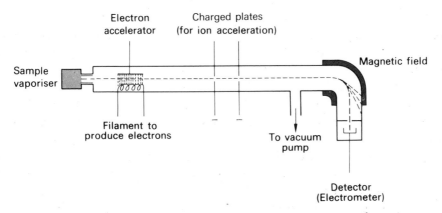

Figure 1.4   Mass spectrometer

with high-energy electrons and some collisions will result in electron loss from the sample with the formation of positive ions. The ions are then accelerated by passage through slits in two plates to which an electric field is applied. At this stage, the positive ions have similar velocities but different charge/mass ratios. They now pass through a magnetic field and are deflected by it, the deflection produced by a given magnetic field depending on the charge/mass ratio of the ion. The magnetic field is gradually adjusted so that ions with different charge/mass ratios are focused, in turn, on the detector. The detector consists of a conductor connected to an electrometer. The relative abundance of each type of ion affects the charge on the electrometer. The mass of an ion and its relative abundance are indicated by a peak on a chart.

Although passing an electric discharge through different gases at low pressure always gives identical negatively charged rays, the positively charged rays are different. The value of e/m for these rays is much smaller than the value for electrons but the largest value is obtained when hydrogen is the source material. Hence it is assumed that the hydrogen nucleus contains a single positive particle, the proton.

c. Neutrons

The existence of neutrons was suggested as a result of experiments done by Rutherford, Geiger, and Marsden using the apparatus shown schematically in Figure 1.5.

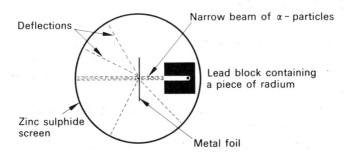

Figure 1.5   Apparatus used in the discovery of neutrons

A piece of radium is placed in a small hole in a lead block to give a narrow beam of α-particles (page 7) and this beam is directed at a very thin piece of gold foil (about 10 000 atoms thick). The foil and lead block are situated in an evacuated vessel with a zinc sulphide screen which emits a flash of light every time an α-particle strikes it, the flashes being detected and counted with the aid of a microscope. It is found that most of the α-particles pass straight through the foil with little or no deviation but a few are deflected, sometimes through angles which approach $180°$.

At the time, atoms were thought by Thomson to be uniform-density balls containing electrons embedded in a sphere of positive charge and, from the physical properties, it was inferred that in solids the atoms were close together. If this were the case, however, an α-particle with sufficient kinetic energy to force its way through 10 000 atoms would undergo little or no deflection. The fact that some large deflections occurred refuted Thomson's concept. Rutherford suggested that atoms, in fact, occupy a spherical volume with a radius of about 10^{-10} m, the electrons moving round a central positively charged nucleus, which contains most of the mass of the atom, and which has a radius of about 10^{-14} m. An atom on this hypothesis is therefore mostly space. Hence α-particles will readily push past the electrons and will be deflected to an appreciable extent only when they approach the nucleus of an atom. Not only are most nuclei bigger than α-particles but they are also positively charged and so they exert a considerable repulsive force on them when they come close together; large deflections are therefore possible.

Rutherford found that he could determine the charge on the nucleus from the pattern made by the deflected α-particles, since the greater the charge on the nucleus, the greater the deflections. This proved to be an accurate method after various improvements had been made to the apparatus. Rutherford then observed that there was no correlation between the actual mass of an atom (as determined by chemical means or by the mass spectrograph) and the total mass of the protons and electrons present and so he suggested the existence of a neutral particle in the nucleus.

In 1930, Bothe and Becker discovered a new type of 'radiation' by irradiating beryllium with α-particles. Two years later Chadwick analysed this 'new radiation' and found that its properties were consistent with it being neutral particles, i.e. neutrons.

RADIOACTIVITY

In 1896, Becquerel found that uranium salts could discharge an electroscope and affect a photographic plate. Two years later Mme Curie and Schmidt found that thorium compounds behaved similarly. Compounds such as this are said to be *radioactive.*

Many naturally occurring radioactive substances are now known. All isotopes with atomic number greater than 83 are radioactive but many radioactive isotopes of the lighter elements can be made artificially by bombarding various elements with α-particles, neutrons, protons, or deuterons ($^2_1$H), etc.

There are three methods of detecting radioactivity.

(a) The action on a photographic plate.

(b) The fluorescence produced in some compounds, e.g. zinc silicate, zinc sulphide, etc.

(c) By its ionisation of gases.

The last method is the most important, the presence of ionised gases being demonstrated by the fact that they discharge a charged gold leaf electroscope.

By nearly surrounding a piece of radium with sheets of metal foil and using strong magnetic or electric fields and an electroscope, it is possible to show that the radium emits three types of rays (Figure 1.6).

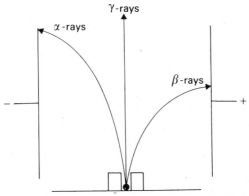

Figure 1.6 Behaviour of α-, β-, and γ-rays in magnetic and electric fields. Note that this diagram greatly exaggerates the deflection of the α-rays.

(i) α-Rays

These are fast moving positively charged particles having a mass of four and two units of positive charge, i.e. helium nuclei. They have a powerful ionising effect on any gases they pass through but they are readily absorbed by, for example, about 7cm of air or a sheet of paper. The actual path of an α-particle can be seen by the Wilson cloud chamber method in which an α-particle is allowed to pass through a gas supersaturated with water vapour. The α-particle ionises the gas and the ions act as centres for condensing the water vapour so that a vapour trail can be seen. The path of an α-particle is affected by a magnetic field.

(ii) β-Rays

These are fast moving electrons with velocities near to that of light. They have considerably more penetrating power than α-particles but are much less effective at ionising gases due to their smaller mass and kinetic energy. The path of β-rays is greatly affected by a magnetic field.

(iii) γ-Rays

These are electromagnetic radiations with the speed of light and identical with very short wavelength X-rays. They are very penetrating and will, for example, penetrate several cm of lead, etc. Their path is unaffected by magnetic fields. They are emitted in conjunction with α- or β-rays and are merely a means of getting rid of excess energy from a nucleus formed in an excited state. Since γ-rays are uncharged, they do not readily cause ionisation on passing through matter.

Reasons for radioactivity

Rutherford and Soddy (1903) suggested that the nuclei of radioactive elements are unstable and that spontaneous disintegrations take place to give the nuclei of different elements. The reason for the instability becomes apparent when the actual masses of isotopes, relative to $^{12}_{6}C$, are compared with the expected masses from the constituent particles. The expected masses may be calculated given that the masses of protons, neutrons, and electrons are 1.007276, 1.008665, and 0.000549 atomic mass units respectively.

Example   Comparison of the actual and calculated masses of the radioactive $^{13}_{8}O$ and the stable $^{16}_{8}O$ isotopes:

Expected mass of $^{13}_{8}O = (8 \times 1.007276) + (8 \times 0.000549) + (5 \times 1.008665) = 13.105925$ a.m.u.
Actual mass as shown by mass spectroscopy $= 13.024799$ a.m.u.
Difference, i.e. expected mass – actual mass $=\ \ 0.081126$ a.m.u.
Expected mass of $^{16}_{8}O = (8 \times 1.007276) + (8 \times 0.000549) + (8 \times 1.008665) = 16.131920$ a.m.u.
Actual mass as shown by mass spectroscopy $= 15.994915$ a.m.u.
Difference $=\ \ 0.137005$ a.m.u.

It is seen that there is a net loss in mass on forming the isotopes from their constituent particles. This loss in mass is known as the *mass defect* and corresponds to the energy that would be required to split the atom into its constituent particles (energy is related to mass according to Einstein's equation $E = mc^2$—see page 14). The energy which appears to have been released in the formation of the atom is known as the *binding energy*. From the example above it is evident that stable isotopes have larger mass defects and binding energies than radioactive ones.

Consequences of radioactive decay

If an atom emits an α-particle, the resultant atom will have an atomic number of two less than the original and a relative atomic mass of four less. On the other hand, if an atom emits a β-particle, the resultant atom will have an atomic number one higher but the same relative atomic mass because, in the process, a neutron disintegrates to give an electron and a proton:

$$^{1}_{0}n \rightarrow\ ^{1}_{1}p +\ ^{0}_{-1}e$$

It often happens that the isotope of the new element formed is radioactive and so the radioactive decay continues until a stable nucleus is formed. For example, uranium, thorium, and actinium all disintegrate to give isotopes of lead as the final product.

A plot of neutrons against protons for the stable isotopes of the various elements gives a graph of the form shown in Figure 1.7. Any isotopes not lying in the stability region are radioactive and undergo decay until an atom is formed which is in the stable band. It is seen that, up to about atomic number twenty, the stable atoms have a proton : neutron ratio of 1 : 1 but, above this point, the neutrons predominate. There is no definite way of predicting how a radioactive atom will decay but some generalisations may be made. Naturally occurring

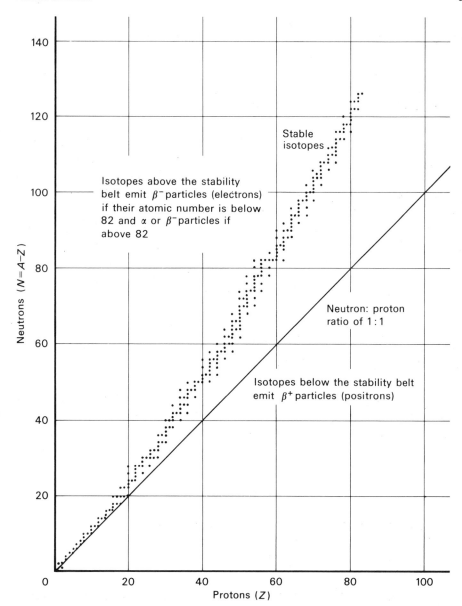

Figure 1.7   Plot of neutrons against protons for stable isotopes

isotopes with atomic number less than 82 do not show α-activity but emit β-particles if they have more neutrons than the stability band. Above atomic number 82, α- and/or β-activity is possible. An element having fewer neutrons than the stable isotopes of that element will rectify this by undergoing *positron* (a positively charged electron, e^+) emission:

$$_1^1p \rightarrow {}_0^1n + {}_1^0e$$

Half-lives of radioactive isotopes

The rate of decay of a radioactive isotope is characteristic of the isotope and is usually expressed in terms of its *half-life*, $t_{\frac{1}{2}}$. The half-life is the time taken for half the number of atoms to disintegrate or the time taken for the intensity of radiation to fall to half its original value. The half-lives of radioactive isotopes vary from very small fractions of a second to millions of years.

NUCLEAR TRANSFORMATIONS

Many nuclear transformations are possible by bombarding various elements with α-particles, neutrons, protons, and deuterons (^2_1H), etc. Neutrons (from an atomic pile) are the most useful bombarding agent because they are neutral and so are not repelled by the positively charged nuclei of the elements. New elements, the transuranic elements ($Z > 92$), have been made by these reactions.

Some examples of nuclear transformations are given below. It should be noted that these equations must balance just like normal chemical equations.

$$^9_4Be + {}^4_2He \rightarrow {}^{12}_6C + {}^1_0n$$
$$^{23}_{11}Na + {}^1_0n \rightarrow {}^{23}_{10}Ne + {}^1_1H$$
$$^{14}_7N + {}^2_1H \rightarrow {}^{15}_8O + {}^1_0n$$
$$^9_4Be + {}^1_1H \rightarrow {}^{10}_5B + \gamma$$
$$^{242}_{96}Cm + {}^4_2He \rightarrow {}^{244}_{98}Cf + 2{}^1_0n$$

Plutonium, for use in bombs and reactors, is made in atomic piles by the following reactions:

$$^{238}_{92}U + {}^1_0n \rightarrow {}^{239}_{92}U \rightarrow {}^{239}_{93}Np + {}^{\ 0}_{-1}e$$
$$\downarrow$$
$$^{239}_{94}Pu + {}^{\ 0}_{-1}e$$

ATOMIC SPECTRUM OF HYDROGEN

1. Absorption spectrum

A metal can absorb all frequencies of ultraviolet light above a certain characteristic frequency and this results in emission of electrons. This absorption is known as non-quantised absorption, and the effect as the *photoelectric effect*. Below the particular frequency mentioned above, no electrons are emitted, and the energy is absorbed, not continuously, but only in multiples of a certain unit which is called a *quantum* or a *photon*. Quantised absorption is then said to occur. The size of the quantum of energy is proportional to the frequency (velocity/wavelength) of radiation and, therefore, inversely proportional to the wavelength. This may be expressed mathematically as $E = h\nu$ where E is the quantum of energy, h is the Planck constant, and ν is the frequency. The smaller the wavelength, the greater will be the frequency, and the greater the energy associated with each quantum.

A diagram showing the frequencies of light absorbed is known as an absorption spectrum. In the simplest case, atomic hydrogen, the absorption takes the form shown in Figure 1.8(a), each line representing an absorbed frequency. Figure 1.8(b) shows the range of the electromagnetic spectrum.

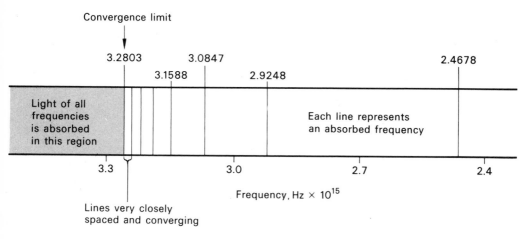

Figure 1.8(a)   Atomic absorption spectrum of hydrogen

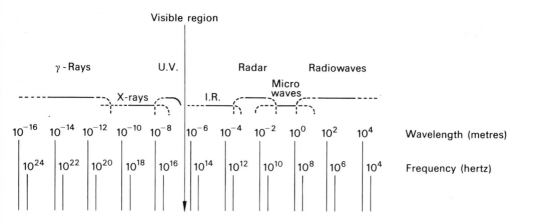

Figure 1.8(b)   The electromagnetic spectrum

2. Emission spectrum

If atoms are excited (i.e. given excess energy) by high temperature or electric discharge etc., they can emit radiation and this also is quantised. Excited hydrogen atoms emit red light and resolution of the radiation into its component frequencies gives the emission spectrum of atomic hydrogen as shown in Figure 1.9. (The absorption spectrum is given as well for comparison.)

Explanation of spectra

The above spectra occur as a result of electronic transitions within the hydrogen atom. As an electron obtains more energy it is promoted to a higher energy level in the atom, that is, further from the nucleus. Since the spectrum of hydrogen atoms is always the same and the energy (energy is related to frequency by the equation $E = h\nu$—see above) absorbed or emitted is sharp and definite, it follows that the energy levels available to an electron must also be precise.

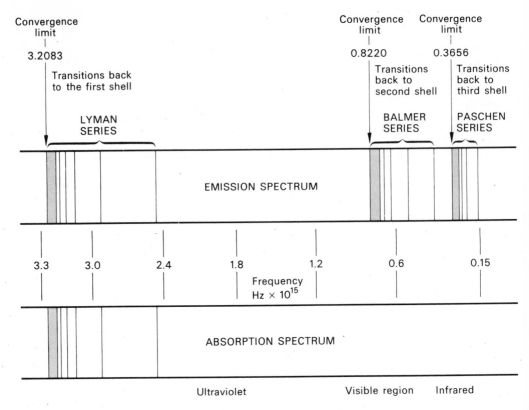

Figure 1.9   Comparison of emission and absorption spectra of atomic hydrogen. Shaded areas denote regions of convergent very closely spaced lines.

Rydberg found by trial and error that the frequencies in the ultraviolet part of the spectrum could, in fact, be calculated by the expression

$$v = R \left(\frac{1}{n_1^2} - \frac{1}{n_2^2} \right)$$

where v = frequency, R = Rydberg constant, and n_1 and n_2 are whole numbers. $R \approx 3.290 \times 10^{15}$ hertz, the limit of the lines in the ultraviolet spectrum. All the lines in the spectrum can be predicted by letting $n_1 = 1$ while n_2 is in turn 2, 3, 4, etc. up to infinity, then letting $n_1 = 2$ and $n_2 = 3$ to infinity, etc.

The energy absorbed for a particular line in the absorption spectrum is utilised in promoting the electron to a higher energy level. Similarly, each line in the emission spectrum represents energy released as the electron falls back from higher to lower energy levels.

Normally the electron in the hydrogen atom occupies the lowest possible energy level and it is said to be in the *ground state*. If a photon is absorbed, the electron is promoted to one of the higher energy levels as illustrated in Figure 1.10. If the photon has sufficiently high energy, the electron may be completely freed from the influence of the nucleus, i.e. ionisation occurs. Once this happens the electron has kinetic energy only and, since this is not quantised, light of all frequencies is then absorbed (see Figure 1.8(a)).

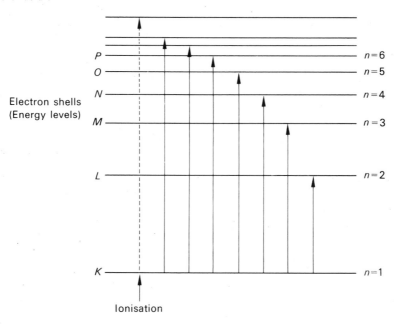

Figure 1.10 Atomic absorption spectrum of hydrogen—electronic transitions

If a hydrogen atom is previously excited, the electron can occupy states other than the ground state. It will then gradually emit energy $(E_{initial} - E_{final} = h\nu)$ and fall back to the ground state (Figure 1.11). If the electron falls back directly to the ground state, the reverse of the absorption spectrum occurs and the Lyman series in the ultraviolet region is produced, whilst fall back to the second shell results in the Balmer series. Similarly, fall back to the third, fourth, and fifth shells gives the Paschen, Brackett, and Pfund series respectively.

The alternative ways of representing the shells or energy levels are shown in Figures 1.10 and 1.11. For example, the first shell is denoted by the symbol K or by a quantum number, n, of one.

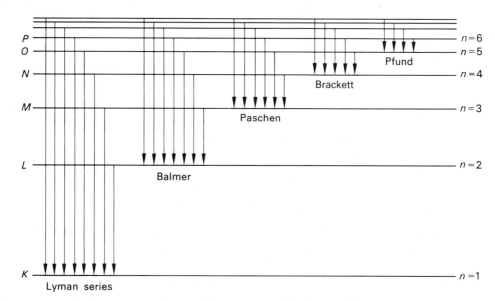

Figure 1.11 Emission spectrum of atomic hydrogen—electronic transitions

WAVE AND PARTICLE NATURE OF ELECTRONS

The next advance towards understanding atomic structure was due to de Broglie. He suggested that all particles in motion have an associated wave character, and he proposed the relationship

$$\lambda = \frac{h}{mv}$$

where λ = wavelength, h = the Planck constant, m = mass and v = velocity. It is apparent from this that the wavelength of a fast moving particle will be significant only if the mass of the particle is very small, i.e. if it has atomic or sub-atomic dimensions. The validity of the de Broglie ideas was confirmed when it was found that a beam of electrons, just like light, could undergo diffraction and interference effects.

The dual wave and particulate nature of electrons is reflected in the Heisenberg uncertainty principle. According to this, it is impossible to know the exact position and velocity of an electron at a given time; instead it is necessary to refer to the probability of finding an electron at a given position. The current wave mechanical or quantum mechanical description of electrons is based on the Schrödinger wave equation but, since the solution of this equation involves very advanced mathematics, it will not be dealt with here. However, the results derived from this equation are very important as described below.

QUANTUM NUMBERS AND ORBITALS

Electrons move round the nucleus of an atom in definite energy levels and these energy levels are identified by numbers called *quantum numbers*. As will be seen later, these energy levels can be sub-divided and so the numbers referred to above are called *principal quantum numbers*. Increase in principal quantum number, n, indicates higher energy and consequently increase in distance from the nucleus. (There are three subsidiary quantum numbers: (i) l, the *azimuthal quantum number* which is an integer and may have any positive value less than n, including zero, (ii) m, the *magnetic quantum number* which also is an integer, and may have any positive or negative value equal to or less than 1, including zero and (iii) s, the *spin quantum number* which may have the values of $+\frac{1}{2}$ or $-\frac{1}{2}$. For the present purposes, however, it is not necessary to discuss these.)

It is not possible to predict the exact position of an electron at a given time in view of its wavelike properties, but quantum mechanics does indicate the regions where there is the maximum probability of finding the electron. These regions are known as *orbitals*. The quantum theory indicates that for each value of the principal quantum number, n, there are n^2 different orbitals. These orbitals differ in their shapes.

When a hydrogen atom is at its lowest energy level (i.e. in the ground state), $n = 1$ and so there is 1 (i.e. 1^2) orbital possible. This orbital is spherically symmetrical about the nucleus and may be compared to a solid ball. It is known as an *s* orbital and, since it is the lowest energy orbital $(n = 1)$, it is designated a 1*s* orbital. As the distance from the nucleus increases so the probability of finding the electron decreases as illustrated in Figure 1.12. An alternative way of representing this is shown in Figure 1.13. Thus, the electron is considered to be distributed as a diffuse negative charge cloud and the diagram endeavours to show the distribution of the charge density.

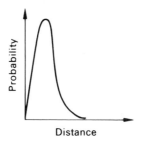

Figure 1.12　Radial probability distribution diagram for 1*s* electrons

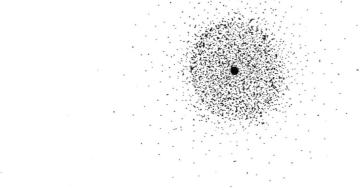

Figure 1.13　Density of the electron charge cloud in a 1*s* orbital

When *n* has a value of two for a hydrogen atom, there are 4 (i.e. 2^2) possible orbitals with the same energy. These orbitals comprise a 2*s* orbital, which like all *s* orbitals is spherically symmetrical about the nucleus, and three 2*p* orbitals. The 2*s* orbital differs from the 1*s* orbital only in that it has more energy and so the electron has a greater probability of being found further away from the nucleus. This may be illustrated by the probability distribution diagram, Figure 1.14.

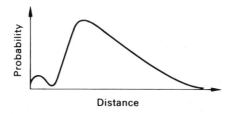

Figure 1.14　Radial probability distribution diagram for 2*s* electrons

The three 2p orbitals are dumb-bell shaped and are situated along the three co-ordinate axes (Figure 1.15).

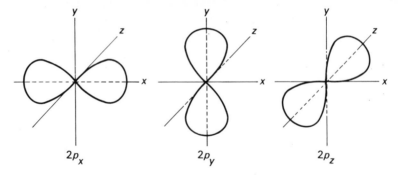

Figure 1.15   The three 2p orbitals

It is seen that there is zero probability of finding a p electron at the nucleus.

When $n = 3$ for a hydrogen atom there are 9 (i.e. 3^2) possible orbitals with the same energy and these are made up of one 3s orbital, three 3p orbitals, and five 3d orbitals. The s and p orbitals are a similar shape as before but further from the nucleus. The shape of the d orbitals is more complicated and will not be discussed here.

When $n = 4$ for a hydrogen atom there are 16 (i.e. 4^2) possible orbitals and these are comprised of one 4s, three 4p, five 4d, and seven 4f orbitals. The shape of f orbitals is also complicated and beyond the scope of this text.

The facts above have all been concerned with the hydrogen atom. However, all atoms give line spectra although generally they are much more complicated. The lines in the spectra are the result of electronic transitions between definite energy levels. The orbitals differ from those in hydrogen in that sub-orbitals in a main shell do not all have the same energy. The relative order of energy levels is indicated by Figure 1.16. The order is given by passing down each diagonal as far as it will go and then continuing at the top of the next one.

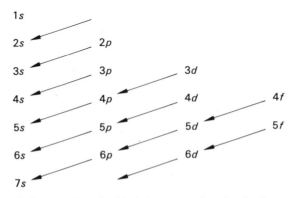

Figure 1.16   The relative energies of orbitals in atoms other than hydrogen

Each orbital may hold up to two electrons and so the number of electrons possible in each shell is:

$$1s^2 \qquad\qquad = 2$$
$$2s^2\,2p^6 \qquad\qquad = 8$$
$$3s^2\,3p^6\,3d^{10} \qquad = 18$$
$$4s^2 4p^6 4d^{10} 4f^{14} = 32$$

The superscripts above indicate the maximum number of electrons possible in the orbitals.

The two electrons in an orbital must spin in opposite directions (each electron in an orbital may be regarded as spinning around its own axis) and they are said to have *paired spins.*

BUILDING UP THE PERIODIC TABLE

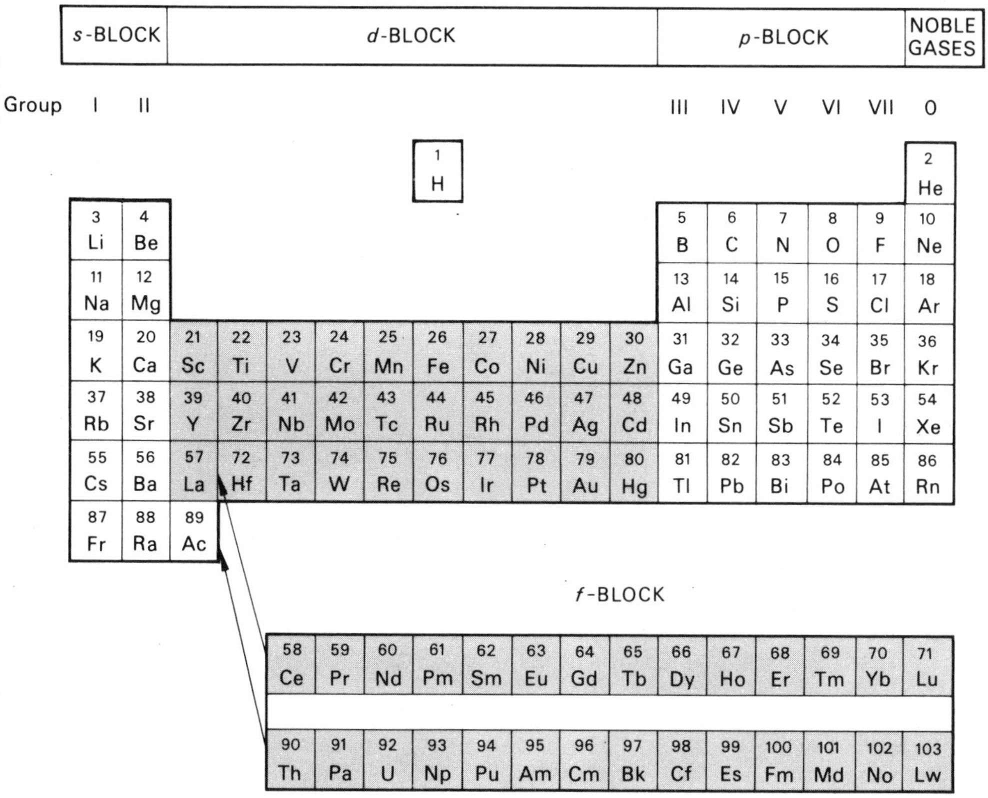

Figure 1.17   The long form of the periodic table. Shaded areas indicate transition elements

Knowing the orbitals possible and their relative energies, it is possible to build up the periodic table based on electronic structures. The following rules must be applied.

(a)   The Aufbau principle: electrons enter the lowest energy orbital available.

(b)   Pauli's exclusion principle: no orbital can accommodate more than two electrons.

(c)   Hund's rule: orbitals of the same energy are occupied singly before pairing occurs. (This is because electrons repel each other and the atom would have a higher total energy if Hund's rule did not apply.)

The electronic structures of the elements up to krypton are given in Table 1.2. The elements scandium to zinc are known as *transition elements* and they constitute the first transition series. Transition elements are elements which are formed by filling up an inner electron shell after an outer shell has been started.

Table 1.2. Electronic structures of elements with atomic number 1–36

Element	Symbol	Atomic Number	Electrons/ shell	Detailed electronic structure
Hydrogen	H	1	1	$1s^1$
Helium	He	2	2	$1s^2$
Lithium	Li	3	2.1	$1s^2\,2s^1$
Beryllium	Be	4	2.2	$1s^2\,2s^2$
Boron	B	5	2.3	$1s^2\,2s^2\,2p^1$
Carbon	C	6	2.4	$1s^2\,2s^2\,2p^2$
Nitrogen	N	7	2.5	$1s^2\,2s^2\,2p^3$
Oxygen	O	8	2.6	$1s^2\,2s^2\,2p^4$
Fluorine	F	9	2.7	$1s^2\,2s^2\,2p^5$
Neon	Ne	10	2.8	$1s^2\,2s^2\,2p^6$
Sodium	Na	11	2.8.1	$1s^2\,2s^2\,2p^6\,3s^1$
Magnesium	Mg	12	2.8.2	$1s^2\,2s^2\,2p^6\,3s^2$
Aluminium	Al	13	2.8.3	$1s^2\,2s^2\,2p^6\,3s^2\,3p^1$
Silicon	Si	14	2.8.4	$1s^2\,2s^2\,2p^6\,3s^2\,3p^2$
Phosphorus	P	15	2.8.5	$1s^2\,2s^2\,2p^6\,3s^2\,3p^3$
Sulphur	S	16	2.8.6	$1s^2\,2s^2\,2p^6\,3s^2\,3p^4$
Chlorine	Cl	17	2.8.7	$1s^2\,2s^2\,2p^6\,3s^2\,3p^5$
Argon	Ar	18	2.8.8	$1s^2\,2s^2\,2p^6\,3s^2\,3p^6$
Potassium	K	19	2.8.8.1	$1s^2\,2s^2\,2p^6\,3s^2\,3p^6\,4s^1$

(The $4s$ level has less energy than the $3d$ and so it is filled first.)

Calcium	Ca	20	2.8.8.2	$1s^2\,2s^2\,2p^6\,3s^2\,3p^6\,4s^2$
Scandium	Sc	21	2.8.9.2	$1s^2\,2s^2\,2p^6\,3s^2\,3p^6\,3d^1\,4s^2$

(The $4p$ level has more energy than the $3d$ and so Sc has the structure 2.8.9.2 not 2.8.8.3.)

Titanium	Ti	22	2.8.10.2	$1s^2\,2s^2\,2p^6\,3s^2\,3p^6\,3d^2\,4s^2$
Vanadium	V	23	2.8.11.2	$1s^2\,2s^2\,2p^6\,3s^2\,3p^6\,3d^3\,4s^2$
Chromium	Cr	24	2.8.13.1	$1s^2\,2s^2\,2p^6\,3s^2\,3p^6\,3d^5\,4s^1$

(Full and exactly half full sub-shells exhibit increased stability i.e. the above configuration for chromium is more stable than

$$1s^2\,2s^2\,2p^6\,3s^2\,3p^6\,3d^4\,4s^2$$
(see page 241.)

Manganese	Mn	25	2.8.13.2	$1s^2\,2s^2\,2p^6\,3s^2\,3p^6\,3d^5\,4s^2$
Iron	Fe	26	2.8.14.2	$1s^2\,2s^2\,2p^6\,3s^2\,3p^6\,3d^6\,4s^2$
Cobalt	Co	27	2.8.15.2	$1s^2\,2s^2\,2p^6\,3s^2\,3p^6\,3d^7\,4s^2$
Nickel	Ni	28	2.8.16.2	$1s^2\,2s^2\,2p^6\,3s^2\,3p^6\,3d^8\,4s^2$
Copper	Cu	29	2.8.18.1	$1s^2\,2s^2\,2p^6\,3s^2\,3p^6\,3d^{10}\,4s^1$

(The configuration of copper again reflects the enhanced stability of an exactly half-full sub-shell.)

Zinc	Zn	30	2.8.18.2	$1s^2\,2s^2\,2p^6\,3s^2\,3p^6\,3d^{10}\,4s^2$
Gallium	Ga	31	2.8.18.3	$1s^2\,2s^2\,2p^6\,3s^2\,3p^6\,3d^{10}\,4s^2\,4p^1$

(The $4p$ is the next lowest energy orbital which is vacant and so this is now filled up from gallium onwards.)

Germanium	Ge	32	2.8.18.4	$1s^2\,2s^2\,2p^6\,3s^2\,3p^6\,3d^{10}\,4s^2\,4p^2$
Arsenic	As	33	2.8.18.5	$1s^2\,2s^2\,2p^6\,3s^2\,3p^6\,3d^{10}\,4s^2\,4p^3$
Selenium	Se	34	2.8.18.6	$1s^2\,2s^2\,2p^6\,3s^2\,3p^6\,3d^{10}\,4s^2\,4p^4$
Bromine	Br	35	2.8.18.7	$1s^2\,2s^2\,2p^6\,3s^2\,3p^6\,3d^{10}\,4s^2\,4p^5$
Krypton	Kr	36	2.8.18.8	$1s^2\,2s^2\,2p^6\,3s^2\,3p^6\,3d^{10}\,4s^2\,4p^6$

In the periodic table (Figure 1.17) the elements are arranged in such a way that all those in a vertical column (known as a group) have similar electronic structures — for example, lithium, sodium, potassium, etc. in the first group all have one electron in the outer s orbital. All the elements in a group exhibit a common valency (and often others as well) since the valency of an element is the same as the number of electrons in the outer shell or eight minus the number if there are more than four. On travelling horizontally across the table (i.e. across a period) the number of electrons in the outer shell gradually increases from one to eight, ignoring the transition elements (shaded in Figure 1.17).

Questions

1.1 Outline the principles of relative atomic mass determination by the mass spectrometer.

The mass spectrum of an element, A, contained four lines at mass/charge of 54, 56, 57, and 58 with relative intensities of $5.84 : 91.68 : 2.17 : 0.31$ respectively. Explain these data and calculate the relative atomic mass of A.

1.2 Given that protons, neutrons, and electrons have masses of 1.007276, 1.008665, and 0.000549 a.m.u. respectively, predict which isotope, $^{56}_{26}Fe$ or $^{55}_{26}Fe$, is most likely to be radioactive. The actual masses of the isotopes are $^{56}_{26}Fe = 55.938299$ a.m.u.   and $^{55}_{26}Fe = 54.938299$ a.m.u.

1.3 Write an essay on radioactivity. Include in your answer reference to the following points, giving experimental details where appropriate.

(a) The nature and origins of radioactive disintegrations.

(b) The importance of this topic for an understanding of atomic structure.

[J.M.B.]

1.4 Identify X in the following nuclear reactions:

$$^{107}_{47}Ag + ^{1}_{0}n \rightarrow X + ^{0}_{-1}e$$
$$^{2}_{1}H + X \rightarrow ^{4}_{2}He + ^{1}_{0}n$$
$$^{12}_{4}Be \rightarrow ^{0}_{-1}e + \gamma + X$$
$$^{12}_{6}C + 2(^{1}_{0}n) \rightarrow ^{14}_{7}N + X$$
$$^{19}_{9}F + X \rightarrow ^{20}_{10}Ne + ^{0}_{-1}e$$

1.5 Insert the missing data in the following decay chain for $^{232}_{90}Th$:

$$^{232}_{90}Th \xrightarrow{\alpha} ? \xrightarrow{?} {}^{228}_{89}Ac \xrightarrow{\beta} ? \xrightarrow{4\alpha} ? \xrightarrow{?} {}^{212}_{83}Bi \xrightarrow{\alpha} ? \xrightarrow{\beta} ?$$

1.6 Calculate the frequencies and hence the wavelengths of the first four lines in the Lyman series of the atomic absorption spectrum of hydrogen. The Rydberg constant is 3.290×10^{15} Hz (hertz $= s^{-1}$), and the speed of light is 2.998×10^8 ms^{-1}.

1.7 Calculate the energy required to ionise a mole of hydrogen atoms given that the convergence limit of the lines in the Lyman series of the absorption spectrum of atomic hydrogen is 3.28×10^{15} Hz (hertz $= s^{-1}$), the Avogadro constant is 6.02×10^{23} mol^{-1}, and Planck's constant, h, is 6.62×10^{-34} Js.

1.8 Give the detailed electronic structures of the elements with atomic numbers 9, 14, 18, 29, and 35. State the relationship between electronic structure and valency and hence predict the valency of each element. Which of the elements would be expected to exhibit similar chemical properties?

1.9 (a) What are α, β, and γ emissions and how do they differ in their penetrating power and their behaviour in a magnetic field? Explain the meaning of the two numbers before the symbol for uranium and identify P, Q, R, and S in the following equations.

$$^{234}_{92}U \rightarrow \quad \alpha \quad + P$$
$$^{239}_{92}U \rightarrow \quad \beta^- \quad + Q$$
$$^{235}_{92}U \rightarrow \quad \gamma \quad + R$$
$$^{2}_{1}H + ^{238}_{92}U \rightarrow ^{239}_{92}U + S$$

(b) Deduce the nature of X in the following reaction.

$$^{235}_{92}U + ^{1}_{0}n \rightarrow ^{95}_{42}Mo + ^{139}_{57}La + 2^{1}_{0}n + 7X$$

There is approximately 0.1% less mass on the right-hand side of the equation than on the left-hand side. What is the significance of this?   [J.M.B.]

2 States of Matter

A given substance may exist as a gas, liquid, or solid or in two or all three of these states simultaneously depending upon the physical conditions. It is one of the aims of the chemist to understand the bulk properties of these states in terms of the constituent atoms or molecules. In this respect, gases are the easiest to study and in fact Boyle's law dates back to 1662.

1 THE GASEOUS STATE

The volumes of liquids and solids are affected to a small extent by changes in temperature and pressure, the volume change being characteristic of the liquid or solid concerned. On the other hand, the volume of gases is affected to a much greater degree by changes in temperature or pressure and all gases are affected by approximately the same amount. This observation led to the gas laws.

Boyle's law

In 1662, Boyle found that the volume of a fixed mass of gas is inversely proportional to the pressure provided that the temperature is kept constant.

This may be expressed mathematically as $V \propto 1/P$ or $PV = \text{constant}$ where P and V represent pressure and volume respectively.

A plot of volume against the reciprocal of the pressure will therefore give a straight line graph passing through the origin (Figure 2.1).

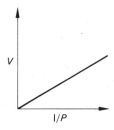

Figure 2.1   Graphical representation of Boyle's law

Charles' law

Charles' law is, in fact, based on observations made by Gay-Lussac and Charles. Gay-Lussac found that gases expand or contract by $\frac{1}{273}$ of their volume at 0 °C for every °C that the temperature is raised or lowered respectively. If the temperature is lowered by 273 °C, then the volume will theoretically be zero (i.e. the molecules will be completely at rest) and so −273 °C is known as *absolute zero* or 0 K (kelvin). 0 °C is therefore 273 K.

Volume – temperature relationships of gases were also studied by Charles and he observed that a number of gases had the same volume coefficient of thermal expansion. Combination of Charles' and Gay-Lussac's findings gives Charles' law: the volume of a fixed mass of gas is directly proportional to the kelvin temperature provided that the pressure is kept constant.

The mathematical expression is $V \propto T$ or $V/T = \text{constant}$ where $V = \text{volume}$ and $T = \text{kelvin temperature}$.

The ideal gas equation

Boyle's law and Charles' law may be combined.

Thus, since $V \propto 1/P$ and $V \propto T$, $V \propto T/P$

This may be written $V = \dfrac{kT}{P}$ or $\dfrac{PV}{T} = k$

$$(2.1)$$

where k is a constant for a particular volume of a particular gas. However, if one mole of a gas is considered, the same number of molecules are always involved and then $PV/T = R$, where R is called the *universal gas constant* — it is a constant for all gases. If n moles of gas are considered, the equation becomes $PV = nRT$ and this is known as the *ideal gas equation*.

The value of R in SI units may be calculated as follows.

For 1 mole, $n = 1$ and so $R = PV/T$

Expressing P as force/area gives:

$$R = \frac{\text{force}}{\text{area}} \times \text{volume} \times \frac{1}{T} = \text{force} \times \text{length} \times \frac{1}{T}$$

But force $\times$ length $=$ work or energy and the SI unit for this is the joule (J) — defined as the work done when a force of 1 newton (N) acts over a distance of 1 metre(m).

The units of R must therefore be expressed in joules per mole per degree absolute, that is, $\text{J mol}^{-1}\text{K}^{-1}$.

Substituting the values for P and V (in SI units) and T in the equation $R = PV/T$ will give the numerical value for R.

$P = $ standard pressure, 1 atmosphere. In SI terms it is the force exerted by a 0.76 m column of mercury at 0 °C on an area of 1 m². Since mercury has a density of $13\,600\,\text{kg m}^{-3}$ the column has a mass of $0.76 \times 1 \times 13\,600$ kg. The acceleration due to gravity is $9.81\,\text{m s}^{-2}$ and so the pressure in SI units $= 0.76 \times 13\,600 \times 9.81\,\text{N m}^{-2} = 101\,325\,\text{N m}^{-2}$.

$V = $ volume of 1 mole of gas at s.t.p. $= 0.0224\,\text{m}^3$.

$T = $ standard temperature $= 273$ K

$$R = \frac{PV}{T} = \frac{101\,325 \times 0.0224}{273} = 8.31\ \text{J mol}^{-1}\text{K}^{-1}$$

The gas equation may be used to derive another very useful equation. Thus, if the volume of one mole of gas is V_1 at pressure P_1 and kelvin temperature

T_1 and at temperature T_2 and pressure P_2 its volume is V_2 then:

$$\frac{P_1 V_1}{T_1} = R \quad \text{and} \quad \frac{P_2 V_2}{T_2} = R$$

$\therefore$

$$\frac{P_1 V_1}{T_1} = \frac{P_2 V_2}{T_2} \tag{2.2}$$

Dalton's law of partial pressures

This states that the total pressure of a mixture of gases is the sum of the pressures each gas would exert if it alone occupied the whole volume of the mixture at the same temperature.

The pressure each constituent gas exerts is called its *partial pressure*.

$$\text{Partial pressure} \quad P_A = \frac{\text{number moles of gas A}}{\text{total number moles in the mixture}} \times P$$

where P = total pressure.

The gas laws may be illustrated by the examples below.

Example   An enclosed vessel contains 2.8 g of nitrogen and 14.2 g of chlorine at atmospheric pressure and 0 °C. What will be the partial pressure of the nitrogen if the temperature is raised to 180 °C? A_r: N = 14,   Cl = 35.5.

$$\frac{P_1 V_1}{T_1} = \frac{P_2 V_2}{T_2} \quad \text{but the volume is constant and so} \quad \frac{P_1}{T_1} = \frac{P_2}{T_2}$$

$$\frac{760}{273} = \frac{P_2}{453} \quad \therefore P_2 = \frac{760 \times 453}{273} = 1261 \text{ mm}$$

The partial pressure of the nitrogen,

$$P_{N_2} = \frac{\text{moles of } N_2}{\text{moles of } N_2 + Cl_2} \times \text{total pressure}$$

$$= \frac{\dfrac{2.8}{28}}{\dfrac{2.8}{28} + \dfrac{14.2}{71}} \times 1261$$

$$= \frac{0.1}{0.3} \times 1261 = 420.3 \text{ mm}$$

Example   9 moles of a gas occupy 1800 cm³ at 113 atmospheres pressure and 27 °C. Calculate the pressure this gas would exert under the same conditions if it were an 'ideal gas'.

Comment on the results.

$$R = 8.31 \text{ J mol}^{-1}\text{K}^{-1}, \quad 1 \text{ atmosphere} = 1.013 \times 10^5 \text{ N m}^{-2}$$

This calculation may be performed in two ways:

(a)
$$PV = nRT$$

$\therefore$
$$P = \frac{nRT}{V} \quad \text{where } P \text{ will be in N m}^{-2},$$

$$R = 8.31 \text{ J mol}^{-1}\text{K}^{-1}$$

$$T = 300 \text{ K}, \ V = \frac{1800}{10^6} \text{ m}^3$$

$\therefore$
$$P = \frac{9 \times 8.31 \times 300 \times 10^6}{1800} \text{ N m}^{-2}$$

$$= \frac{9 \times 8.31 \times 300 \times 10^6}{1800 \times 1.013 \times 10^5} \text{ atmospheres}$$

$$= 123 \text{ atmospheres}$$

(b) 9 moles of an ideal gas occupy 9×22.4 litres at s.t.p. and so its pressure, when it occupies 1.8 litres at 27 °C, may be calculated:

$$\frac{P_1 V_1}{T_1} = \frac{P_2 V_2}{T_2} \qquad \frac{1 \times 9 \times 22.4}{273} = \frac{P_2 \times 1.8}{300}$$

$\therefore$
$$P_2 = \frac{9 \times 22.4 \times 300}{273 \times 1.8} = 123 \text{ atmospheres}$$

The pressure of the real gas is less than that of an ideal one because, at these high pressures, the molecules are close together and intermolecular attractions become appreciable. The number of molecules, and hence the pressure, is effectively reduced.

Real and ideal gases, and the limitations of the gas laws, are discussed in more detail later in this chapter.

Determination of relative molecular masses of gases and vapours

Relative molecular masses can be determined by mass spectroscopy as described in the previous chapter. Alternatively, they can be obtained for gases or vapours from a knowledge of their relative densities (vapour densities), since relative molecular mass is related to relative density as illustrated below.

Hydrogen is the lightest gas known and the densities of all other gases are compared with it, that is:

Relative density =

$$\frac{\text{mass of 1 volume of gas}}{\text{mass of 1 volume of hydrogen at the same temperature and pressure}}$$

But, according to Avogadro's law, equal volumes of gases at the same temperature and pressure contain equal numbers of molecules.

$\therefore$ Relative density $= \dfrac{\text{mass of } n \text{ molecules of gas}}{\text{mass of } n \text{ molecules of hydrogen}}$

$= \dfrac{\text{mass of 1 molecule of gas}}{\text{mass of 1 molecule of hydrogen}}$

$= \dfrac{\text{mass of 1 molecule of gas}}{\text{mass of 2 atoms of hydrogen}}$

But relative molecular mass $= \dfrac{\text{mass of 1 molecule of gas}}{\text{mass of 1 atom of hydrogen}}$

$\therefore$ Relative density $= \frac{1}{2}$ relative molecular mass
or relative molecular mass $= 2 \times$ relative density.

This expression is based on the old relative atomic mass scale of $H = 1$ and not on the present scale of $^{12}_{6}C = 12.0000$. This is, however, unimportant since this error is considerably less than the accuracy of this method of relative molecular mass determination. Also, accurate relative molecular mass determinations are not always necessary, for example, the molecular formula of a compound is readily apparent if the empirical formula and approximate relative molecular mass are known.

Victor Meyer's method of relative molecular mass determination

This method is based on the determination of the volume occupied by a given mass of the gas.

The apparatus is as shown in Figure 2.2. The liquid D in the outer jacket is chosen so that its boiling point is about 30 °C above the boiling point of the

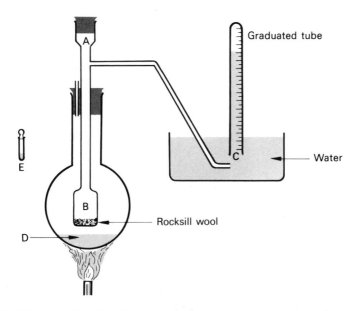

Figure 2.2   Apparatus for Victor Meyer's method of relative molecular mass
            determination

liquid whose relative molecular mass is being determined. D is boiled and the apparatus is left until no more air is displaced through C. Some of the liquid under test is weighed into the stoppered tube E. The bung in A is removed, the tube E containing the liquid is dropped in, and the bung quickly replaced. The graduated tube, filled with water, is then placed in position over C. The liquid in E vaporises, blows out the stopper, and displaces some of the air in B. The displaced air is collected in the graduated tube. When no more bubbles are expelled through C, the levels of water in the graduated tube and the surrounding bath are equalised so that the pressure in the tube is the same as atmospheric pressure. The volume is then read off and the atmospheric pressure and room temperature are noted.

Note the following:

(a) The temperature of the vapour round B is not required since the displaced air is measured at room temperature in the graduated tube.

(b) The amount of sample used in B must be such that its vapour does not reach the side-arm because it could then condense on the cooler parts of the tube. Diffusion will be slow since the tubes are narrow.

(c) The error in this method of relative molecular mass determination is normally about 2 per cent.

(d) The method has the advantage of needing only a small amount of sample and it is quick.

Calculation   In a Victor Meyer determination of the relative molecular mass of benzene, the heating vessel was maintained at 120 °C. A mass of 0.1528 g of benzene was used and the volume of displaced air collected over water at 15 °C, was 48.0 cm³. The barometric pressure was 743 mm mercury. Calculate the relative molecular mass of benzene. The vapour pressure of water at   15 °C = 13 mm   mercury,   molar   volume = 22.4 litres   at   s.t.p.,   1 atmosphere = $1.013 \times 10^5 \, \mathrm{N\,m^{-2}}$, and   R = 8.31 J mol⁻¹K⁻¹.
The calculation may be done in two ways.

(a) Actual pressure of displaced air = atmospheric pressure – vapour pressure of water at 15 °C = 743 – 13 = 730 mm Hg

Correcting the volume of displaced air to s.t.p.:

$$\frac{P_1 V_1}{T_1} = \frac{P_2 V_2}{T_2} \qquad \frac{730 \times 48}{288} = \frac{760 \times V_2}{273}$$

$$V_2 = \frac{730 \times 48 \times 273}{288 \times 760} = 43.7 \text{ cm}^3$$

$$\text{Density of benzene} = \frac{0.1528}{43.7} \text{ g cm}^{-3}$$

$$\text{Density of hydrogen} = \frac{2}{22\,400} \left(\frac{M}{\text{Molar vol.}} \right)$$

∴ $$\text{Relative density} = \frac{0.1528}{43.7} \Bigg/ \frac{2}{22\,400}$$

$$= 39.1$$

$$\text{Relative molecular mass} = 2 \times \text{relative density}$$
$$= 78.2$$

(b)
$$PV = nRT \quad \text{but} \quad n = \frac{W}{M} \left(\frac{\text{mass}}{\text{relative molecular mass}} \right)$$

$\therefore$
$$PV = \frac{WRT}{M} \quad \text{or} \quad M = \frac{WRT}{PV}$$

All figures must be in SI units, i.e.

$\therefore$
$$W = 0.1528 \text{ g} \qquad R = 8.31 \text{ J mol}^{-1}\text{K}^{-1}$$
$$V = \frac{48}{10^6} \text{ m}^3 \qquad P = \frac{730}{760} \times 1.013 \times 10^5 \text{ Nm}^{-2}$$
$$T = 288 \text{ K} \qquad M = \text{g relative molecular mass}$$
$$M = \frac{0.1528 \times 8.31 \times 288 \times 760 \times 10^6}{730 \times 1.013 \times 10^5 \times 48}$$
$$= 78.2$$

$\therefore$ Relative molecular mass = 78.2

The Dumas method of relative molecular mass determination

In this method, the mass of a given volume of vapour is found.

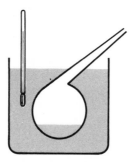

Figure 2.3   Apparatus for the Dumas method of relative molecular mass determination

The Dumas bulb is weighed and the room temperature and atmospheric pressure are noted. The bulb is warmed and the tip is placed under the surface of the test liquid which is sucked into the bulb as it cools. The bulb is about one quarter filled by this method and then it is placed in the heating bath containing a boiling liquid (often water). The liquid in the bulb vaporises and air and excess vapour are blown out. When none of the liquid is left and the vapour has had time to reach the same temperature as the bath liquid, the end of the bulb is sealed in a flame and the temperature of the bath is noted. The bulb is dried, allowed to cool, and weighed.

The tip of the bulb is then broken off under a large volume of gas-free water (water which has been boiled for some time). If the experiment has been a success, the bulb will fill completely with water — the vapour will condense to a very small volume of liquid and there will be no air present. The bulb full of water and the broken tip are weighed and so the volume of the bulb is known since 1g of water occupies 1 cm^3.

Again the method is not very accurate because:

(a)   it is difficult to displace all the air in the bulb by vapour,

(b)   the glass expands in the hot bath and so the volume of vapour at elevated temperatures is greater than that at room temperature.

This method has the disadvantage of requiring a fairly large volume of liquid compared with Victor Meyer's method.

Calculation   A Dumas bulb with a capacity of 248 cm^3 was sealed full of ethanol vapour at 99 °C. The mass of the bulb and vapour was 53.6712 g, whilst the mass of the bulb full of air was 53.6039 g. If the room temperature and pressure were 15 °C and 750 mm, calculate the relative molecular mass of ethanol. A litre of air at s.t.p. weighs 1.293 g.

The volume of air in the bulb must be converted to s.t.p. before its mass can be found.

$$\frac{P_1 V_1}{T_1} = \frac{P_2 V_2}{T_2} \qquad \frac{750 \times 248}{288} = \frac{760 \times V_2}{273}$$

$$\therefore \qquad V_2 = \frac{750 \times 248 \times 273}{288 \times 760} = 232.1 \text{ cm}^3$$

$$\text{The mass of air in the bulb} = \frac{232.1 \times 1.293}{1000}$$

$$= 0.3001 \text{ g}$$

$$\text{Mass of empty bulb} = 53.6039 - 0.3001$$

$$= 53.3038 \text{ g}$$

$$\text{Mass of ethanol vapour} = 53.6712 - 53.3038$$

$$= 0.3674 \text{ g}$$

Hence 0.3674 g of ethanol vapour occupy 248 cm^3 at 750 mm and 99 °C. Correcting this volume to s.t.p. gives:

$$\frac{750 \times 248}{372} = \frac{760 \times V_2}{273}$$

$$\text{or } V_2 = \frac{750 \times 248 \times 273}{372 \times 760} = 179.7 \text{ cm}^3$$

$$\text{Density of vapour} = \frac{0.3674}{179.7}$$

$$\text{Density of hydrogen} = \frac{2}{22\,400}$$

$$\text{Relative density of ethanol} = \frac{0.3674}{179.7} \times \frac{22\,400}{2}$$

$$= 22.89$$

Relative molecular mass $= 2 \times$ relative density $= 45.8$

N.B. This calculation may also be worked out using the gas equation in a similar manner to that shown for the Victor Meyer calculation.

Kinetic theory of gases

In the gaseous state under normal conditions, molecules are so far apart as to have little effect on one another. They are in a state of rapid random motion and travel in straight lines until they collide with other molecules or with the walls of the containing vessel. There is no limit to the volume of a gas and, since the molecules are moving in all directions, their ability to diffuse is readily understood. The pressure exerted by a gas is the result of continual collisions of the molecules on the walls of the container. It is apparent that the collisions of molecules do not result in a lowering of the total kinetic energy of the gas since the temperature of a gas in a container does not fall with time. The conclusion from this is that all the collisions are perfectly elastic.

It is possible to derive, from the kinetic theory of gases, a quantitative relationship between pressure, volume, and the velocity of molecules. This is

$$PV = \tfrac{1}{3}mn\overline{c^2} \tag{2.3}$$

where $P =$ pressure, $V =$ volume, $m =$ mass of molecule, $n =$ the number of molecules and $\overline{c^2} =$ mean square velocity. (The meaning of $\overline{c^2}$ may be clarified by considering three molecules with velocities of 1, 2 and 3. The average velocity is 2 but $\overline{c^2} = \tfrac{1}{3} (1^2 + 2^2 + 3^2) = 4.67$. Root mean square velocity, $\sqrt{(\overline{c^2})}$, is therefore $\sqrt{4.67} = 2.16$.) Note that velocity is the rate of change of the position of a body in a particular direction, i.e. it is a vector quantity. The average velocity of the molecules of a gas is therefore zero, since they are moving in all directions. However, squaring the velocities eliminates negative signs, and so the root mean square velocity is not zero.

Equation (2.3) holds for *ideal gases*, that is, gases to which the following assumptions apply.

(a) They consist of a large number of molecules all of which are in a state of rapid random motion.

(b) The size of molecules is negligible compared to the distance they travel between collisions (known as the mean free path).

(c) The molecules are so far apart that the attractive forces between them are negligible.

(d) There is no kinetic energy loss in a collision.

(e) The molecules can convert heat energy to kinetic energy.

It is possible to derive all the gas laws from the above equation thus showing its validity.

(i) Boyle's law

$PV = \frac{1}{3}mn\overline{c^2}$ but for a given mass of gas at constant temperature m, n, and $\overline{c^2}$ are all constant and so $PV = $ constant.

(ii) Charles' law

$$PV = \frac{1}{3}mn\overline{c^2} \quad \text{or} \quad V = \frac{1}{3}\frac{mn\overline{c^2}}{P}$$

But, for a given mass of gas, m and n are constant and $\overline{c^2}$ is directly proportional to the absolute temperature and so, at constant pressure, $V \propto T$.

(iii) Dalton's law

Consider a mixture of two gases with pressures, volumes, masses, numbers of molecules, and mean square velocities of P_1, V_1, m_1, n_1, and $\overline{c_1^2}$ and P_2, V_2, m_2, n_2, and $\overline{c_2^2}$ respectively.

$$P_1V_1 = \frac{1}{3}m_1n_1\overline{c_1^2} \quad \text{and} \quad P_2V_2 = \frac{1}{3}m_2n_2\overline{c_2^2}$$

If the gases are in the same vessel $V_1 = V_2$ and, since the temperature is the same, the kinetic energy of the molecules of each gas is the same, i.e:

$$\frac{1}{2}m_1\overline{c_1^2} = \frac{1}{2}m_2\overline{c_2^2} \quad \text{and so} \quad m_1\overline{c_1^2} = m_2\overline{c_2^2}$$

$\therefore \quad P_1 \propto n_1 \quad \text{and} \quad P_2 \propto n_2$

Hence,
$$\frac{P_1}{P_2} = \frac{n_1}{n_2} \quad \text{or} \quad P_1 = \frac{n_1}{n_2} \times P_2$$

The total pressure $P = P_1 + P_2 = \left(\frac{n_1}{n_2} \times P_2\right) + P_2$

$\therefore \qquad P = P_2\left(\frac{n_1 + n_2}{n_2}\right) \quad \text{giving} \quad P_2 = P\left(\frac{n_2}{n_1 + n_2}\right)$

i.e. partial pressure = total pressure × the mole fraction of that gas.

(iv) Graham's law

This states that the rate of diffusion of a gas is inversely proportional to the square root of its density.

$$\text{Now } PV = \frac{1}{3}mn\overline{c^2} \quad \text{or} \quad \sqrt{(\overline{c^2})} = \sqrt{\frac{3PV}{mn}}$$

$$\text{but, density, } d = \frac{mn}{V} \quad \text{and so} \quad \sqrt{(\overline{c^2})} = \sqrt{\frac{3P}{d}}$$

Now the rate of diffusion of a gas is directly proportional to the root mean square velocity of its molecules – the faster the molecules move the faster they diffuse;

i.e. $\sqrt{(\overline{c^2})} \propto r = \sqrt{\frac{k}{d}}$ at a given pressure, where $k = $ constant and

$r = $ rate of diffusion

or
$$r \propto \sqrt{\frac{1}{d}}$$

(v) Avogadro's law

Consider two gases with pressures, volumes, numbers of molecules, masses, and mean square velocities of P_1, V_1, n_1, m_1, and $\overline{c_1^2}$ and P_2, V_2, n_2, m_2, and $\overline{c_2^2}$ respectively, then:

$$P_1V_1 = \tfrac{1}{3}m_1 n_1 \overline{c_1^2} \quad \text{and} \quad P_2V_2 = \tfrac{1}{3}m_2 n_2 \overline{c_2^2}$$

$$\therefore \qquad n_1 = \frac{3P_1 V_1}{m_1 \overline{c_1^2}} \quad \text{and} \quad n_2 = \frac{3P_2 V_2}{m_2 \overline{c_2^2}}$$

If equal volumes at the same temperature and pressure are considered, then $V_1 = V_2$ and $P_1 = P_2$. If the temperature is the same in each case, then their kinetic energies are the same, i.e.

$$\tfrac{1}{2}m_1 \overline{c_1^2} = \tfrac{1}{2}m_2 \overline{c_2^2} \quad \text{and so} \quad m_1 \overline{c_1^2} = m_2 \overline{c_2^2}$$

$$\therefore \qquad \frac{3P_1 V_1}{m_1 \overline{c_1^2}} = \frac{3P_2 V_2}{m_2 \overline{c_2^2}}$$

Hence, $n_1 = n_2$ as stated by Avogadro's law.

Energy–temperature relationship in gaseous molecules

From the kinetic theory, $PV = \tfrac{1}{3} mn\overline{c^2}$ or for 1 mole $PV = \tfrac{1}{3} mN\overline{c^2}$ where N is the Avogadro constant $(6.02 \times 10^{23} \text{ mol}^{-1})$, i.e. the number of molecules of the gas in 1 mole.

Now according to the ideal gas equation, $PV = RT$ for 1 mole.

$$\therefore \qquad \tfrac{1}{3} mN\overline{c^2} = RT \text{ mol}^{-1} \qquad (2.4)$$

Since the kinetic energy of a molecule $= \tfrac{1}{2}m\overline{c^2}$, the kinetic energy for 1 mole $= N \times \tfrac{1}{2} m\overline{c^2}$ or $2\,K.E. = Nm\overline{c^2}$,

Substituting $2\,K.E.$ for $Nm\overline{c^2}$ in equation (2.4) gives

$$\tfrac{1}{3} \times 2K.E. = RT$$
$$\text{i.e. } K.E. = \tfrac{3}{2}RT \text{ mol}^{-1}$$
$$\text{or } K.E. = \tfrac{3}{2}kT \text{ molecule}^{-1} \qquad (2.5)$$

Monatomic molecules can have only one form of motion associated with them, that is, translational motion or simple motion of the molecules through space; their kinetic energy is simply translational energy. With di- and polyatomic molecules, however, rotations and vibrations occur as well as translational motion.

Consider, for example, a molecule of carbon dioxide; its translational motion can be illustrated as in Figure 2.4.

Its rotational and vibrational motion is comparable to what would happen if three balls of similar mass, held together by springs, were set vibrating and then thrown end over end through space (Figure 2.5).

The kinetic energy of carbon dioxide is therefore distributed between three kinds of motion.

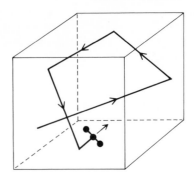

Figure 2.4   Translational motion of a carbon dioxide molecule

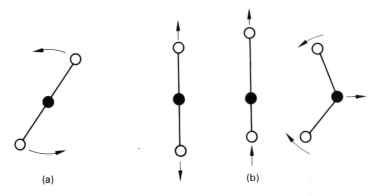

(a)   (b)

Figure 2.5   (a) Rotational and (b) vibrational motions of a triatomic molecule

Graham's law as evidence of translational energy

According to Graham's law, the rate of diffusion of a gas is inversely proportional to the square root of its density, i.e.

$$R \propto \sqrt{\frac{1}{d}} . \text{ For two gases } \frac{R_1}{R_2} = \sqrt{\frac{d_2}{d_1}} \qquad (2.6)$$

This relationship may be used to determine the relative molecular mass of a gas.

Example   100 cm³ of hydrogen chloride were found to diffuse through a porous plug in 43.0 seconds whilst the same volume of a nitrogen compound was found to diffuse in 29.3 seconds under the same conditions. Calculate the relative molecular mass of the nitrogen compound   H = 1,   Cl = 35.5.

From Graham's law $\dfrac{R_1}{R_2} = \sqrt{\dfrac{d_2}{d_1}}$

Now density of a gas $= \dfrac{\text{relative molecular mass}}{\text{molar volume}}$

and so the density is directly proportional to the relative molecular mass.

$\therefore$ $\dfrac{R_1}{R_2} = \sqrt{\dfrac{M_2}{M_1}}$

where $R_1 = \dfrac{100}{43.0}\ cm^3 s^{-1}$ $R_2 = \dfrac{100}{29.3}\ cm^3 s^{-1}$ and $M_1 = 36.5$

$$\therefore \qquad \frac{100}{43} \times \frac{29.3}{100} = \sqrt{\frac{M_2}{36.5}}$$

$$M_2 = \left(\frac{29.3}{43.0}\right)^2 \times 36.5 = 16.9$$

Since relative molecular mass is directly proportional to density, it follows that a light gas will diffuse more rapidly than a heavy one. The diffusion must be due to translational motion and it may be demonstrated using the apparatus shown in Figure 2.6.

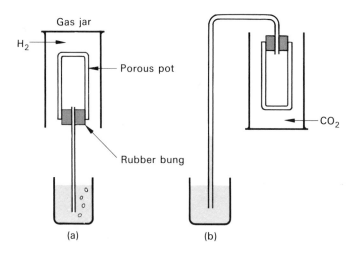

Figure 2.6 Apparatus for demonstrating diffusion of (a) light and (b) heavy gases

If a gas jar of hydrogen is inverted over the porous pot, a stream of bubbles issues from the tube in the water because fast moving hydrogen diffuses in faster than air can diffuse out and so the pressure rises. If the gas jar is removed, the level of the water in the tube rises because the hydrogen in the pot diffuses out faster than air diffuses in, thus creating a partial vacuum.

If the pot is placed in a jar of carbon dioxide, the pressure falls and water is sucked up the tube because air diffuses out more quickly than carbon dioxide can diffuse in. If the jar is removed, the pressure in the pot increases because air diffuses in faster than carbon dioxide can diffuse out.

Molecular velocities

From the kinetic theory of gases, $PV = \frac{1}{3} mN\overline{c^2}$ for one mole whilst according to the ideal gas equation, $PV = RT$ for one mole.

$$\therefore \qquad \tfrac{1}{3} Nm\overline{c^2} = RT$$

or $\qquad \sqrt{(\overline{c^2})} = \sqrt{\dfrac{3RT}{Nm}}$ but Nm = relative molecular mass and so

$$\sqrt{(\overline{c^2})} = \sqrt{\frac{3RT}{M}} \qquad\qquad (2.7)$$

where $\sqrt{(\overline{c^2})}$ is the root mean square velocity of the molecules. It is therefore possible to work out approximate speeds of molecules. Using SI units, the speeds will be in $m\,s^{-1}$.

Calculation Find the approximate average speed of oxygen molecules at 20 °C. $R = 8.31\ J\,mol^{-1}K^{-1}.$

Relative atomic mass of oxygen $= 16$

$$\sqrt{(\overline{c^2})} = \sqrt{\frac{3RT}{M}} = \sqrt{\frac{3 \times 8.31 \times 293}{32 \times 10^{-3}}} \quad \text{(The } 10^{-3} \text{ converts g to kg)}$$

$$= \quad 473\ m\,s^{-1}$$

$1\ m\,s^{-1} \approx 2.24$ m.p.h. and so the speed is approximately 1000 m.p.h.

Distribution of molecular velocities

The molecules of a gas have a wide range of velocities because of their frequent collisions. A good indication of the range of velocities can, in fact, be obtained from the Zartmann experiment using the apparatus in Figure 2.7.

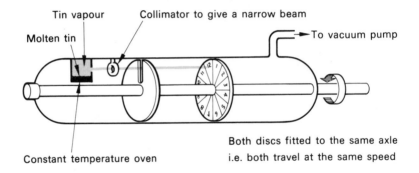

Figure 2.7   Apparatus for the Zartmann experiment

A narrow beam of tin molecules is admitted through the slit in the first disc once every revolution. They then travel at their own individual velocities until they hit the second disc and condense (tin gas to tin solid). Since the second disc is also rotating, they will spread out over the disc, the fastest ones condensing on the low numbered segments and the slower ones on the high numbered segments because the disc will have rotated further before they strike it. A layer of tin molecules will gradually build up on the second disc as the slit lets in a burst once a revolution (Figure 2.8). The mass of tin on the various segments of the disc will be directly proportional to the distribution of velocities. From the rate of rotation of the discs and the distance between them, it is possible to calculate the velocity of the molecules hitting any particular segment of the disc. It should be noted that the velocities of a collimated beam of molecules are measured here but nevertheless it will give a good indication of the general velocity distribution.

A plot of number of molecules (determined from the mass of tin deposited) against molecular velocity takes the form shown in Figure 2.9.

Some molecules have very low or very high velocities but most have intermediate velocities. The second curve shows the effect of raising the temperature.

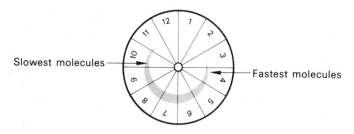

Figure 2.8 Distribution of tin in the Zartmann experiment

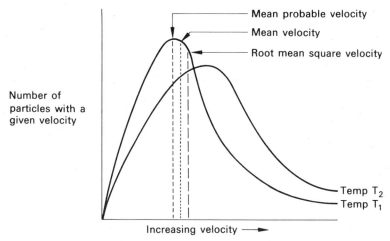

Figure 2.9 Effect of temperature on the distribution of molecular velocities

There is a general shift to higher velocities and a considerable increase in the number of molecules having very high velocities. The flattening of the peak indicates a wider distribution of velocities.

Real or non-ideal gases

It was stated above that the kinetic theory of gases, and hence the gas laws and gas equation, are concerned with ideal gases. Two of the major assumptions for an ideal gas are that its molecules are so far apart that intermolecular attractions are negligible and that the molecules travel a very large distance between collisions compared with their own size.

This situation will occur with gases at high temperature and/or low pressures. However, with *real gases* — that is, gases under normal conditions — this requirement is not strictly attained and deviations of up to about 2 per cent are observed. This point is illustrated by the pressure-volume relationships given in Table 2.1.

As the pressure is increased, or as the temperature is decreased, the molecules are brought closer together. Hence the deviations become greater because the molecules have definite attractions for each other and the size of the molecule is not negligible compared to the distance between collisions. The nature of the deviation depends upon how close the temperature is to the point of liquefaction.

Evidence for the mutual attraction of molecules is provided by the Joule–Thomson effect. A stream of gas at constant pressure is passed through a

Table 2.1.   Pressure–volume relationships for 1 mole of ammonia at 25 °C

Pressure/ atmosphere	Volume/litre	$P \times V$
0.250	97.72	24.43
0.500	48.80	24.40
0.750	32.45	24.34
1.000	24.34	24.34
3.000	8.05	24.15
8.000	2.925	23.40
9.800	2.360	23.10 condensation starting
9.800	0.020	0.20 condensation complete

tube, containing a porous plug, into a low pressure region so that no work is done against the atmosphere. The gas is cooler on leaving the plug than on entering, provided that the initial temperature is below a certain temperature known as the inversion temperature. The molecules come closer together on passing through the plug and the fall in temperature reflects the energy required to overcome the intermolecular attractions when the gas expands on the other side. If molecules did not attract one another, liquefaction would not be expected. These weak intermolecular forces may be due to *van der Waals forces, dipole–dipole attractions,* or *hydrogen bonding.*

Van der Waals forces arise from induced fluctuating dipoles in atoms and molecules. Thus, when one atom or molecule approaches another, the electrons of one or both are temporarily displaced owing to their mutual repulsion and polarisation occurs as illustrated in Figure 2.10. This then leads to a degree of attraction between the slightly positive end of A and the outer electrons of B. As the atoms or molecules move on, their electron clouds return to normal until their next encounter. It is apparent that van der Waals forces will increase in importance as the number of electrons in the atoms and molecules increases. The magnitude of these forces obviously varies greatly from substance to substance but an indication of the energy involved is given on page 184.

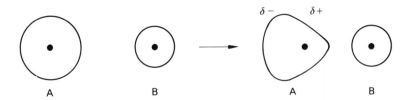

Figure 2.10   Temporary polarisation of atoms or molecules on their close approach

Dipole–dipole attractions occur between molecules which are permanently polarised, for example propanone (acetone):

$$\begin{array}{l} CH_3 \\ \!\!\!\!\!\!\!\searrow \overset{\delta+}{C} = \overset{\delta-}{O} \\ CH_3 \end{array}$$ (The delta sign indicates that it is a small charge.)

The negative end of one molecule will therefore attract the positive end of the next. The effect of this is evident when the boiling point of propanone is compared with that of butane, a compound which has the same relative molecular mass but is not polarised:

$$CH_3 \cdot CO \cdot CH_3, \text{ b.p. } 50 \,°C; \quad CH_3 \cdot CH_2 \cdot CH_2 \cdot CH_3, \text{ b.p. } 0 \,°C$$

Hydrogen bonds occur between molecules which contain a highly electro-negative atom linked covalently to a hydrogen atom. For example, water exists as clusters of molecules owing to hydrogen bonding.

These intermolecular attractions occur because the oxygen–hydrogen bonds are highly polarised, the hydrogen atoms having a high charge density as a result of their very small size. Hydrogen bonds are, in fact, a special case of dipole–dipole attractions.

At very close quarters the attractive forces give way to repulsive forces. (These are indicated by molecules having fairly definite collision diameters.) However, the repulsive forces act only over a short distance compared with the attractive forces.

Van der Waals' equation

Van der Waals tried to allow for the deviations from ideal behaviour by proposing a modified version of the ideal gas equation:

$$\left(P + \frac{a}{V^2}\right)(V - b) = RT \quad \text{for 1 mole of gas}$$

He arrived at this expression as outlined below.

The pressure of a real gas in a container is lower than that of an ideal gas. This is because molecules attract one another and those near the walls of the container tend to be pulled inwards due to there being no molecules outside them. The attractive force inwards is proportional to the number of molecules in the container, i.e. to the density. Also, the number of molecules colliding with the walls of the container at a given instant is proportional to the density of the gas. Hence the total inward force (which consequently reduces the pressure) is proportional to the square of the density of the gas. Now density is inversely proportional to volume and so the reduction in pressure arising from the inward force can be represented by the term a/V^2 where a is a constant for a particular gas at a particular temperature. This value must be added to the pressure to give the pressure an ideal gas would give.

The volume of a gas needs to be corrected to allow for the volume which the molecules themselves occupy. Van der Waals represented the 'true' volume as $(V - b)$ where b is a constant for each gas, its magnitude depending on the size of the molecules.

There are more accurate equations than the van der Waals equation but they are much more complicated and beyond the scope of this text.

Liquefaction of gases

The early work on liquefaction was done on carbon dioxide by Andrews in the eighteen sixties. He studied the effect of changing the pressure on the volume of a fixed mass of the gas at given temperatures. The results are shown in Figure 2.11.

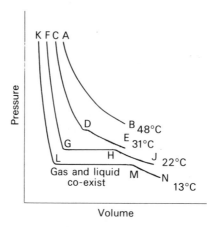

Figure 2.11   Isothermals for carbon dioxide

Each line on the graph shows the variation of volume with pressure at the temperature indicated and is known as an isothermal. At 48 °C approximately ideal behaviour occurs and so PV = constant.   At 13 °C, however, section NM is the expected curve but along ML a large decrease in volume occurs for a very small change in pressure, i.e. the gas is liquefying. Section LK shows the very small contraction in volume as the pressure is increased on the liquid carbon dioxide. As the temperature is raised the horizontal section where liquefaction occurs gets shorter and at 31 °C it is just a point. Above this temperature, liquefaction does not occur whatever the pressure. The progressively shorter horizontal sections in the isothermals up to 31 °C indicate the increasing reluctance of the gas to liquefy even though it is highly compressed.

The maximum temperature at which a gas can be liquefied is known as the *critical temperature* and the pressure required at this temperature is called the *critical pressure*. The volume occupied by one mole of a substance at its critical temperature and pressure is known as the *critical volume*.

Carbon dioxide, sulphur dioxide, and chlorine have critical temperatures above room temperature and so they can be liquefied by pressure alone. However, nitrogen, hydrogen, and oxygen have very low critical temperatures and so they have to be cooled considerably before liquefaction can be carried out.

2.   THE LIQUID STATE

If gases are cooled, their kinetic energy, and hence velocity, falls. If the cooling is continued the intermolecular attractions become progressively more important and liquefaction eventually occurs. The molecules in the liquid move throughout its bulk but they are obviously more restricted than in the vapour state. Further cooling will finally result in solidification and the molecules can then vibrate only about certain fixed positions. Molecules in liquids are therefore in an intermediate state between the order in the crystalline state and the disorder of the gaseous state.

Evidence for the continuous motion and frequent collisions in the liquid state is furnished by the Brownian movement. Colloidal particles (insoluble particles in suspension, their size being about 10^{-9} to 2×10^{-7} m whereas molecules have sizes up to about 1×10^{-10} m) observed through an ultramicroscope are seen to be in a state of continual random motion owing to bombardment with solvent molecules. Larger particles would remain almost stationary since simultaneous collisions from all sides would cancel themselves out.

A molecule in the middle of the liquid bulk will be subject to equal molecular attractions from all sides. However, a molecule near the surface will be attracted by greater forces from below than above and the net effect of this will be a pull inwards, thus accounting for the surface tension of liquids.

A few of the molecules in the liquid will have very fast speeds and, if they are near the surface, they may overcome the attractive forces and escape as vapour, i.e. evaporation will occur. If no heat is supplied, the temperature will fall since the high kinetic energy molecules are escaping. However, the vapour is not at a higher temperature than the liquid since the molecules lose a lot of their energy in overcoming the attractive forces. The average kinetic energy of the vapour molecules is the same as that of the liquid.

3. THE SOLID STATE

The arrangements of particles in solid structures are determined by X-ray diffraction and to a lesser extent by electron diffraction. These processes are based on the fact that both X-rays and electrons have wavelengths which are of the same order as interatomic distances (10^{-10} m) and so crystals may be used as diffraction gratings for them. The fundamentals of X-ray and electron diffraction are discussed below.

X-ray diffraction

X-rays are produced when cathode rays fall on metals. The effect of atomic-sized particles on them is similar to the effect produced when a handful of pebbles is thrown into a pond. In the pond a series of ripples is produced and the crests of some of them will meet and reinforce each other while some crests will meet troughs and cancel each other out. In the same way, some of the X-ray waves are reinforced and some cancelled out (Figure 2.12).

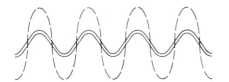

Wave crests meeting to give a reinforced wave

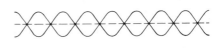

Out of phase waves cancelling each other out

Figure 2.12   Interaction of X-rays

In crystals there are regular layers of atoms or ions and it is possible to calculate the conditions under which reinforcement will occur if a beam of X-rays strikes them. When X-rays are being diffracted by the layers they behave as if they are being reflected. In Figure 2.13, two waves, which are in phase, are shown striking the first and second layers of the crystal.

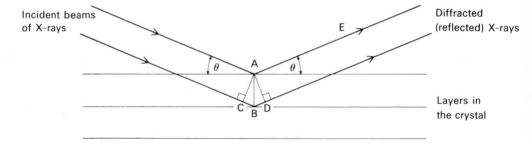

Figure 2.13   The diffraction of X-rays from the planes of a crystal lattice

Obviously, the lower wave travels a greater distance than the upper one. Now, if the diffracted waves are still to be in phase, the extra distance travelled by the lower wave (i.e. CB + BD) must be a whole number of wavelengths.

∴   For reinforcement $CB + BD = n\lambda$
     where n = any integer and λ = wavelength.
     But $CB = AB \sin \theta$ and $BD = AB \sin \theta$
∴ $AB \sin \theta + AB \sin \theta = n\lambda$

If d is used to replace AB and represent the distance between layers, then:

$$2d \sin \theta = n\lambda \qquad (2.8)$$

This equation is known as the Bragg equation.

For a given crystal face, d is fixed and, if X-rays of a definite wavelength are used, maximum reinforcement will occur only for certain values of θ corresponding to $n = 1, 2, 3,$ etc. If these angles are found experimentally, the distance d can be calculated.

The diffracted X-rays are detected by the diffraction pattern they produce on a photographic plate or by their ionising effect on a gas such as bromoethane (ethyl bromide). In the latter method, the ionised gas will pass an electric current, the strength of the current being proportional to the intensity of the X-rays.

The diffraction or scattering of X-rays when they fall on a crystal is due to the electrons around the atoms or ions, the greater number of electrons the atom or ion has, the greater the scattering. For this reason, hydrogen atoms are difficult to locate exactly.

The spots in the photograph occur in a regular pattern and so it is easy to determine the size of the unit cell (i.e. the repeating unit) in the crystal from the distance apart of the relevant spots. However, it is much more difficult to determine the exact position of the atoms or ions since the photograph is a two-dimensional picture of a three-dimensional structure. Complex calculations are required using not only the distance apart of the spots but also their relative intensities and this often necessitates the use of computers.

X-ray diffraction is restricted to solids but single crystals or fine powder may be used. It can give the arrangement of molecules or ions in the crystal and the distance between them. Bond lengths, with the exception of those involving hydrogen (very low electron density), can be measured to within $\pm 1 \times 10^{-3}$ nm (1nm = 10^{-9} m).

Electron diffraction

This is most appropriate to gases. If a beam of electrons is passed through a gas at low pressure and then allowed to fall on a photographic plate, a series of concentric rings is formed on the plate (Figure 2.14). The beam is diffracted by the atoms in the molecules of the gas, the pattern obtained depending on the molecular structure. The rings in the diffraction pattern correspond to interatomic distances in the molecules.

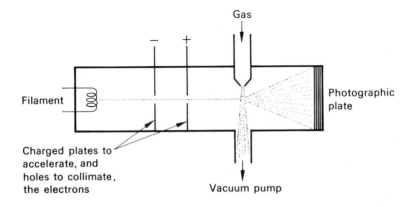

Figure 2.14   The basis of electron diffraction

The diffraction pattern is compared with that calculated for a likely structure. The postulated structure is modified until the calculated and observed patterns agree.

Electrons have poor powers of penetration and so, with solids, the incident beam must just skim the surface.

Electron diffraction yields information about bond angles and bond lengths, the latter being determined to within $\pm 3 \times 10^{-4}$ nm except where bonds to hydrogen atoms are concerned. Hydrogen atoms can nevertheless be located more accurately by this method than by X-ray diffraction, the error being within $\pm 10^{-3}$ nm.

Structure of metal crystals

Atoms and ions in solid structures may be considered as hard incompressible spheres. Metallic crystals generally consist of a three-dimensional arrangement of spheres occupying the minimum volume and so the spheres (atoms) are said to be close packed. Close packing in a layer takes the form shown in Figure 2.15(a) not (b) since in the former 60.4 per cent of the available space is utilised whereas in the latter only 52.4 per cent is used.

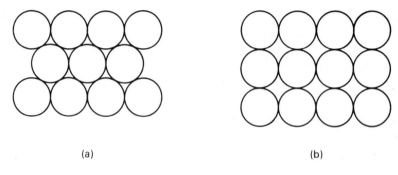

(a)   (b)

Figure 2.15   Efficient and inefficient packing of spheres

Figure 2.16 shows the first layer, A, of a metallic crystal and how the second layer, B, fits into the hollows of A.

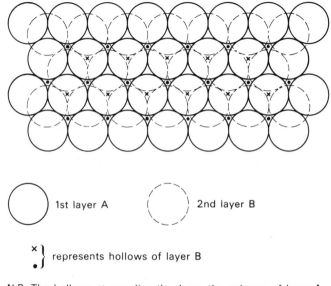

1st layer A   2nd layer B

$\left. \begin{array}{c} \times \\ \bullet \end{array} \right\}$ represents hollows of layer B

N.B. The hollows × are directly above the spheres of layer A.
      The hollows • are directly above hollows in A

Figure 2.16   Packing arrangements in metallic lattices

It is seen that there are two types of hollow in layer B, one type being marked by crosses and the other type by dots. Those marked by a cross are directly above spheres in row A and so a third row filling these positions would be identical to A. The fourth row is formed by repeating row B and so there is a repeating pattern ABAB; this is known as *hexagonal close packing*.

Figure 2.17 shows the *unit cell* of hexagonal close packing derived from three rows of spheres in ABAB sequence. (Just as patterned wallpaper is made by repeating a unit pattern two-dimensionally so crystals are made up by repeating the unit cell almost an infinite number of times three-dimensionally.)

As stated above, Figure 2.16 shows that there are two types of hollow formed by row B. As well as those marked by crosses and already discussed, there are those marked by dots which are over hollows of row A. If the third row, C, takes up these positions then an ABCABC repeating pattern is obtained and this is known as *cubic close packing*. The unit cell derived from four layers of cubic

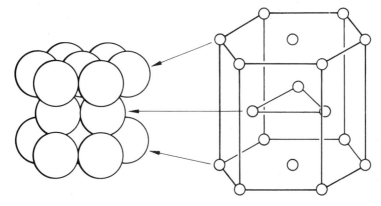

Figure 2.17 Hexagonal close packing, unit cell

close packing is shown in Figure 2.18; it is often known as a *face centred cube* since there is a sphere in the centre of each face of the cube. Figure 2.19 attempts to show how the unit cell is situated in four layers of the lattice.

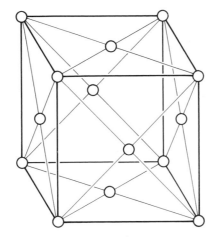

Figure 2.18 Cubic close packing, unit cell (face centred cube) — exploded view

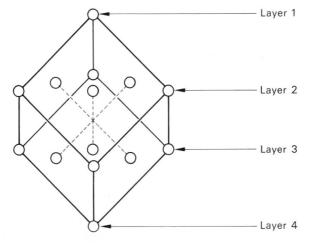

Figure 2.19 The position of the face centred cube in the cubic close packing arrangement

The following points may be made regarding close packed structures.

(a)   Any particular sphere is in contact with three spheres in the layer above and three in the layer below as well as the six spheres in the same layer. It therefore has twelve nearest neighbours and it is said to have a *co-ordination number* of twelve.

(b)   In both hexagonal and cubic close packing, 74% of the available volume is occupied.

(c)   Irregular arrangements of close packed layers such as AB AB ABC AB are not encountered.

A further, less efficient, mode of packing is encountered particularly with the alkali metals. This is known as *body centred cubic packing* and it utilises only 68% of the available space. Compact and exploded views of the unit cell are shown in Figure 2.20. In this structure, each sphere has eight nearest neighbours and so the co-ordination number is said to be eight.

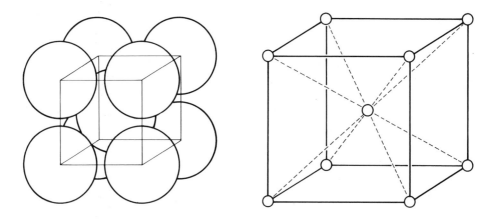

Figure 2.20   Body centred cubic packing, unit cell – compact and exploded views

Almost all metals crystallise into one or more of these systems. Some examples are:

Hexagonal close packing	Mg Zn Ni
Cubic close packing	Cu Ag Au Al
Body centred cubic packing	Li Na K Rb Cs

The type of packing a metal crystal adopts is important since it affects such physical properties as melting point, boiling point, hardness, density, etc. Three of the most malleable metals known are copper, silver, and gold and they are all cubic close packed. The malleability results from the fact that cubic close packed crystals contain more planes of atoms to allow deformation.

The type of crystal structure may alter with temperature, for example, iron is body centred cubic packed up to 906 °C but cubic close packed between 906   and   1401 °C.

Ionic packing

Three main factors determine the type of packing in ionic crystal lattices.

(a) Close packing considerations. As with metals, the spheres (in this case ions) attempt to form a condensed structure. Many ionic crystals may be described as close packed but the oppositely charged ions must be considered separately. The sodium chloride lattice may be regarded as two inter-penetrating cubic close packed lattices, one of Na^+ ions and the other of Cl^- ions. Figure 2.21 illustrates the sodium chloride lattice and a more obvious description is apparent, that is, a cubic lattice containing alternate Na^+ and Cl^- ions.

   Both the Na^+ and Cl^- ions have six oppositely charged ions as their nearest neighbours and so the co-ordination is referred to as $6:6$.

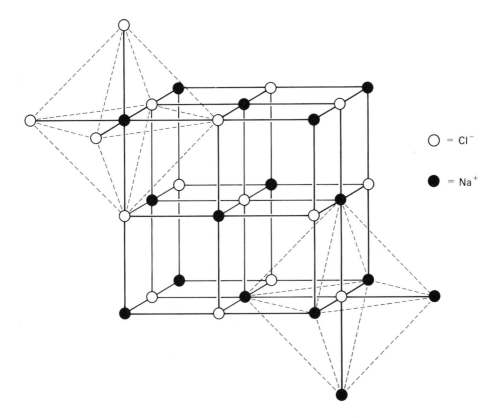

$\bigcirc = Cl^-$

$\bullet = Na^+$

Figure 2.21   Sodium chloride lattice showing the 6 : 6 co-ordination

(b) The relative size of the ions. The ionic radii of Na^+, Cs^+, and Cl^- are 0.098, 0.167, and 0.181 nm respectively and so it is apparent (Figure 2.22) that, if caesium chloride had the same type of lattice as sodium chloride, the packing would be much less efficient.

Caesium chloride therefore adopts a more compact structure (Figure 2.23) but even this is not close packed. Each ion has eight oppositely charged ions as nearest neighbours and so the co-ordination is $8:8$. The lattice is not called a body centred cubic lattice because not all the ions in the cube are the same.

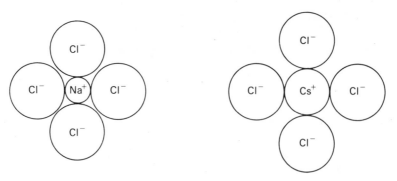

Figure 2.22   An illustration of the wastage of space which would occur if caesium chloride had the sodium chloride type of lattice

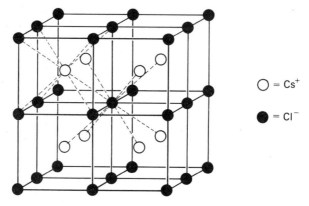

Figure 2.23   Caesium chloride lattice showing the 8 : 8 co-ordination

The radius ratios for the sodium and caesium chloride lattices are discussed on the next page.

(c)   The degree of covalent character in the bond. Some compounds have bonds which are intermediate in character and two important types will be considered.

(i)   Binary 1 : 1 type compounds, e.g. zinc sulphide. This compound is said to be polymorphic which means that it exists in more than one crystalline form. (Allotropy refers only to elements.) In fact, two forms exist and these are known as zinc blende and wurtzite. In both of these forms, each zinc ion is surrounded tetrahedrally by four sulphide ions and *vice versa* (Figure 2.24) and so 4 : 4 co-ordination occurs.

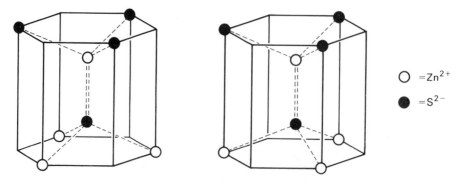

Figure 2.24   Zinc blende and wurtzite lattices respectively

(ii)   Binary 1 : 2 type compounds, e.g. calcium fluoride (fluorite). In this structure, each calcium ion is surrounded by eight fluoride ions at the corners of a cube and each fluoride ion is surrounded tetrahedrally by four calcium ions and so 8 : 4 co-ordination occurs (Figure 2.25).

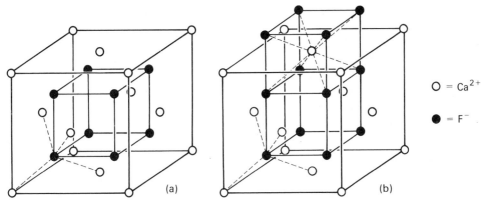

$\bigcirc = Ca^{2+}$

$\bullet = F^-$

Figure 2.25   The calcium fluoride lattice: (a) the face centred cube of Ca^{2+} ions with the interlocking simple cube of F^- ions and (b) an additional layer of F^- ions to illustrate the co-ordination of the Ca^{2+} as well as the F^- ions

Radius ratios

Consider an ionic substance A^+B^- with the sodium chloride type of lattice. In any layer of the lattice, each positive ion is surrounded by four negative ions as illustrated in Figure 2.26.

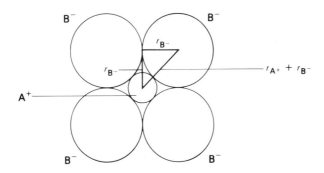

Figure 2.26   Calculation of limiting radius ratio in sodium chloride type lattices

The structure will be at its most stable when the cation fits exactly between the anions and the radius ratio for this limiting case can be calculated as follows.

From the diagram,
$$\cos 45° = \frac{r_{B^-}}{r_{A^+} + r_{B^-}}$$

$$0.7072 = \frac{r_{B^-}}{r_{A^+} + r_{B^-}}$$

$$0.7072\, r_{A^+} + 0.7072 r_{B^-} = r_{B^-}$$

∴
$$\frac{r_{A^+}}{r_{B^-}} = \frac{0.2928}{0.7072} = 0.414$$

As the ratio increases, i.e. as r_{A^+} gets larger in comparison with r_{B^-}, the repulsive forces between the electrons of the ions gradually outweigh the attractive forces and so the stability decreases. When the r_{A^+}/r_{B^-} ratio reaches 0.732, the sodium chloride type lattice with co-ordination number six gives way to the caesium chloride type lattice with co-ordination number eight since the ions are then 'in contact' again. The limiting value of 0.732 for the caesium chloride lattice is obtained from trigonometrical calculations.

Calculation of the Avogadro constant from unit cell measurements

X-ray diffraction can be used to determine the dimensions, and the number of ions, in the unit cell of a crystal. From this and a knowledge of the density of a solid, it is possible to calculate a value for the Avogadro constant, N, as shown in the example below with sodium chloride.

The unit cell of sodium chloride is a cube with sides of 5.641×10^{-10} m. Reference to Figure 2.21 shows that the cell has:

$$\tfrac{1}{8} \text{ share of each } Na^+ \text{ at the corners of the cube} = 1 \; Na^+$$

$$\tfrac{1}{2} \text{ share of each } Na^+ \text{ in the faces of the cube} = 3 \; Na^+$$

$$\tfrac{1}{4} \text{ share of each } Cl^- \text{ along the edges of the cube} = 3 \; Cl^-$$

$$1 \text{ unshared } Cl^- \text{ in the centre of the cube} = 1 \; Cl^-$$

The unit cell therefore contains the equivalent of four Na^+ and four Cl^- ions.

$$\text{Now,} \quad \text{density} = \frac{\text{relative molecular mass}}{\text{volume of 1 mole}}$$

where density $= 2.165 \times 10^6$ g m^{-3}, relative molecular mass $= 58.4428$, and volume of 1 mole $= N \times \dfrac{(5.641 \times 10^{-10})^3}{4}$

$$\therefore \qquad 2.165 \times 10^6 = \frac{58.4428 \times 4}{N \times (5.641 \times 10^{-10})^3}$$

$$N = \frac{58.4428 \times 4}{(5.641 \times 10^{-10})^3 \times 2.165 \times 10^6} = 6.015 \times 10^{23}$$

Hydrogen bonding in ionic crystals

Hydrogen bonding plays an important part in the lattice structure of salt hydrates such as gypsum, $CaSO_4 \cdot 2H_2O$. Gypsum has a somewhat complicated structure in which layers of calcium sulphate(VI) are held together by hydrogen bonds with the water molecules (Figure 2.27). Consequently, the crystals cleave readily along the layers because the hydrogen bonds are much weaker than the ionic attractions. On the other hand, anhydrite, $CaSO_4$, is resistant to cleavage because it has no hydrogen bonds or inherently weak planes.

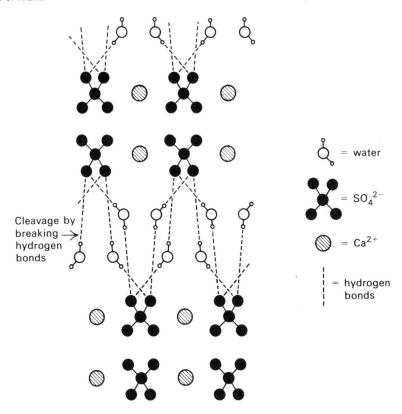

Figure 2.27   Diagrammatic representation of the gypsum, $CaSO_4 \cdot 2H_2O$, lattice

Van der Waals forces in ionic crystals

Van der Waals forces make a contribution to the stability of the lattices of compounds containing polyatomic ions. Calcite, $CaCO_3$, anhydrite, $CaSO_4$, and sodium nitrate(V) are examples where this occurs. These compounds may be regarded as having distorted sodium chloride type lattices. In the case of calcite, for example, Ca^{2+} ions take the place of Na^+ ions while CO_3^{2-} ions replace the Cl^- ions. The three oxygens in a carbonate ion are situated triangularly round the carbon and, because of the space occupied by these ions, the cubic unit cell of sodium chloride expands in a horizontal direction for calcite.

Macromolecular structures

Carbon exists in two allotropic forms — that is, diamond and graphite — and these are examples of giant three-dimensional and two-dimensional networks respectively. In diamond (Figure 2.28(a)), each carbon atom is sp^3 hybridised and is joined by covalent bonds to four others, the four bonds pointing towards the corners of a regular tetrahedron. In graphite, each carbon atom is sp^2 hybridised and a planar network of hexagons results (Figure 2.28(b)). The unpaired $2p$ orbital on each carbon atom overlaps with those on adjacent carbon atoms to form π-bonds and so these electrons are delocalised over the whole layer. (Before the technicalities of the bonding in diamond and graphite can be understood, it will be necessary to read pages 287–9.) The layers of hexagons are held together by weak van de Waals forces.

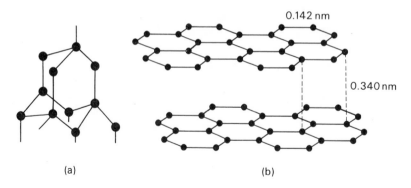

Figure 2.28   The crystal lattices of (a) diamond, and (b) graphite

Molecular crystals

Each covalent molecule in a molecular crystal is independent of the other molecules except for weak intermolecular forces. The sort of lattice formed is largely dependent on the molecular shape.

The iodine lattice (Figure 2.29) is held together by van der Waals forces as is the naphthalene lattice (Figure 2.30).

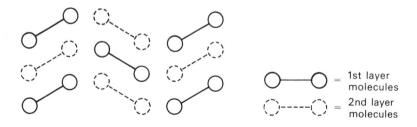

Figure 2.29   Iodine lattice – plan view of the unit cell

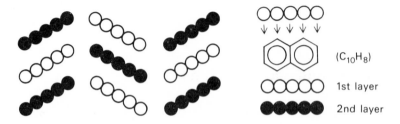

Figure 2.30   Naphthalene lattice

Ice, on the other hand, is held together by hydrogen bonds; its structure is variable under certain conditions but it tends to have the wurtzite structure (page 46). The unit cell of the wurtzite form is shown in Figure 2.31. The orientation of the water molecules is arbitrary but there is one hydrogen along each oxygen–oxygen axis. Since the hydrogen bond length is 0.177 nm compared to the 0.096 nm for an O—H bond, a very open structure results and consequently ice has a low density.

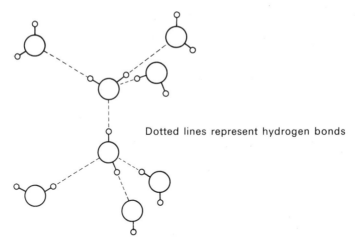

Dotted lines represent hydrogen bonds

Figure 2.31   The ice lattice

Relationship between physical properties and crystal structure

Metallic crystals

Metals are malleable and ductile due to the mobility of the valency electrons. Thus, one plane of metal atoms may slip over another but the electrons can move easily between the planes to maintain the bonding. The strength of metallic bonds increases as the number of valency electrons and the nuclear charge increases. This may be illustrated by the enthalpies (latent heats) of fusion of say sodium, magnesium, and aluminium which are 2.60, 8.95, and 10.75 kJ mol^{-1} respectively.

Ionic crystals

Ionic crystals are generally brittle because any movement of the planes of ions will result in repulsion between them (Figure 2.32). The crystals are hard and have high melting and boiling points because many electrostatic attractions must be overcome to break down the lattice. The higher the charge on the ions, the higher the melting and boiling points; for example, Na_2O, MgO, and Al_2O_3 have melting points of 920, 2900, and 2980 °C respectively.

Attraction of planes   Repulsion

Figure 2.32   The reason for the brittleness of ionic crystals

Macromolecular crystals

Diamond is very hard because each atom is held firmly in position by covalent bonds to four other atoms. It follows that its melting and boiling point will also be high because many covalent bonds must be broken to destroy the crystal lattice.

Graphite has a high melting and boiling point for similar reasons. It has high electrical conductivity along the layers, due to the delocalised electrons, but poor conductivity in a direction perpendicular to the layers because electrons cannot pass between them. The layers are bonded together only by weak van der Waals forces and so graphite is soft and slippery.

Molecular crystals

These are soft, low melting, and low boiling crystals because the molecules are held together only by weak attractions such as van der Waals forces or hydrogen bonding.

Questions

2.1   Calculate the volume occupied by a mixture of 4 g methane and 35.5 g chlorine at 750 mm and 18 °C.

2.2   (a)   Outline the assumptions made in deriving the equation $PV = RT$ for a mole of an ideal gas. Explain why a gas such as chlorine does not strictly obey the equation. When are the deviations at a minimum?

    (b)   When 5.35 g ammonium chloride is heated to 350 °C it vaporises and occupies a volume of 10.5 litres at 740 mm pressure. What may be deduced from this information?

2.3   State Dalton's law which enables one to calculate the total pressure for a mixture of gases. Explain how this law is consistent with the kinetic theory of gases.

A mixture of nitrogen and hydrogen, in the ratio of one mole of nitrogen to three moles of hydrogen, was partially converted into ammonia, so that the final product was a mixture of all these three gases. The mixture was found to have a density of 0.497 g litre^{-1} at 25 °C and 1.00 atm. What would be the mass of gas in 22.4 litres at s.t.p.? Calculate the percentage composition by volume of this gaseous mixture.

[J.M.B.]

2.4   When 0.1204 g bromobutane were volatilised in a Victor Meyer apparatus, 22.1 cm^3 of air was displaced. The atmospheric pressure and room temperature were 745 mm and 17 °C respectively. Calculate the relative molecular mass of bromobutane. The vapour pressure of water is 15 mm at 17 °C.

2.5   When 0.1357 g of an alkene was vaporised in a Victor Meyer apparatus, 35.8 cm^3 of air at 12 °C and 756 mm was displaced. Calculate the relative molecular mass of the alkene. The vapour pressure of water is 11 mm at 12 °C.

2.6   The following results were obtained in a relative molecular mass determination of trichloromethane by the Dumas method:

Weight of bulb full of air	53.200 g
Weight of bulb and vapour	53.806 g
Temperature of sealing	100 °C
Atmospheric pressure	753 mm
Room temperature	19 °C
Capacity of bulb	237.0 cm^3

Calculate the relative molecular mass of trichloromethane from these results. A litre of air at s.t.p. weighs 1.293 g.

2.7 (a) Using the molecular kinetic theory explain the following:

  (i) a substance is usually less compressible in the solid state than in the liquid state,

  (ii) a solid usually expands slightly on melting,

  (iii) a substance is much more compressible in the gaseous state than in the liquid state,

  (iv) a volatile liquid cools when a stream of inert gas is passed through it.

(b) The following table gives the vapour pressure of propanone at various temperatures:

Vapour pressure/mm Hg	40	100	400	760
Temperature T/K	263.8	280.9	312.7	329.7

Make a suitable plot of lg (v.p. in mm Hg) against $10^3/T$.

Use your graph to find the vapour pressure of propanone at 25 °C.   [J.M.B.]

2.8 The kinetic theory leads to the equation

$$PV = \tfrac{1}{3} nm\overline{c^2} \quad \text{for an ideal gas.}$$

(a) What do the letters P, V, n, m and $\overline{c^2}$ in this equation stand for? give one consistent set of units in which they may be expressed.

(b) Draw a rough sketch to show how the speeds of the molecules of a gas are distributed at some temperature, T_1. On the same sketch, show the distribution of the speeds at some higher temperature, T_2. Discuss briefly the bearing this shift in distribution has on the rates of reactions of gases. [The final part of this will require a knowledge of Chapter 8.]

(c) State Avogadro's law and show that it necessarily follows for an ideal gas from the equation above.

(d) Real gases deviate to a greater or lesser extent from 'ideal' behaviour. Discuss qualitatively those assumptions made in deducing the ideal gas equation which are not true for real gases and the effect they have on the behaviour of the latter.

Under what conditions does the behaviour of a gas like carbon dioxide tend to that of an ideal gas?   [University of London]

2.9 (a) State the assumptions of the kinetic theory of gases which correctly describe an ideal gas.

(b) Which of these assumptions are invalid for a non-ideal gas?

(c) Under what conditions may it be said that all gases depart from ideal behaviour?

(d) Describe, using the molecular kinetic theory, what happens when iodine sublimes.

(e) Explain why the molar enthalpies of vaporisation of liquids are very similar to enthalpies of sublimation but both are appreciably larger than the enthalpies of fusion.

2.10 Give a concise description of an experiment which demonstrates the phenomenon of gaseous diffusion.

State Graham's law of gaseous diffusion.

A specimen of N_2O_4, partially dissociated into NO_2, was found to diffuse at exactly one-fifth of the speed of hydrogen under the same conditions. Calculate (a) the relative density of the gas mixture, (b) the degree of dissociation of the dinitrogen tetroxide, and (c) the percentage by volume of dinitrogen tetroxide in the gas mixture. (It may be assumed that the composition of the gas mixture remains constant during the measurement of the rate of diffusion.)   [J.M.B.]

2.11 State Graham's law of diffusion of gases. A gas-jar filled with nitrogen is placed on top of a gas-jar filled with chlorine, the two gases being separated by a partition. Describe and explain what happens when the partition is gently removed. How does this behaviour differ from that of a similar experiment where the jar of chlorine is above the jar of nitrogen? Give reasons for the differences.

A certain volume of sulphur dioxide diffuses through a porous plug in 2 min. 40 sec. The same volume of a gaseous compound of bromine diffuses under the same conditions in 3 min. Deduce what bromine compound is being used. Suggest TWO tests which could be used to confirm this conclusion.

[J.M.B.]

2.12 (a) State what is understood by the term 'diffusion', and explain how Graham's law can be used to measure the relative values of the mean molecular velocities in two gases.

(b) Sketch a graph, with appropriately labelled axes, illustrating the distribution of molecular velocities in a gas. Describe how this distribution changes with an increase in temperature.

(c) Describe the differences between molecular motion in gases and liquids. Show how the distribution of molecular velocities in a liquid can be used to account for the observed variation of the vapour pressure of a liquid with temperature.

[J.M.B.]

2.13 Discuss the assumptions on which the kinetic theory of ideal gases is based and state the equation which governs the behaviour of such gases. The van der Waals equation for one mole of a gas,

$$(P + \frac{a}{V^2})(V - b) = RT \qquad (a \text{ and } b \quad \text{are constants})$$

is often used to describe the behaviour of real gases. Suggest reasons for the terms a/V^2 and b in this equation.

Explain why, under the same conditions, different gases diffuse at different rates. A wad saturated with aqueous ammonia is placed at one end of a narrow tube 1 m long and one saturated with concentrated hydrochloric acid at the other. Describe and explain what happens.   [J.M.B.]

2.14 Describe how X-rays can be used to show that there is order in the crystalline state. Explain how the order arises in crystals of (a) sodium chloride, (b) naphthalene, (c) diamond, and (d) a metal; show the effect upon the physical properties in each case.   [J.M.B.]

2.15 Describe and illustrate the crystal lattices of (a) naphthalene, (b) a metal, (c) sodium chloride, and (d) graphite.

In each case, explain the type of bonding involved in the lattice and the effect this has on
(i) the melting point,
(ii) the hardness, and
(iii) the ability to conduct heat and electricity.

2.16 Briefly outline the principles involved in one of the methods for determining the arrangement of atoms or ions in solid structures. Describe and illustrate two types of lattice adopted by metals.

2.17 What is meant by (a) a unit cell, and (b) co-ordination number?
Illustrate your answer by reference to a specific example.
Briefly describe the three types of intermolecular attractions which may be encountered in molecular crystals.

2.18 (a) Explain how X-rays are diffracted by crystals. What information about the crystal can be obtained from X-ray diffraction? Why cannot light be used for the same purpose?
The first reflection from a face of a crystal of iron occurs at an angle of 15° 35′ between the face and the X-rays. The wavelength of the X-rays responsible for this reflection was 1.54×10^{-10} m. Calculate the spacing of the planes responsible for this.

(b) The diagram shows a cross-section of a crystal of sodium chloride.

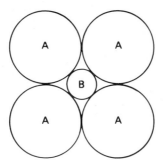

Identify the positive and negative ions giving the reasons for your choice. What is the co-ordination number of the ions in this crystal? As the value of

$$\frac{\text{radius of positive ion } (r_+)}{\text{radius of negative ion } (r_-)}$$

is reduced, the crystals which have this type of structure become more stable until a limiting value is reached. Explain the increase in stability and draw the cross-section of the limiting case. Calculate the value of r_+/r_- for the limiting case. [J.M.B.]

2.19   Describe the principles of a method which is used to determine the arrangement of atoms in a crystal of a metal.

Explain what is meant by *face centred cubic, body centred cubic,* and *hexagonal close packing.*

Suggest briefly how measurements of interionic distances in a crystal of sodium chloride could be used to find the Avogadro constant.

[University of London]

3 Electronic Theory and Chemical Bonding

The plot of ionisation energy (see page 163) atomic number (Figure 3.1) shows that the noble gases have particularly stable structures.

This stability of structures with eight electrons in the outer shell, or two in the case of the first shell, is confirmed by the successive ionisation energies of potassium (Figure 3.2). It is not surprising, therefore, that atoms attempt to

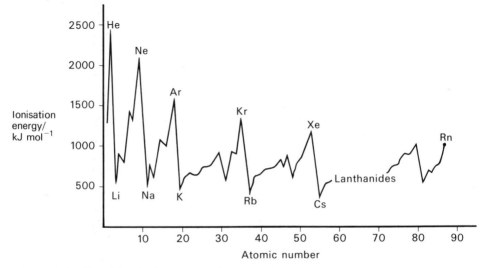

Figure 3.1 Plot of first ionisation energies of the elements

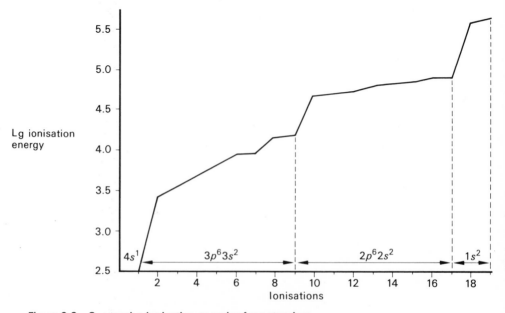

Figure 3.2 Successive ionisation energies for potassium

achieve the noble gas structure. This can be done by the formation of ions or covalent bonds.

Formation of ions

Atoms become ions by gaining or losing electrons and so electrons are transferred until the outer electron shell is identical to that of the nearest noble gas, e.g.

$$\text{Li} \ + \ \text{Cl} \longrightarrow \text{Li}^+ \ + \ \text{Cl}^-$$
$$\begin{array}{cccc} 2.1 & 2.8.7 & 2 & 2.8.8 \\ & & \text{Helium structure} & \text{Argon structure} \end{array}$$

$$\text{Mg} \ + \ 2\,\text{F} \longrightarrow \text{Mg}^{2+} \ + \ 2\,\text{F}^-$$
$$\begin{array}{cccc} 2.8.2 & 2.7 & 2.8 & 2.8 \\ & & \text{Neon structure} & \text{Neon structure} \end{array}$$

However, not all cations have the noble gas configuration, for example, the electronic configuration of the Cu^{2+} ion is $1s^2\,2s^2\,2p^6\,3s^2\,3p^6\,3d^9$.

The ions are held together by electrostatic attraction, the lattice energy (see Born–Haber cycles, Chapter 4) of the crystal being a measure of the strength of this attractive force. It is evident from the discussion on ionic lattices (Chapter 2) that ionic solids are not merely an assemblage of ion pairs. In sodium chloride, for example, a Na^+ ion is no more or less bonded to a particular Cl^- ion than it is to any of its other five nearest neighbours.

The tendency to form ions is greatest in Groups I, II, VI, and VII since atoms will form ionic compounds only if the energy evolved in the formation of the ionic crystal lattice exceeds the energy required to form the ions. So much energy is required to remove or add three electrons that Group III and V atoms do not often form M^{3+} or M^{3-} ions respectively. Group IV elements do not form simple M^{4+} or M^{4-} ions although M^{4+} can occur in complex ions, e.g. H_2PbCl_6 (see Chapter 10).

Ionic radii are discussed in Chapter 9.

Covalent bonds

Atoms may also attain the noble gas structure by sharing electrons to give covalent bonds, for example:

$$2\text{F} \longrightarrow \text{F—F}$$
$$\begin{array}{ccc} 2.7 & 2.8 & 2.8 \end{array}$$

Each atom in the molecule has, in the outer shell, six electrons to itself and a share of the two electrons in the bond between them, i.e. they have the neon electronic structure.

Similarly, hydrogen chloride is covalent:

$$\text{H} \ + \ \text{Cl} \longrightarrow \text{H} \text{———————} \text{Cl}$$
$$\begin{array}{cccc} 1 & 2.8.7 & 2 & 2.8.8 \\ & & \text{Helium structure} & \text{Argon structure} \end{array}$$

The atoms in the molecule are held together by the attractive force of the two nuclei for the negative charge cloud (the bonding electrons) between them.

Generally, atoms share electrons so as to achieve the stability of the noble gas electronic structure but they are not always successful. For example, in boron trifluoride the boron has only six electrons in its outer shell:

$$B + 3F \longrightarrow B \underset{\diagdown F}{\overset{\diagup F}{\displaystyle{-}F}}$$

$$2.3 \quad 2.7 \qquad 2.6 \; 2.8$$

Covalent radii are discussed in Chapter 9.

It has been stated that covalent bonds are formed simply by sharing electrons but a brief look at this from the orbital point of view has distinct advantages. For instance, it can explain the reactions of unsaturated organic compounds, the formation of PF_5 and SF_6, etc. and also the existence of odd electron molecules.

From an orbital viewpoint, covalent bonds are formed when atomic orbitals of different atoms overlap and, since two atomic orbitals are involved, it is not surprising that two molecular orbitals result.

Considering the case of two hydrogen atoms forming a hydrogen molecule, it is seen that this is extremely exothermic (enthalpies or heats of reactions are discussed in Chapter 4):

$$H + H \rightarrow H_2; \qquad \Delta H = -436 \; kJ \, mol^{-1}$$

This indicates that one of the molecular orbitals is more stable (i.e. has lower energy) than the atomic orbitals of the hydrogen atoms and so the other must be less stable by a corresponding amount. This is illustrated by Figure 3.3.

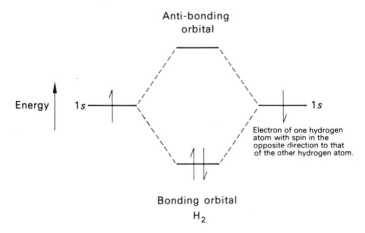

Figure 3.3   Formation of two molecular orbitals from two atomic orbitals

The two electrons in the bonding orbital are equivalent to a shared pair of electrons, i.e. a covalent bond. The *bond order* is said to be one. The correlation between the shapes of the various orbitals is illustrated in Figure 3.4. The bonding molecular orbital corresponds to an increase in electron density between the two nuclei thus reducing the electrostatic repulsion between them. The reverse is true for the anti-bonding molecular orbital.

A similar diagram may be used to illustrate the bonding in any simple system which just involves $1s$ orbitals, for example, the H_2^+ ion. The situation here is represented by Figure 3.5.

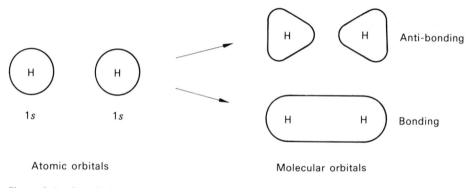

Figure 3.4 Correlation between atomic and molecular orbitals

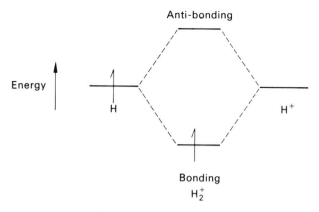

Figure 3.5 Formation of the H_2^+ ion

Again the atomic orbitals have overlapped to give a bonding and an anti-bonding molecular orbital. The fact that there is only one electron in the bonding orbital is unimportant; the bond order is simply a half. Since the ion contains one unpaired electron, it is paramagnetic. (A spinning electron behaves as a small magnet. Thus, a substance containing unpaired electrons behaves as if it contained many small magnets and consequently it is weakly attracted by a magnetic field. This phenomenon is known as *paramagnetism*. If two electrons are paired, the effect of one is cancelled out by that of the other. Hence, substances in which all electrons are paired are not attracted by magnetic fields and are said to be *diamagnetic*. A diamagnetic substance is, in fact, weakly repelled by a magnetic field owing to a very small induced magnetic moment which acts in the opposite direction to the applied field.)

For a hypothetical He_2 molecule the situation would be represented by Figure 3.6. In this case, any stability gained by the bonding orbital is cancelled out by the electrons having also to utilise the anti-bonding orbital. He_2 molecules, therefore, do not exist because their formation is energetically unfavourable.

Returning to the case of hydrogen, it is apparent that the ionisation energy (page 163) for a hydrogen molecule is greater than that for a hydrogen atom (Figure 3.7), the difference being the stabilisation energy of the bonding molecular orbital.

The structures of organic compounds in terms of orbital overlap are discussed in Chapter 14.

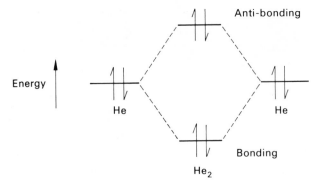

Figure 3.6   Location of electrons in a hypothetical He_2 molecule

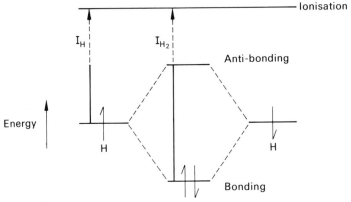

Figure 3.7   Ionisation energy of a hydrogen atom and molecule

Co-ordinate (dative covalent) bonds

These are a special case of covalent bond, the distinguishing feature being that both electrons for the bond come from the same atom. Although the mode of formation is different, covalent and co-ordinate bonds are identical once they have been formed. Examples are

$$H-\overset{..}{\underset{..}{O}}-H \; + \; H^+ \longrightarrow H-\overset{\overset{\textstyle H}{|}}{\underset{..}{O}}{}^+-H$$

$$H-\overset{\overset{\textstyle H}{|}}{\underset{\underset{\textstyle H}{|}}{N}}: \quad + \; BF_3 \longrightarrow H-\overset{\overset{\textstyle H}{|}}{\underset{\underset{\textstyle H}{|}}{N}}{}^+-\bar{B}F_3$$

Co-ordinate bonds are sometimes indicated by an arrow to show that they have been obtained differently, the arrow head pointing away from the donor atom; for example, $H_3N \longrightarrow BF_3$.

Similarities between ionic and covalent bonding

The main difference between ionic and covalent bonds is the distribution of the electrons. This may be illustrated by X-ray diffraction of solid compounds, the results of which can be obtained as a straightforward series of light and dark patches on a photographic plate or as an electron density contour map showing the intensity of the electron cloud round the atoms involved. In the case of ionic solids, e.g. sodium chloride, the map takes the form shown in Figure 3.8(a) and it is seen that the electron intensity falls to zero between the ions.

Theoretical calculations for the hydrogen molecule ion, H_2^+ (chosen because of its simplicity), give a map of the form shown in Figure 3.8(b). It is seen here that the electron density is highest between the two atoms and decreases outwards. In other words, the electrons in a covalent bond are concentrated along a line joining the two nuclei.

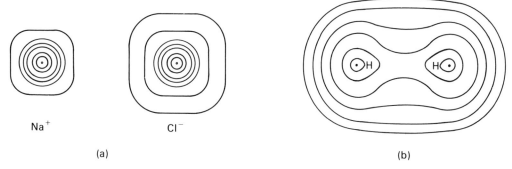

Na$^+$ Cl$^-$

(a)   (b)

In each case contour values decrease from the centre outwards

Figure 3.8   Electron density map for (a) sodium chloride and (b) the H_2^+ molecule ion

In ionic compounds the oppositely charged ions attract one another whereas, in covalent compounds, the positively charged nuclei attract the electrons shared by the atoms concerned. A comparison (Table 3.1) of bond enthalpies (energies) of formation of covalent compounds and the lattice energies of simple ionic compounds shows that they are of the same order of magnitude (bond and lattice energies are discussed in Chapter 4):

Table 3.1.   Comparison of bond enthalpies of formation and lattice energies

Bond enthalpies of formation/kJ mol^{-1}		Lattice energies/kJ mol^{-1}	
C–H	−413	NaCl	−781
O–H	−464	KBr	−679
C–C	−346	KI	−643
N–H	−389		

Intermediate types of bond

It is now necessary to consider whether bonds are purely ionic or purely covalent or if there are intermediate types of bonding. In fact, intermediate types of bond may occur from two processes.

a.   Polarisation of ions

By assuming that ions are separate spherical particles with their charge distributed uniformly round them, it is possible to calculate a theoretical value for the lattice energy of an ionic compound. In the case of the sodium and potassium halides the assumptions made in these calculations are obviously justified since the theoretically calculated lattice energies agree to within 2% (Table 3.2) of the experimental figures obtained from Born–Haber cycles.

Table 3.2.   Comparison of calculated and experimental values of lattice energy

Compound	Theoretical value/$kJ\,mol^{-1}$	Experimental value/$kJ\,mol^{-1}$
NaCl	−766	−781
NaI	−686	−699
KBr	−667	−679
KI	−631	−643

However, discrepancies arise in some cases, the silver halides and zinc sulphide being examples as shown by Table 3.3.

Table 3.3.   Discrepancies between calculated and experimental values of lattice energy

Compound	Theoretical value/$kJ\,mol^{-1}$	Experimental value/$kJ\,mol^{-1}$
AgF	−870	−943
AgCl	−769	−890
AgBr	−759	−877
AgI	−736	−867
ZnS	−3427	−3565

In these cases, it is apparent that the idea of a completely ionic bond is incorrect.

Further evidence is obtained from spectroscopic studies of the alkali metal halide vapours which show them to contain diatomic molecules with internuclear distances less than in the solid state where they are ionic, e.g.

Internuclear distances/nm

	Solid	Vapour
LiBr	0.275	0.217
LiI	0.300	0.239

Since the bonds are shorter in the vapour state they are obviously stronger and so the electron density between the nuclei is greater. The electron cloud of one or both of the ions must have been distorted; this may be represented diagramatically as in Figure 3.9.

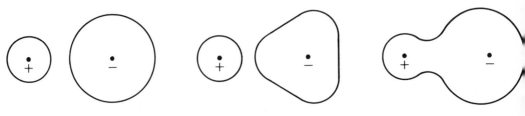

Pure ionic pair   Mild distortion or   Considerable covalent
                                    polarisation   character

Figure 3.9   Progressive polarisation in ionic compounds

The factors leading to polarisation, and so covalent character, in ionic compounds are summarised by Fajan's rules. These state that a compound will have appreciable covalent character if:

(i) either the anion or cation is highly charged (this will make the anion highly polarisable and the cation highly polarising),

(ii) the cation is small (the charge density, and hence the polarising power, will be high), and

(iii) the anion is large (the electrons are far from the nucleus and so less under its control).

Increasing polarisation of ionic bonds can readily explain a series of melting points such as:

NaCl   808 °C, $MgCl_2$ 714 °C, $AlCl_3$ 180 °C, and $SiCl_4$ – 70 °C

b.   Polarisation of covalent bonds — bond polarisation

As seen above, polarisation of ions results in some ionic compounds having a degree of covalent character and so it would also seem feasible that polarisation of covalent bonds could give them some ionic character.

The fact that polarisation of covalent bonds does occur is due to atoms having differing electronegativities or tendencies to attract electrons to themselves. It is difficult to give satisfactory numerical values to electronegatives but those derived by Pauling are often quoted. He utilises a scale from zero to four, the four being for the most electronegative element, i.e. fluorine. The values for some elements are given below.

H						
2.1						
Li	Be	B	C	N	O	F
1.0	1.5	2.0	2.5	3.0	3.5	4.0
Na	Mg	Al	Si	P	S	Cl
0.9	1.2	1.5	1.8	2.1	2.5	3.0
K						Br
0.8						2.8

The trends are clear: electronegativity increases on going from left to right across the periodic table and decreases down a group. Now, if two atoms of the same element are joined by a covalent bond, the electrons in the bond will be shared equally between them. However, if different atoms are linked the electrons will be displaced towards the atom with the highest electronegativity, that is, the bond is polarised as in, for example, hydrogen chloride $H^{\delta+}—Cl^{\delta-}$ (The δ sign — the Greek letter delta — is used in chemistry to indicate a small value, in this instance a small charge.)

It is apparent that bonds between elements with a large difference in electronegativity will be highly polarised, i.e. predominantly ionic, whilst bonds between elements with similar electronegativities will be only slightly polarised, i.e. predominantly covalent. It should be noted that pure ionic or covalent bonds are extreme cases; there are very many compounds with intermediate types of bonding.

The phenomenon of polarisation of covalent bonds may be readily demonstrated with the apparatus shown in Figure 3.10.

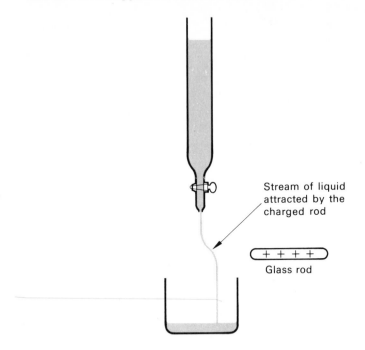

Figure 3.10   Demonstration of polarisation of covalent bonds

The glass rod acquires a positive charge by rubbing it with poly(ethene) sheet. Alternatively, an ebonite rod may be used; it becomes negatively charged when rubbed with fur. The results obtained with a series of liquids are given in Table 3.4; they are independent of the sign of the charge on the rod. The liquids which are deflected all have the common factor of being polarised and in each case a small highly electronegative atom such as F, Cl, O, or N is involved. Identical results are obtained with both rods because the charge on the rod attracts the oppositely charged end of the dipole in the liquid molecules. (A dipole is a separation of two opposite charges by a small distance.) No deflection occurs with tetrachloromethane because the molecule is symmetrical and the four dipoles cancel each other out. Benzene does not possess a permanent dipole but its π electrons (see the structure of benzene, page 306) are readily displaced and so, in an electrostatic field, it has a temporary induced dipole which subsequently causes some deflection.

Table 3.4.   Behaviour of various liquid streams towards a charged rod

Compound	$CHCl_3$ Trichloromethane	CCl_4 Tetrachloromethane	$CH_3 \cdot (CH_2)_4 \cdot CH_3$ Hexane	$CH_3 \cdot CH_2 \cdot CH_2 \cdot CH_2OH$ Butan-1-ol
Attraction	Yes	No	No	Yes

Compound	$CH_3 - \overset{\overset{\textstyle O}{\|\|}}{C} - CH_3$ Propanone	H_2O Water	C_6H_6 Benzene	$C_6H_5 \cdot CN$ Benzenecarbonitrile
Attraction	Yes	Yes	Slight	Yes

The above experiment shows qualitatively that some covalent molecules are polarised but quantitative evidence is also available. Thus, non-symmetrical polar molecules have a dipole moment, symbol p or μ, the value of which is the product of the charge (note that the fractional charges, $\delta +$ and $\delta -$ are equal in magnitude and of opposite sign) and the distance of separation. The SI unit for dipole moment is the coulomb metre, $C\,m$, and the values for some hydrides are given in Table 3.5. The fall in dipole moment down each series of hydrides reflects the decreasing electronegativity of the atoms as the Group is descended.

Table 3.5. Dipole moments of some hydrides

Molecule	p/Cm	Molecule	p/Cm	Molecule	p/Cm
NH_3	1.48	H_2O	1.84	HF	1.91
PH_3	0.55	H_2S	0.92	HCl	1.05
AsH_3	0.16	H_2Se	0.40	HBr	0.80
SbH_3	0.12	H_2Te	0.20	HI	0.42

Many molecules have zero dipole moment as a result of symmetry, examples being CCl_4, CO_2, and BF_3 (the dipole moment of a molecule is the vector sum of the individual bond moments).

Metallic bonds

Practically all metals are silvery white in colour and are bright and shiny because they reflect all frequencies of light. Light absorption and reflection are related to energy levels and so the inference is that all metals are utilising similar energy levels. Two other features that metal atoms have in common are low ionisation energies and vacant valency orbitals; these characteristics are essential for metallic bonds.

Metallic bonds are in fact rather complicated but it is thought that, whilst the atoms are held firmly in the lattice, the valency electrons move in orbitals enveloping the whole lattice. The electrons are, therefore, delocalised throughout the lattice similar in fact to the delocalisation which occurs in benzene (page 306) only more so. It is this delocalisation which results in the ability of metals to conduct electricity and heat. The malleability shows that the atoms can occupy various places within the structure without destroying the lattice. The lustre of metals is due to the fact that the delocalised electrons are in energy levels which are close together, and these can absorb light of almost all frequencies, most of it being immediately re-emitted, i.e. reflected.

Multiple bonds and delocalisation of electrons

From valency considerations, nitric(V) acid may be expected to have the structure

$$H-O-\overset{+}{N}\underset{\underset{O}{\diagdown}}{\overset{\diagup O}{\diagup}}$$

However, electron diffraction studies show that the two bonds from the nitrogen to the oxygen atoms are equal in length and so identical in character. Hence the molecule is more correctly represented as a *resonance hybrid* (note the use of the double-headed arrow);

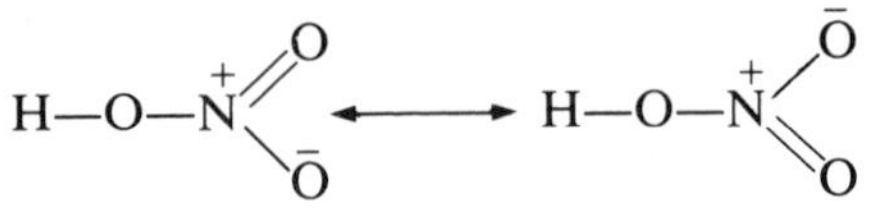

It should be understood that this does not imply continual interconversion between two forms but rather that the true structure cannot be represented by a single classical type structure. The true structure is intermediate between the two extremes shown, i.e. both bonds have partial double bond character. In other words, the two electrons initially written as making the second bond are, in fact, delocalised between the three atoms.

The three nitrogen-oxygen bonds in nitrate(V) ions all have the same length and so the electron, gained from the hydrogen of the acid, must be delocalised between the three oxygens:

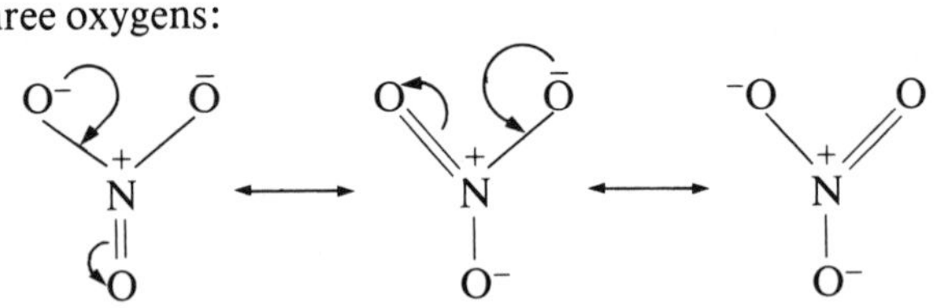

Similar delocalisation occurs in sulphate(VI) ions, all the bond lengths being the same and intermediate between those of sulphur-oxygen single and double bonds:

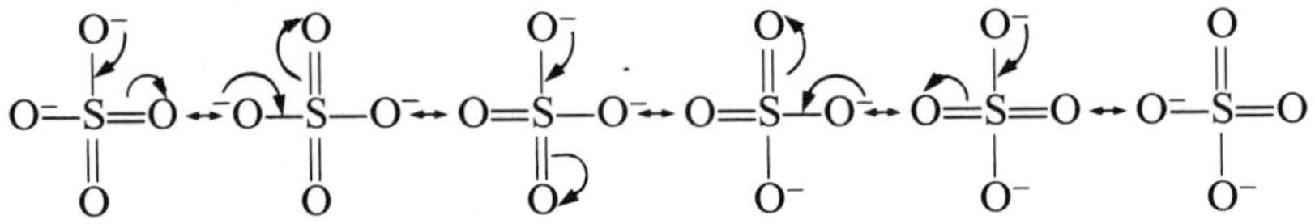

Delocalisation and the structure of organic compounds, in terms of molecular orbitals, is discussed on pages 286–9 and page 306.

Shapes of molecules

The shape of a molecule will be determined by the electron pairs in the valency shells of the atoms. These electron pairs will repel one another and try to get as far apart as possible. The shapes of simple molecules may therefore be predicted as in Figure 3.11 and these predictions are confirmed by X-ray and electron diffraction and by spectroscopic measurements.

Distortions from the expected shape

Not all electron pairs are equivalent since they may be bound to different atoms (with the result that the bond lengths are different), they may be unshared (i.e. lone pairs) or they may be involved in multiple bonds. All these factors lead to distortion of the basic shape of the molecule.

Number of electron pairs	Shape	Bond angles	Example
2	Linear	180°	$BeCl_2$
3	Trigonal	120°	BF_3
4	Tetrahedral	109.5°	CH_4
5	Trigonal bipyramidal	120° and 90°	PF_5
6	Octahedral	90°	SF_6

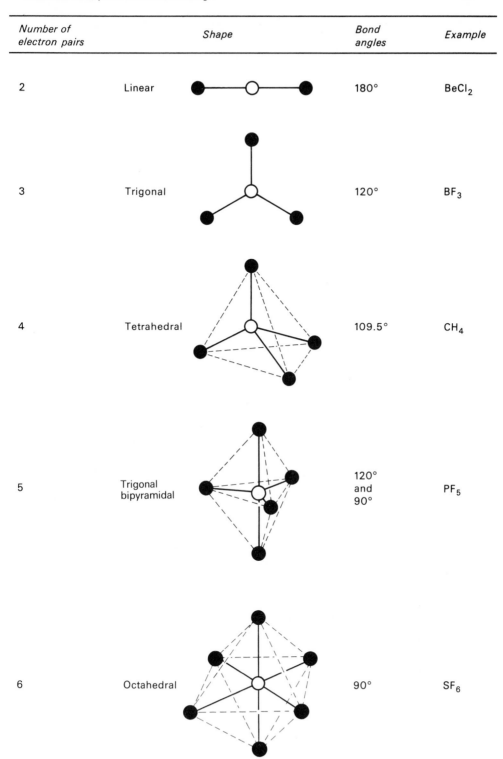

Figure 3.11   Shapes adopted by molecules owing to mutual repulsion of electron pairs

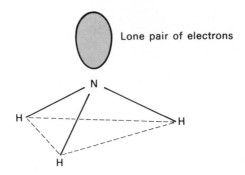

Figure 3.12   Shape of the ammonia molecule

Ammonia, for example, would be expected to be based on the tetrahedron structure since there are four electron pairs in the outer shell of the nitrogen. In fact, the H—N—H bond angles are 107.0° instead of the expected 109.5°. The deviation arises because one of the electron pairs is a lone pair, and this is closer to the nucleus than the bonding pairs, and so it repels them more strongly. It should be noted that, although the ammonia shape is based on the tetrahedron, the molecule is trigonal pyrimidal (Figure 3.12). This is not a rigid structure, however, since the molecule is continually inverting. The structure of the water molecule is based on the tetrahedron but it is more distorted since it contains two lone pairs. The bond angles are consequently reduced to 104.5°.

Questions

3.1   Discuss the three types of bonding present in ammonium chloride and the similarity between them.
  Explain why it is energetically favourable for hydrogen atoms to form H_2 molecules but unfavourable for helium to form He_2 molecules.

3.2   Write a short account on the role of the noble gases in the development of chemical theory.
  Arrange aluminium bromide, magnesium fluoride, and sodium iodide in order of increasing covalent character. Justify your arrangement.

3.3   (a)   Describe in detail the bonding which occurs in (i) calcium hydride and (ii) methane.

  (b)   Hydrogen chloride may be written as $H^{\delta+}-Cl^{\delta-}$. Discuss the bonding in this molecule and account for (i) the charge separation and (ii) the changed situation when the gas is dissolved in water.

3.4   How can the shapes of simple molecules be explained in terms of electron pair repulsions? Your answer should be illustrated by reference to five specific molecules with different shapes.
  Explain how the presence of lone pairs of electrons affect the shape of (a) ammonia and (b) water molecules.

3.5   Predict, with reasons, the shapes of the following ions:

$$NH_2^-, \quad PCl_4^+, \quad NO_3^-, \quad ICl_4^-, \quad SO_4^{2-}, \quad AlF_6^{3-}, \quad \text{and} \quad CO_3^{2-}.$$

4 Energetics

Practically all reactions, whether on a laboratory or industrial scale, involve energy changes and a study of these changes is known as *energetics*.

Before energy changes can be compared they must obviously be measured under the same conditions. Generally, an *enthalpy change* is considered: this is the heat that would be exchanged with the surroundings if the temperature and pressure of the system were the same before and after the reaction, i.e.

enthalpy change or heat of reaction =

enthalpy of products at – enthalpy of reactants at
temperature, t, and   temperature, t, and
atmospheric pressure   atmospheric pressure

or $\qquad\qquad\qquad \Delta H = H_2 - H_1$

The enthalpy change is usually determined by insulating the system from its surroundings and allowing the heat of reaction to alter the system's temperature. The amount of heat that has to be put into or removed from the system to restore it to its initial temperature (i.e. the enthalpy change) is then calculated.

Enthalpy change refers to the energy change with the surroundings at constant pressure and so includes any work done against the atmosphere — as in the case, for example, if a gas is evolved. This work will not arise if the reaction is done at constant volume; the energy change under these conditions is known as the *internal energy change*, ΔU. Internal energy change is less useful than enthalpy change since reactions are not generally performed under conditions of constant volume.

In exothermic reactions, heat is lost by the system and the enthalpy change is given a negative sign whereas, in endothermic reactions, the system gains heat from the surroundings and so the enthalpy change has a positive sign.

The first law of thermodynamics

This resembles the law of conservation of mass and may be stated as follows: energy cannot be created or destroyed although it may be changed from one form into another.

A change in the energy of a system may be brought about by the system gaining or losing heat, doing mechanical work, or having work done upon it. The symbols used for these values are q and w respectively.

The heat absorbed by a system may be utilised in two ways: (a) to increase the internal energy, U, of the system, and (b) to enable the system to carry out mechanical work on its surroundings. This may be expressed mathematically as

$$q = \Delta U + w \qquad\qquad (4.1)$$

where ΔU is the change in the internal energy. (It should be noted that q is the actual heat whereas ΔH is the heat change which would occur if the initial and final conditions were the same.) As with enthalpy, a positive sign for q indicates heat absorbed by the system. Work done by the system on the surroundings is represented as $+w$ and work done on the system by the surroundings as $-w$. Although ΔU can be estimated by physical means, absolute values of internal energy cannot be measured since it involves very complex features such as atomic and molecular crystalline forces, etc.

The work term, w, obviously involves movement and its magnitude can be assessed by considering a gas expanding in a cylinder and piston arrangement. The external pressure, P, on the piston is equal to the force divided by the area, A. Therefore

$$\text{force} = P \times A$$

and so

$$\text{work done in expansion} = P \times A \times \text{distance travelled}$$

or

$$w = P\Delta V$$

where ΔV is the change in the volume. The work term will be significant only when ΔV is large or P is large. In reactions involving gases, ΔV is often large and so w is appreciable, but liquids and solids seldom expand significantly and so the work term is negligible.

For 1 mole of gas, $P\Delta V = w = RT$

Therefore at 298 K, $\quad w = \dfrac{8.31 \times 298}{1000} = 2.5 \text{ kJ mol}^{-1}$

Changes in systems may be effected under three types of conditions.

a. Constant volume

Here $\quad \Delta V = O$
Therefore $\quad w = P\Delta V = O$
Equation (4.1) then becomes

$$q_v = \Delta U \tag{4.2}$$

the suffix $_v$ indicating constant volume conditions. This means that any heat changes are reflected as changes in internal energy.

b. Constant pressure

In most chemical reactions the process is carried out in an open beaker on the bench, the pressure remaining constant and mechanical work being done by the gas or other system expanding against the atmospheric pressure. Here $w = P\Delta V$ and equation (4.1) becomes

$$q_p = \Delta U + P\Delta V \tag{4.3}$$

the suffix p indicating constant pressure. This equation may be rewritten

$$q_p = (U_{final} - U_{initial}) + P(V_{final} - V_{initial})$$

Collecting like terms gives

$$q_p = (U_{final} + PV_{final}) - (U_{initial} + PV_{initial}) \tag{4.4}$$

and so q_p is the difference between two similar terms. The value $U + PV$ is a measure of the energy possessed by the system and is the enthalpy or heat content, H. Equation (4.4) now becomes

$$q_p = H_{final} - H_{initial} = \Delta H$$

Substituting this value for q_p in equation (4.3) gives

$$\Delta H = \Delta U + P\Delta V \qquad (4.5)$$

c. Adiabatic changes

Here no heat enters or leaves the system. This arrangement is not common but it is encountered in the bomb calorimeter. Since under these conditions $q = 0$, equation (4.1) becomes

$$\Delta U = -w \qquad (4.6)$$

Enthalpies or heats of formation, combustion, and atomisation

Consider the reaction

$$C_{(graphite)} + O_2(g) \rightarrow CO_2(g)$$

where (g) indicates the gaseous state. The enthalpy change, ΔH, is equal to -393.5 kJ mol^{-1} and this applies to the amounts shown in the equation, that is one mole each of carbon, oxygen, and carbon dioxide. Usually, *standard enthalpy changes* are quoted and these refer to standard pressure and some temperature which is generally 25 °C, i.e. 298 K; these are written as $\Delta H^{\ominus}_{298}$. The substances must be in the normal physical state applicable to these conditions and so the example above concerns the most stable allotrope of carbon, gaseous oxygen, and gaseous carbon dioxide. If solutions are being used, they must be at unit activity (for most practical purposes this approximates to one mole per litre). Obviously, if the reaction does not occur under these conditions, $\Delta H^{\ominus}_{298}$ cannot be measured directly but it can be calculated from the enthalpy change at a temperature at which the reaction does take place. (This calculation is, however, beyond the scope of this text.)

The -393.5 kJ mol^{-1} in the reaction above refers to the enthalpy change when one mole of carbon dioxide is formed in the standard state from the elements in the standard state and so this is known as the *standard enthalpy (heat) of formation* of carbon dioxide. It is represented as $\Delta H^{\ominus}_{f,298}[CO_2(g)]$. It follows from this that the enthalpy of formation of an element in its standard state must be taken as zero.

The *standard enthalpy (heat) of combustion* of a substance is represented as $\Delta H^{\ominus}_{c,298}$ and refers to the enthalpy change when one mole of it undergoes complete combustion in the standard state. In the combustion of organic compounds, for example, all the carbon would have to be converted to carbon dioxide not to the monoxide, e.g.

$$CH_3 \cdot CH_2 OH(l) + 3O_2(g) \rightarrow 2CO_2(g) + 3H_2 O(l); \quad \Delta H^{\ominus}_{c,298} = -1366.7 \text{ kJ mol}^{-1}$$

where (1) stands for the liquid state.

The *standard enthalpy (heat) of atomisation* of an element is represented as $\Delta H^{\ominus}_{at,298}$ and refers to the enthalpy change when one mole of gaseous atoms is formed from the element in the standard state, e.g.

$$\tfrac{1}{2} Cl_2(g) \rightarrow Cl(g); \quad \Delta H^{\ominus}_{at,298[Cl_2(g)]} = 121.1 \text{ kJ mol}^{-1}$$

N.B.   Absolute enthalpies cannot be determined and so elements in the standard state are regarded as having zero enthalpy. Enthalpy changes are, however, readily obtained.

Hess's law

If a set of reactants can be converted to the same products by more than one route, it is found that the total energy involved in each series of reactions is the same. This leads to Hess's law: *the heat change in a reaction depends only on the initial and final states and is independent of the route followed.* The value of this law lies in the fact that it enables changes, which cannot be measured directly, to be calculated if related facts are known.

Enthalpies (heats) of formation can be calculated from an energy cycle involving the enthalpy (heat) of combustion of the compound and of the constituent elements. This calculation is necessary in the determination of the standard enthalpy (heat) of formation of carbon monoxide, the direct measurement being impracticable since some carbon dioxide is always formed when carbon is burnt in a limited supply of air. Some typical calculations are given below and it should be stressed that the approach to these is always the same, i.e:

(a)   write down the equation for which the enthalpy change is required,

(b)   complete the triangular energy cycle by adding the equations for the information given,

(c)   apply Hess's law and solve the resultant equation in which there is one unknown.

Example   Calculate the standard enthalpy (heat) of formation of carbon monoxide given that the standard enthalpies (heats) of combustion of graphite and carbon monoxide are -393.5 and $-283.0 \text{ kJ mol}^{-1}$ respectively.

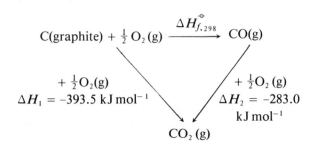

According to Hess's law, $\Delta H_1 = \Delta H^{\ominus}_{f,298} + \Delta H_2$ (the carbon dioxide has been made in two ways).

Therefore
$$\Delta H^{\ominus}_{f,298} = \Delta H_1 - \Delta H_2$$
$$= -393.5 - (-283.0)$$
$$= -393.5 + 283.0$$
$$= -110.5 \text{ kJ mol}^{-1}$$

Example Calculate the enthalpy (heat) of formation of methane given that the enthalpies (heats) of combustion of carbon, hydrogen, and methane are -393.5, -285.7, and -890.4 kJ mol^{-1} respectively.

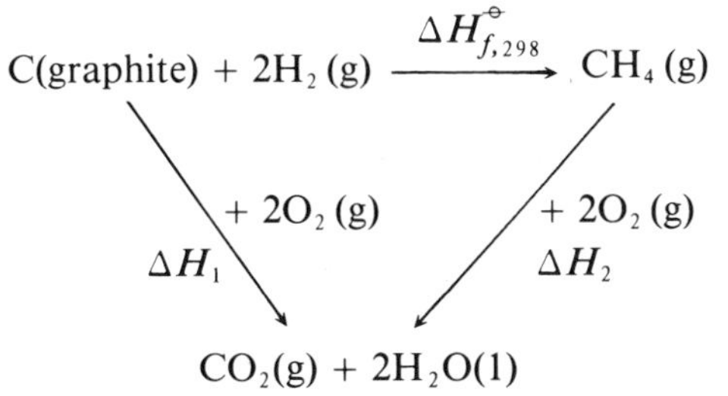

$$C(\text{graphite}) + 2H_2(g) \xrightarrow{\Delta H^{\ominus}_{f,298}} CH_4(g)$$

$$+ 2O_2(g) \qquad + 2O_2(g)$$
$$\Delta H_1 \qquad \qquad \Delta H_2$$

$$CO_2(g) + 2H_2O(l)$$

$$\Delta H_1 = \Delta H^{\ominus}_{c,298 \, [C \, (\text{graphite})]} + 2 \times \Delta H^{\ominus}_{c,298[H_2 \, (g)]}$$

$$= -393.5 + 2(-285.7)$$

$$= -964.9 \text{ kJ}$$

$$\Delta H_2 = \Delta H^{\ominus}_{c,298[CH_4 \, (g)]} = -890.4 \text{ kJ}$$

Now
$$\Delta H_1 = \Delta H^{\ominus}_{f,298 \, [CH_4 \, (g)]} + \Delta H_2$$

or
$$\Delta H^{\ominus}_{f,298 \, [CH_4 \, (g)]} = \Delta H_1 - \Delta H_2$$

$$= -964.9 - (-890.4)$$

$$= -74.5 \text{ kJ mol}^{-1}$$

Enthalpies (heats) of formation can be used to calculate the enthalpy change in a given reaction. The standard enthalpy of reaction, $\Delta H^{\ominus}_{rctn,298}$, may be defined as the enthalpy change when the moles of reactants in the equation react at a temperature of 298 K and a pressure of 1 atmosphere.

Example Calculate the enthalpy (heat) change for the reaction

$$NH_3(g) + HCl(g) \rightarrow NH_4Cl(s)$$

given that the standard enthalpies (heats) of formation of ammonia, hydrogen chloride, and ammonium chloride are -46.0, -92.3, and -315.5 kJ mol^{-1} respectively.

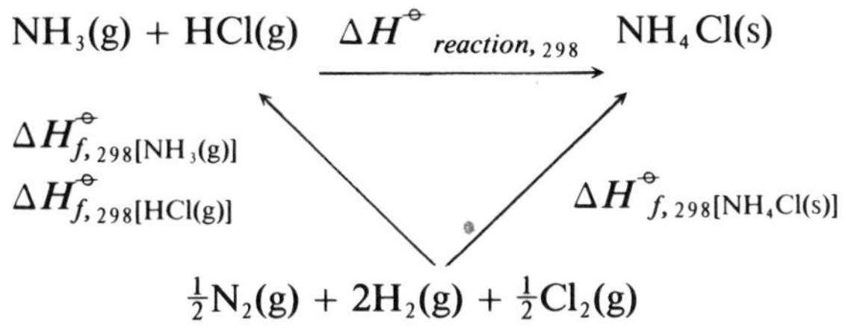

$$NH_3(g) + HCl(g) \xrightarrow{\Delta H^{\ominus}_{reaction, 298}} NH_4Cl(s)$$

$$\Delta H^{\ominus}_{f,298[NH_3(g)]}$$
$$\Delta H^{\ominus}_{f,298[HCl(g)]} \qquad \qquad \Delta H^{\ominus}_{f,298[NH_4Cl(s)]}$$

$$\tfrac{1}{2}N_2(g) + 2H_2(g) + \tfrac{1}{2}Cl_2(g)$$

The total enthalpy change must be the same regardless of the route followed.

Therefore:

$$\Delta H_{f,298[NH_4Cl(s)]}^{\ominus} = \Delta H_{f,298[NH_3(g)]}^{\ominus} + \Delta H_{f,298[HCl(g)]}^{\ominus} + \Delta H_{reaction,298}^{\ominus}$$

i.e. $-315.5 = -46.0 + (-92.3) + \Delta H_{reaction,298}^{\ominus}$

or $\Delta H_{reaction,298}^{\ominus} = -315.5 + 46.0 + 92.3$

$$= -177.2 \text{ kJ}$$

Bond dissociation enthalpies (energies)

During a reaction the number of bonds and their type may alter and so it is natural to attempt to allot to particular bonds specific contributions to the overall energy changes. The feasibility of this may be illustrated by comparing the standard enthalpies (heats) of combustion of a series of compounds, for example, the alkanes. Each member of the series differs from adjacent members by a CH_2 unit and the figures in Table 4.1 illustrate that $\Delta H_{c,298}^{\ominus}$ for the series also changes by a fairly regular amount.

Table 4.1. Standard enthalpies (heats) of combustion of some alkanes

Compound	Formula	$\Delta H_{c,298}^{\ominus}$ /kJ mol^{-1}	Difference from previous members of the series
Methane	CH_4	-890.4	
Ethane	C_2H_6	-1559.8	-669.4
Propane	C_3H_8	-2220.0	-660.2
Butane	C_4H_{10}	-2877.1	-657.1
Pentane	C_5H_{12}	-3509.4	-632.3
Hexane	C_6H_{14}	-4194.7	-685.3
Heptane	C_7H_{16}	-4853.5	-658.8
Octane	C_8H_{18}	-5512.5	-659.0

It is apparent that each C—C and C—H bond has a definite amount of energy associated with it and, taking the simplest compound, methane, as an example, it is reasonable to suggest that each C—H bond makes a quarter contribution to the total energy required to break the molecule into its constituent atoms:

$$\text{H—C—H(g)} \rightarrow C(g) + 4H(g); \quad \Delta H_{298}^{\ominus} = +1662 \text{ kJ mol}^{-1}$$

The figure of 1662 kJ mol^{-1} may be obtained from an energy cycle calculation, i.e.

$$CH_4(g) \xrightarrow{\Delta H_{reaction}^{\ominus}} C(g) + 4H(g)$$

$$\Delta H_f^{\ominus} \qquad \Delta H_{at,[C(s)]}^{\ominus} + 4\Delta H_{at,[H_2(g)]}^{\ominus}$$

$$C(graphite)(s) + 2H_2(g)$$

Now $\Delta H^{\ominus}_{f,[CH_4]} = -75.0 \text{ kJ mol}^{-1}$, $\Delta H^{\ominus}_{at,[C(graphite)]} = 715 \text{ kJ mol}^{-1}$

and $\Delta H^{\ominus}_{at,[H_2]} = 218 \text{ kJ mol}^{-1}$

$\therefore$ $\Delta H^{\ominus}_{f,[CH_4]} + \Delta H^{\ominus}_{reaction,[CH_4(g) \rightarrow C(g) + 4H(g)]} = 4\Delta H^{\ominus}_{at,[H_2]} + \Delta H^{\ominus}_{at,[C]}$

or $\qquad -75.0 + \Delta H^{\ominus}_{reaction,[CH_4(g) \rightarrow C(g) + 4H(g)]} = (4 \times 218) + 715$

and so $\qquad \Delta H^{\ominus}_{reaction,[CH_4(g) \rightarrow C(g) + 4H(g)]} = 872 + 715 + 75$

$$= 1662 \text{ kJ mol}^{-1}$$

(For the reverse reaction, i.e. $C(g) + 4H(g) \rightarrow CH_4(g)$ the sign of $\Delta H^{\ominus}$ must be changed and so the enthalpy (heat) of formation of methane from the atoms is $-1662 \text{ kJ mol}^{-1}$.)

Each C—H bond would therefore be expected to make an energy contribution of $1662/4 = 415.5 \text{ kJ mol}^{-1}$ to the total energy since the four bonds are all equivalent. (It should be noted that a different amount of energy is required for each of the processes

$$CH_4(g) \rightarrow CH_3(g) + H(g)$$
$$CH_3(g) \rightarrow CH_2(g) + H(g)$$
$$CH_2(g) \rightarrow CH(g) + H(g)$$
$$CH(g) \rightarrow C(g) + H(g)$$

because the environment of the carbon is different each time.) Many bond dissociation enthalpies (energies) are known and it is found that the values for a particular bond are influenced to some extent by the nature of the other atoms or groups joined to the atoms involved. Nevertheless, the average bond enthalpy (energy) between two atoms has considerable utility as, for example in the estimation of enthalpies (heats) of formation and as an indication of the strength of bonds.

Some average bond dissociation enthalpies (energies) are listed in Table 4.2 and the increased strength of the bonds with highly electronegative atoms is clearly evident. Note that bond enthalpies are always positive because energy is required to pull the atoms apart.

Table 4.2. Some average bond dissociation enthalpies (energies)

Bond	Average bond enthalpies/$kJ\,mol^{-1}$
C—F	485
C—Cl	339
C—Br	284
C—I	218
C—H	413
C—C	346
C=C	611
C≡C	835
N—H	389
P—H	322
O—H	464
S—H	347

The enthalpy change involved in the atomisation of a compound in the vapour state is approximately equal to the sum of the bond dissociation enthalpies (energies) involved. For example, the atomisation energy of one mole of gaseous 1-bromobutane ($CH_3 \cdot CH_2 \cdot CH_2 \cdot CH_2Br$) is approximately:

$$(3 \times E_{C-C}) + (9 \times E_{C-H}) + (E_{C-Br}) \quad \text{where} \quad E = \text{bond energy}$$

$$= (3 \times 346) + (9 \times 413) + 284$$

$$= 5039 \text{ kJ mol}^{-1}$$

Small discrepancies will obviously arise since average bond enthalpies (energies) are used.

The approximate enthalpy change in a reaction can be calculated as the sum of the energy used in bond breaking and the energy evolved in bond making as illustrated in the example below.

Example   Calculate the approximate enthalpy (heat) change in the reaction

$$CH_2{=}CH_2 + H_2 \rightarrow CH_3 \cdot CH_3$$

Relevant bond dissociation enthalpies (energies) in kJ mol^{-1} are:

$$C{=}C, \quad 611; \quad C{-}H, \quad 413; \quad C{-}C, \quad 346; \quad H{-}H, \quad 436.$$

Energy involved in breaking 1 mole of $C{=}C$ bonds $= 611$ kJ
Energy involved in breaking 4 moles of $C{-}H$ bonds $= 1652$ kJ
Energy involved in breaking 1 mole of $H{-}H$ bonds $= 436$ kJ
Hence the total energy involved in bond breaking is 2699 kJ

Energy involved in forming 1 mole of $C{-}C$ bonds $= -346$ kJ
Energy involved in forming 6 moles of $C{-}H$ bonds $= -2478$ kJ
Hence the total energy involved in bond forming is -2824 kJ
Thus the net energy change, $\Delta H = 2699 - 2824$
$$= -125 \text{ kJ}$$

Enthalpy changes have provided important evidence in the elucidation of the structure of some compounds, for example benzene (see page 305).

Born–Haber cycles

The lattice energy of an ionic crystal is the standard enthalpy (heat) of formation of one mole of the crystal lattice from the gaseous constituent ions.

Lattice energies cannot be determined directly and so they are found from energy cycles, called Born–Haber cycles, similar to those used for calculating enthalpies (heats) of formation. The Born–Haber cycle for sodium chloride is shown in Figure 4.1.

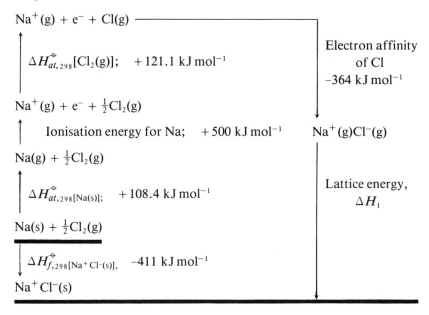

$Na^+(g) + e^- + Cl(g)$ —

$\Delta H^\ominus_{at,298}[Cl_2(g)];$ $+121.1\ kJ\,mol^{-1}$

Electron affinity of Cl $-364\ kJ\,mol^{-1}$

$Na^+(g) + e^- + \tfrac{1}{2}Cl_2(g)$

Ionisation energy for Na; $+500\ kJ\,mol^{-1}$

$Na^+(g)Cl^-(g)$

$Na(g) + \tfrac{1}{2}Cl_2(g)$

$\Delta H^\ominus_{at,298[Na(s)]};$ $+108.4\ kJ\,mol^{-1}$

Lattice energy, ΔH_1

$Na(s) + \tfrac{1}{2}Cl_2(g)$

$\Delta H^\ominus_{f,298[Na^+Cl^-(s)]},$ $-411\ kJ\,mol^{-1}$

$Na^+Cl^-(s)$

Figure 4.1   Born–Haber cycle for sodium chloride

Applying Hess's law:

$$\Delta H^\ominus_{f,298[Na^+Cl^-(s)]} = \Delta H^\ominus_{at,298[Na(s)]} + I_{Na} + \Delta H^\ominus_{at,298[Cl_2(g)]} + E_{Cl} + \Delta H_1$$

i.e. $-411.0 = 108.4 + 500 + 121.1 + (-364) + \Delta H_1$

∴ $\Delta H_1 = -411.0 - 108.4 - 121.1 + 364$

$$= -776.5\ kJ\,mol^{-1}$$

A knowledge of lattice energy is very useful in matters concerning the breakdown and formation of crystal lattices. For example:

(a)   The melting points of ionic solids are dependent on their lattice energies as illustrated by Table 4.3. The greater the lattice energy, the greater the temperature required to break down the lattice.

Table 4.3.   Variation of melting point with lattice energy

	NaF	NaCl	NaBr	NaI
Lattice energy/kJ mol^{-1}	−915	−781	−743	−699
Melting point/°C	995	808	750	662

(b)   The solubility of an ionic solid is usually governed by the lattice and hydration energies (page 80), although the gain in entropy (page 81) is sometimes the predominant factor. In the case of the lithium halides, for example, the solubilities are in accord with the differences between the lattice and hydration energies (Table 4.4). High solubility results when the hydration energy is appreciably greater than the lattice energy.

Table 4.4. Effect of relative values of lattice and hydration energies on solubility

Halide	LiF	LiCl	LiBr
Lattice energy, ΔH_1/kJ mol^{-1}	-1029	-849	-804
Hydration energy, ΔH_h/kJ mol^{-1}	-956	-883	-850
$\Delta H_h - \Delta H_1$	+73	-34	-46
Solubility	Very low	High	Higher

(c) A comparison of calculated and experimental values of lattice energy gives an indication of the degree of covalent character in the compound.

(d) Lattice energies may be utilised to explain the non-existence of some compounds as illustrated in the example below.

Example Calculate the enthalpy (heat) of formation of $CaCl_2$ and of the hypothetical compound, CaCl, given the following data:

$$\Delta H_{at,298[Ca]} = 176.6 \text{ kJ mol}^{-1}$$
$$\Delta H_{at,298[Cl_2]} = 121.1 \text{ kJ mol}^{-1}$$

I_1 and I_2 for calcium = 590   and   1100 kJ mol^{-1} respectively

Electron affinity for chlorine = -364 kJ mol^{-1}
Lattice energy of $CaCl_2$ = -2197 kJ mol^{-1}
Estimated lattice energy of   CaCl = -736 kJ mol^{-1}.

(The lattice energy of Ca^+Cl^- has been estimated from those of Na^+Cl^- and K^+Cl^- because the covalent radii of sodium, calcium, and potassium are 0.157, 0.174, and 0.203 nm respectively.)

Use the results to calculate the enthalpy change for the hypothetical reaction

$$2CaCl \rightarrow Ca + CaCl_2$$

and so compare the relative stabilities of CaCl and $CaCl_2$.

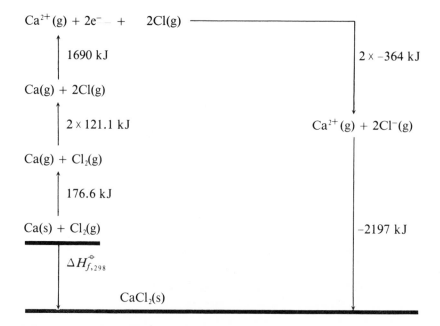

$$\Delta H^{\ominus}_{f,298[CaCl_2]} = 176.6 + 242.2 + 1690 - 728 - 2197$$
$$= -816.2 \text{ kJ mol}^{-1}$$

For Ca^+Cl^-,

$$\Delta H^{\ominus}_{f,298[CaCl]} = 176.6 + 121.1 + 590 - 364 - 736$$
$$= -212.3 \text{ kJ mol}^{-1}$$

$$2CaCl(s) \xrightarrow{\Delta H_{rctn,298}} Ca(s) + CaCl_2(s)$$

$$2 \times \Delta H^{\ominus}_{f,298[CaCl]} \qquad \Delta H^{\ominus}_{f,298[CaCl_2]}$$

$$2Ca(s) + Cl_2(g)$$

$$2 \times \Delta H^{\ominus}_{f,298[CaCl]} + \Delta H^{\ominus}_{rctn,298} = \Delta H^{\ominus}_{f,298[CaCl_2]}$$
$$2 \times -212.3 + \Delta H^{\ominus}_{rctn,298} = -816.2$$
$$\Delta H^{\ominus}_{rctn,298} = -391.6 \text{ kJ}$$

Hence Ca^+Cl^- is stable (ΔH negative) with respect to its elements. It is, however, unstable with respect to $CaCl_2$ because more energy is released when it is converted to this compound.

Solvation

Polar substances are found to dissolve in polar solvents whilst non-polar substances dissolve in non-polar solvents. Iodine, for example, readily dissolves in a solvent such as tetrachloromethane because the van der Waals forces (page 36) holding the iodine molecules together in the lattice are similar in value to those between the solvent molecules and so the iodine and tetrachloromethane molecules attract one another. However, iodine is very sparingly soluble in water because the intermolecular attractions between water molecules (i.e. hydrogen bonding — see page 37) are greater than those between iodine molecules or iodine and water molecules. For a substance to dissolve, the resultant solute–solvent attractions must exceed or be equal to the solute–solute and solvent–solvent attractions.

When an ionic solid is dissolved, it is quite often found that the temperature rises slightly and it is pertinent to ask why this is so. Before a solid can dissolve, it is apparent that the crystal lattice must be broken down and that this will require energy. If the ions are separated to such a distance that there is no longer any interaction between them, as is nearly the case in dilute solutions, this energy will be numerically equal to the lattice energy with the sign reversed. Clearly this endothermic process must be accompanied by an exothermic one and this is, in fact, the solvation of the ions. Thus positive ions attract the negative end of the dipole in several solvent molecules, whilst negative ions attract the positive end of the dipole in a number of solvent molecules (Figure 4.2). The attraction of the ions for the solvent molecules is illustrated by the fact that the volume of a solution may be seen to be less than the combined volume of the solvent plus solute.

If water is the solvent the process of solvation is known as *hydration* and the energy released is the *hydration energy*. Hydration of positive ions may occur by ion–dipole attraction as illustrated in Figure 4.2 or by co-ordination (page 243, etc.).

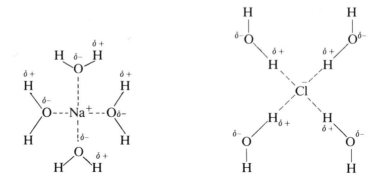

Figure 4.2   Hydration of Na⁺ and Cl⁻ ions

Solvation has the effect of delocalising the charge on the ions, so increasing their stability. The degree of solvation increases as the charge density on the ion increases. A consequence of solvation is that the movement of the ions in solution will be impeded to some extent and this is reflected in highly solvated ions diffusing more slowly than less solvated ones.

The *enthalpy (heat) of solution* is the enthalpy change when one mole of a substance is dissolved in so much water that further dilution produces no detectable heat change. When these conditions prevail, the solution is said to be at infinite dilution. The energy relationships involved in the dissolution of an ionic solid are shown in Figure 4.3, i.e. enthalpy (heat) of solution equals the lattice energy, with the sign reversed, plus the hydration energy. Enthalpies (heats) of solution are generally negative (i.e. exothermic) but positive ones are by no means uncommon (e.g. sodium nitrate(III) (sodium nitrite)). However, the figures involved are always relatively small, that is a few kJ mol⁻¹, compared to lattice energies which are often in excess of 500 kJ mol⁻¹.

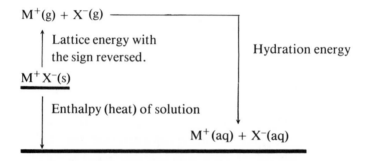

Figure 4.3   Energy cycle relating to enthalpy (heat) of solution; (aq) indicates fully hydrated ions

The second law of thermodynamics — entropy

The first law of thermodynamics (page 69) states that energy cannot be created or destroyed although it can be converted from one form into another. This enables the heat and work involved in a chemical reaction to be calculated but it has to be assumed that the reaction will take place, i.e. that it is *spontaneous*. The second law, however, provides a means of predicting whether a particular reaction is possible, although it does not indicate if the rate is such as to make the reaction feasible.

Most chemical reactions which are spontaneous are exothermic but a negative ΔH is not a sufficient criterion for spontaneity. For example, the thermal decomposition of calcium carbonate,

$$CaCO_3 \rightarrow CaO + CO_2$$

is a spontaneous endothermic process. Clearly, another factor is involved; this is the change in *entropy* and it is denoted by ΔS. A change in entropy is defined as the heat involved in a particular process divided by the temperature at which the action occurs, i.e. $\Delta S = q/T$. Its value depends upon the amounts of reactants considered (i.e. it is an *extensive property*) and so its units are $J K^{-1} mol^{-1}$. (Properties such as density and vapour pressure which are independent of the amount of substance considered are known as *intensive properties*.)

A knowledge of the variation of ΔS for various processes will aid an understanding of what entropy entails. Thus, any reactions in which there is an increase in the number of moles of gas present have high positive values of ΔS (often greater than $100 \ J K^{-1} mol^{-1}$). If there is no change in the number of moles of gas present, ΔS is small (often $\pm 10 \ J K^{-1} mol^{-1}$), whereas a decrease in the number of moles of gas leads to high negative values of ΔS (often more than $-100 \ J K^{-1} mol^{-1}$). These figures may be illustrated by a specific example.

Example Calculate the change in entropy when one mole of water is vaporised at 100 °C. The enthalpy (latent heat) of vaporisation of water is $40\,700 \ J \ mol^{-1}$.

$$\Delta S = \frac{q}{T} = \frac{40\,700}{373} = 109.1 \ J K^{-1} mol^{-1}$$

Entropy may be thought of as a measure of the disorder in a system, and so it is apparent that it will increase as the order of the crystal lattice is replaced by the increasing chaos or randomness of the liquid and gaseous states. Substances at absolute zero are regarded as having zero entropy because they then have maximum order. The entropy of a mole of the substance at any other temperature is known as the molar entropy, *standard molar entropy* being the entropy at 25 °C and 1 atmosphere. The molar entropies of substances increase as the molecular complexity increases because there are then more ways of distributing the energy (e.g. more vibrating bonds) and so more chaos.

According to the second law of thermodynamics, the entropy of the universe as a whole is continually increasing. A simple example of increasing entropy is when two containers of different coloured balls are poured at the same time into a larger third container. The balls mix and, no matter how much this container is shaken, unmixing does not occur even with quite small numbers of balls.

Although the second law of thermodynamics refers to the universe as a whole, it is possible, for most practical purposes, to interpret this as any system which is isolated from its surroundings, for example, a room. It follows therefore that the general condition for a reaction to be spontaneous is that

$$\Delta S_{total} > 0 \qquad (4.7)$$

The ΔS_{total} comprises the change in entropy of the system, i.e. the reaction vessel and contents, plus the change of entropy of the surroundings within the isolated system, i.e. the room:

$$\Delta S_{total} = \Delta S_{system} + \Delta S_{surroundings}$$

If the reaction is exothermic and the heat evolved, $-\Delta H$, is transferred to the surroundings, then the entropy gained by the surroundings is $-\Delta H/T$ if the surroundings are so large that there is negligible change in temperature. Equation (4.7) then becomes

$$\Delta S_{system} - \frac{\Delta H}{T} > 0$$

Multiplying through by T gives

$$T\Delta S - \Delta H > 0$$

or

$$\Delta H - T\Delta S > 0 \qquad (4.8)$$

It follows from equation (4.8) that a reaction will be spontaneous if $\Delta H < T\Delta S$. An exothermic reaction will obviously be spontaneous if it also results in an increase in entropy. It will also be spontaneous if the entropy decreases, provided that the product of $T\Delta S$ is not more negative than ΔH. Endothermic reactions will be spontaneous only if the increase in entropy is such that $T\Delta S$ outweighs ΔH. At normal temperatures, ΔS is not very large compared to ΔH but at high temperatures $T\Delta S$ may become appreciable and become the predominant factor. This is the case in the manufacture of water gas:

$$C + H_2O \rightarrow CO + H_2; \quad \Delta H \text{ positive}$$

Below about 700 °C, ΔH is greater than $T\Delta S$ and so the reaction is not possible but, above this figure, $T\Delta S$ exceeds ΔH and the reaction is spontaneous.

A further example of a spontaneous endothermic process is the dissolution of ammonium nitrate(V) in water. The temperature falls noticeably but the process proceeds because there is a gain in entropy, resulting from the break down of the crystal lattice of the salt, and $T\Delta S$ is greater than ΔH.

Questions

4.1 (a) Define (i) enthalpy (heat) of formation, (ii) exothermic reaction, and (iii) enthalpy of combustion.

(b) When 12 g each of carbon, hydrogen, and methanol are completely burnt in oxygen, 393.5, 1715.4, and 272.4 kJ are evolved respectively. Calculate the enthalpy of formation of methanol and state whether the reaction is endothermic or exothermic.

4.2 (a) What is meant by the following terms: (i) enthalpy (heat) of reaction, (ii) enthalpy of formation, and (iii) enthalpy of combustion?

(b)   Use the data below to calculate the enthalpy of formation of propane.

$$C(graphite) + O_2(g) \rightarrow CO_2(g) \quad \Delta H = -393.5 \text{ kJ mol}^{-1}$$

$$H_2(g) + \tfrac{1}{2}O_2(g) \rightarrow H_2O(l) \quad \Delta H = -285.9 \text{ kJ mol}^{-1}$$

$$C_3H_8 + 5O_2(g) \rightarrow 3CO_2(g) + 4H_2O(l) \quad \Delta H = -2220.0 \text{ kJ mol}^{-1}$$

(c)   The enthalpies of combustion of rhombic and monoclinic sulphur are -296.6 and $-297.0 \text{ kJ mol}^{-1}$ respectively. Why do the values differ and what conclusion may be drawn from them? (A negative sign for ΔH indicates an exothermic reaction.)

4.3   Define the terms enthalpy (heat) of formation, enthalpy of combustion, enthalpy of atomisation, and endothermic compound. Show how Hess's law may be applied in the calculation of the enthalpy of formation of butane, C_4H_{10}, from the following data:

$$\text{enthalpy of combustion of graphite} = -393.5 \text{ kJ mol}^{-1}$$

$$\text{enthalpy of combustion of hydrogen} = -285.9 \text{ kJ mol}^{-1}$$

$$\text{enthalpy of combustion of butane} = -2877.1 \text{ kJ mol}^{-1}$$

In what way is the stability of a substance connected with its enthalpy of formation? (A negative ΔH indicates heat evolved.)

4.4   Calculate the enthalpies (heats) of formation of ethane (C_2H_6) and ethyne (C_2H_2) given that the enthalpies of formation of water and carbon dioxide are -285.9 and $-393.5 \text{ kJ mol}^{-1}$ respectively and the enthalpies of combustion of ethane and ethyne are -1559.8 and $-1299.6 \text{ kJ mol}^{-1}$ respectively. Determine the enthalpy of reaction for

$$C_2H_2 + 2H_2 \rightarrow C_2H_6$$

(Exothermic processes have a negative ΔH.)

4.5   State Hess's law and give two illustrative examples of its use, without including numerical data.
  Calculate the enthalpy (heat) change for the reaction:

$$P_4O_{10}(s) + 6H_2O(l) \rightarrow 4H_3PO_4(s)$$

given that the standard enthalpies of formation of tetraphosphorus decaoxide, water, and phosphoric(V) acid are -2984.0, -285.9, and $-1279.0 \text{ kJ mol}^{-1}$ respectively.
(A negative sign for ΔH indicates an exothermic reaction.)

4.6   Calculate the standard enthalpy of formation of ethanol given that the standard enthalpies of combustion of graphite, hydrogen, and ethanol are -393.5, -285.9, and $-1366.7 \text{ kJ mol}^{-1}$ respectively.

4.7 (a)   Calculate the standard enthalpy change for the reaction

$$2H_2O + 4NO_2 + O_2 \rightarrow 4HNO_3$$

given that the standard enthalpies of formation of water, nitrogen dioxide, and nitric acid are -285.9, 33.2, and $-173.2 \text{ kJ mol}^{-1}$ respectively.

(b). The standard enthalpies of combustion of diamond and graphite are -395.7 and $-393.5 \text{ kJ mol}^{-1}$ respectively. Explain what conclusion may be drawn concerning the relative stability of these allotropes.

4.8 (a)   Calculate the standard enthalpies (heats) of formation of $MgCl_2$ and the hypothetical compound, $MgCl$, from the following data:

$$Mg(s) \rightarrow Mg(g) \qquad \Delta H = 149 \text{ kJ mol}^{-1}$$

$$\tfrac{1}{2}Cl_2(g) \rightarrow Cl(g) \qquad \Delta H = 121 \text{ kJ mol}^{-1}$$

$$Mg(g) \rightarrow Mg^+(g) \qquad \Delta H = 740 \text{ kJ mol}^{-1}$$

$$Mg^+(g) \rightarrow Mg^{2+}(g) \qquad \Delta H = 1500 \text{ kJ mol}^{-1}$$

$$Cl(g) + e^- \rightarrow Cl^-(g) \qquad \Delta H = -364 \text{ kJ mol}^{-1}$$

$$Mg^{2+}(g) + 2Cl^-(g) \rightarrow MgCl_2(s) \quad \Delta H = -2489 \text{ kJ mol}^{-1}$$

$$Mg^+(g) + Cl^-(g) \rightarrow MgCl(s) \quad \Delta H = -815 \text{ kJ mol}^{-1} \text{ (est.)}$$

(b)   From the results of (a), calculate the enthalpy change for the reaction

$$2MgCl \rightarrow Mg + MgCl_2$$

(c)   In the light of your answer to (b), comment on the relative stabilities of MgCl and $MgCl_2$.

4.9 (a)   Calculate the $C-Cl$ bond energy given that the enthalpy (heat) of formation of tetrachloromethane is $-135.5 \, kJ \, mol^{-1}$ and the enthalpies of atomisation of graphite and chlorine are 715.0 and $121.1 \, kJ \, mol^{-1}$ respectively.

(b)   Calculate the approximate enthalpy change in the reaction

$$CH_4 + Br_2 \rightarrow CH_3Br + HBr$$

Relevant mean bond energies in $kJ \, mol^{-1}$ are $C-H$, 413; $C-Br$, 209; $Br-Br$, 193; $H-Br$, 366.

4.10   The enthalpy (heat) of formation of ammonia is $-46.0 \, kJ \, mol^{-1}$ and the bond dissociation energies of hydrogen and nitrogen are 436 and $945 \, kJ \, mol^{-1}$ respectively. Use this data to calculate the average $N-H$ bond energy in ammonia.

4.11   Discuss the factors which affect the spontaneity of reactions.
       For the reaction

$$N_2 + 3H_2 \rightarrow 2NH_3$$

$\Delta H = -92 \, kJ$ and $\Delta S = -97.8 \, J$. Determine the feasibility of the reaction at (a) 400 °C and (b) 700 °C.

5 Phase Equilibria

The term *phase* applies to any part of a system which is physically distinct from the remainder of the system but which is itself homogeneous (i.e. has the same composition throughout). For example, solutions and mixtures of gases consist of a single phase but the gaseous, liquid, and solid forms of a substance constitute three phases.

Phase equilibria of one- and two- component systems will be discussed.

ONE-COMPONENT SYSTEMS

Consider a liquid in a closed container. Some of the molecules near the surface have sufficient energy to escape from the liquid and accumulate as vapour in the space above. The vapour molecules will collide with one another and with the walls of the container and some will collide with the liquid and so return to it. As the vapour pressure builds up, so the number of molecules returning to the liquid will also increase. At any given temperature, the rate of evaporation will be constant and so eventually a state will be reached where the number of molecules leaving the liquid will equal the number returning. The system is then said to be at equilibrium and this is a dynamic not a static equilibrium. If the temperature is raised, then the rate of evaporation increases and so does the rate of return, but to a lesser extent, and so a new equilibrium is set up. It should be noted that the vapour pressure above a liquid at equilibrium in a closed system is independent of the amount of liquid present, its surface area, or the volume of the vapour.

Figure 5.1 illustrates the shape of the plot of vapour pressure against temperature for a single substance. The curves AB and BC represent the saturated vapour pressure curves of the solid and liquid phases respectively. At the point of intersection, B, the solid and liquid are in equilibrium and the line BD shows how this equilibrium point (the freezing point) is affected by applying pressure. The slope of BD indicates that increasing the pressure raises the freezing point of the liquid. In the case of water, however, the line slopes in the opposite direction. Hence the freezing point of water falls as the pressure is increased, the reason being that water, unlike most liquids, expands when it freezes.

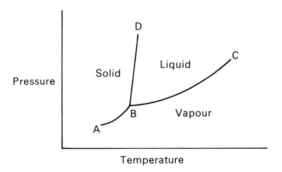

Figure 5.1 Plot of vapour pressure against temperature for a single substance

In the region bounded by ABD, solid is the only stable phase, whilst DBC and ABC show the regions in which liquid and vapour respectively are the only stable phases. Along the lines, two phases are in equilibrium and can co-exist indefinitely. Point B (known as the *triple point*) indicates the only set of conditions under which all three phases are in equilibrium.

When a liquid evaporates, only the high energy molecules escape and, if no heat is supplied, the temperature of the liquid falls. The enthalpy change required to convert one mole of it to vapour with no change in temperature is known as the *enthalpy (heat) of vaporisation or evaporation*.

Some solids, for example iodine, turn to vapour without an intermediate liquid state because the vapour pressure of the liquid form is greater than atmospheric pressure. These solids are said to *sublime*. At any given temperature, such a solid in a closed vessel will reach equilibrium with its vapour. The *enthalpy or heat of sublimation* is the enthalpy change required to sublime one mole of a solid without change in temperature.

If a substance is at its melting point, then solid and liquid are in equilibrium. The *enthalpy or heat of fusion* is the enthalpy change when one mole of a substance is changed from solid to liquid without change in temperature.

BINARY SOLUTIONS

1. Solutions of gases in liquids

The solubility of a gas in a liquid is usually indicated by an *absorption coefficient* which is defined as the number of cm^3 of the gas at s.t.p. which saturate $1\ cm^3$ of the liquid at a given temperature and at one atmosphere pressure.

The solubilities of gases are determined by two main methods.

Pyknometer method

The very soluble gases, for example, hydrogen chloride, sulphur dioxide, and ammonia, react with water giving acidic or alkaline solutions and so the method involves adding a saturated solution of the gas to an excess of standard alkali or acid respectively and back titrating the resulting solution.

A preweighed pyknometer, Figure 5.2, is partially filled with solvent and then placed in a constant temperature bath. When equilibrium has been attained, gas is passed through the apparatus until the solvent is saturated and then the ends are sealed in a flame. The apparatus is allowed to cool and weighed so that the mass of solution is known. One end of the pyknometer is then broken under an excess known volume of standard acid or alkali (depending upon whether the solution is alkaline or acid respectively) and the amount of this reagent used up is determined. From the results, the mass of gas present, and hence the solubility, can be determined.

Ostwald's method

This method is used with sparingly soluble gases and involves the apparatus illustrated in Figure 5.3. The gas burette is filled with mercury, the pipette is filled with the solvent, and the flexible tube C is flushed out with the gas under test. The

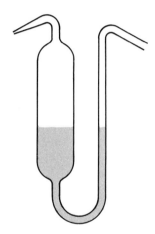

Figure 5.2 A pyknometer

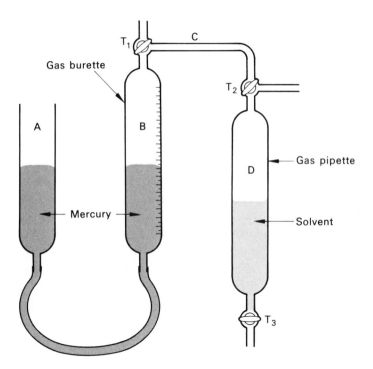

Figure 5.3 Ostwald's apparatus for determining the solubility of sparingly soluble gases

three-way taps, T_1 and T_2, are adjusted and A is lowered so that a sample of the gas is drawn into B; the taps are then closed. The mercury levels in A and B are equalised and the volume in B is noted. Taps T_1, T_2, and T_3 are then adjusted so that about half the solvent in D runs out through T_3 and gas is drawn in from B. T_1, T_2, and T_3 are closed. The solvent which has been run out is weighed and the volume noted so that the volume of gas remaining in D at the end is known. D is then immersed in a constant temperature bath and shaken vigorously. T_1 and T_2 are again opened, so that more gas goes into D, and then closed and the shaking resumed. The above processes are repeated until no more gas dissolves. The mercury levels in A and B are now equalised whilst B is open to D and the volume of gas remaining in B is noted and so is the temperature. If the total mass of solvent D can hold is determined, the mass of solvent remaining in D at the end of the experiment is known. The solubility can therefore be calculated.

Factors affecting solubility of gases

In the majority of cases, the dissolving of a gas in a solvent is an exothermic process and so, at a given pressure, the solubility is decreased by increasing the temperature.

The relationship between pressure and solubility is expressed by *Henry's law*. This states that the mass of gas dissolved by a given volume of solvent, to give a saturated solution, is directly proportional to the pressure of the gas, provided that the temperature remains constant, and there is no reaction between the gas and solvent. Slight deviations from the law will occur because it is being applied to real not ideal gases. The law does not apply to hydrogen chloride in water because reaction occurs, i.e.

$$HCl + H_2\ddot{O} \rightarrow H_3O^+ + Cl^-$$

Ammonia, carbon dioxide, sulphur dioxide, etc. are also exceptions for the same reason. If, however, only the free, i.e. unreacted, gas is considered the results are in agreement with Henry's law.

When a mixture of gases is in equilibrium with a solvent, the solubility of each gas is proportional to its partial pressure.

2. Solutions of solids in liquids

If a non-volatile solute is dissolved in a liquid, it is found that the vapour pressure of the solution is lower than that of the pure solvent. This is because part of the surface of the solution is occupied by solute molecules and so the number of solvent molecules escaping is reduced whilst the number of those returning is unaltered. Knowing the vapour pressures of the solvent and solution at the same temperature, and the concentration of the solution, it is possible to calculate the relative molecular mass of the solute (see below).

Instead of direct measurement of the vapour pressure, it is often more convenient to measure effects which are dependent upon vapour pressure, for example, rise in boiling point, depression of the freezing point, and osmotic pressure. These methods are accurate only if the measurements are made on dilute solutions so that there are not many solute molecules present to interact with those of the solvent. Properties such as the above, which depend on the number of particles of solute in a given volume of solvent, are known as *colligative properties* and are discussed below.

a. Relative molecular mass and lowering of the vapour pressure

The relationship between lowering of the vapour pressure and concentration is given by *Raoult's law* which states that the relative lowering of the vapour pressure is equal to the mole fraction of the solute. This is expressed mathematically as:

$$\frac{p_1-p_2}{p_1} = \frac{n}{N+n} \tag{5.1}$$

where p_1 = vapour pressure of the pure solvent
p_2 = vapour pressure of the dilute solution
N = number of moles of solvent
n = number of moles of solute

If very dilute solutions are used, then n is very small compared to N and so the expression may be reduced to

$$\frac{p_1-p_2}{p_1} = \frac{n}{N} \tag{5.2}$$

The use of the equation is illustrated in the example below. It should be noted that Raoult's law applies to ideal solutions and so when used with real solutions, slight deviations occur. Further, erroneous relative molecular masses will be obtained if the solute dissociates or associates under the conditions of the reaction.

Example Calculate the relative molecular mass of a non-volatile solute if 10 g of it dissolved in 75 g propanone at 25 °C gives a solution with vapour pressure 215.8 mm. The vapour pressure of pure propanone at 25 °C is 229.2 mm mercury. (Propanone = C_3H_6O.)

$$\frac{p_1-p_2}{p_1} = \frac{n}{N+n}$$

$$\frac{229.2-215.8}{229.2} = \frac{\dfrac{10}{M}}{\dfrac{75}{58} + \dfrac{10}{M}}$$

where M = relative molecular mass of the solute

Therefore $M = 125$

b. Elevation of the boiling point

The boiling point of a liquid is the temperature at which its saturated vapour pressure equals atmospheric pressure. If a non-volatile solute is dissolved in the liquid, then its vapour pressure falls and so the solution needs to be heated to a higher temperature before its vapour pressure equals atmospheric pressure, i.e. the boiling point is elevated—see Fig. 5.4.

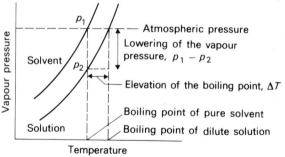

Figure 5.4   Boiling point elevation

The elevation of the boiling point is proportional to the lowering of the vapour pressure provided that only dilute solutions are considered, i.e.

$$\Delta T \propto (p_1 - p_2)$$

where ΔT = elevation of the boiling point and p_1 and p_2 are the vapour pressures of solvent and solution respectively.

Therefore

$$\Delta T = k(p_1 - p_2) \quad \text{where} \quad k = \text{constant} \tag{5.3}$$

Now from Raoult's law

$$\frac{p_1 - p_2}{p_1} = \frac{n}{N} \quad \text{for very dilute solutions}$$

and so

$$p_1 - p_2 = p_1 \frac{n}{N}$$

Substituting this value for $p_1 - p_2$ into equation (5.3) gives:

$$\Delta T = kp_1 \frac{n}{N}$$

But the vapour pressure of any given solvent is constant at a particular temperature and so kp_1 may be replaced by the constant k', i.e.

$$\Delta T = k' \frac{n}{N} \tag{5.4}$$

But $n = \dfrac{W_2}{M_2}$, where W_2 and M_2 = mass and relative molecular mass respectively of the solute, and $N = \dfrac{W_1}{M_1}$, where W_1 and M_1 = mass and relative molecular mass respectively of the solvent.

Substituting these values of n and N in equation (5.4) gives:

$$\Delta T = k' \frac{W_2}{M_2} \times \frac{M_1}{W_1} \tag{5.5}$$

If a fixed mass of 1000 g of solvent is always taken, then M_1/W_1 gives a constant figure for that particular solvent and so equation (5.5) may be reduced to:

$$\Delta T = \frac{KW_2}{M_2} \tag{5.6}$$

where K is known as the *ebullioscopic constant, molecular elevation constant*, or the *boiling point constant*. K is the elevation in boiling point which would theoretically result if one mole of any solute was dissolved in 1000 g solvent. In fact, this solution would be too concentrated for the vapour pressure to be directly proportional . to the concentration. The boiling point elevation is therefore found for a dilute solution and K is found by simple proportion.

The boiling point elevation may be determined by the Landsberger method using the apparatus in Figure 5.5. Solvent is boiled in the conical flask and its vapour passes through the perforated bulb which is immersed in the same solvent in the graduated tube, through the hole at the top, into the outer jacket, and finally to a condenser. After some time, the Beckmann thermometer (a very accurate thermometer reading to 1/1000 °C) shows a constant reading and this is the boiling point of the solvent at the prevailing atmospheric pressure. A weighed pellet of the solute is then added to the graduated tube and the vapour is passed through again until a new higher constant temperature is achieved (the higher temperature can be achieved because the condensing vapour releases its enthalpy or latent heat of vaporisation). The passage of vapour is immediately stopped to avoid further dilution and the volume of liquid in the graduated tube is noted. The relative molecular mass of the solute may be calculated from the observed results as illustrated in the example below.

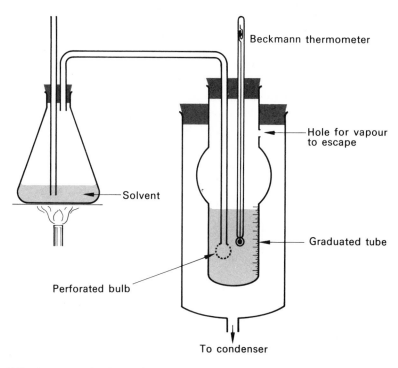

Figure 5.5   Apparatus for determing relative molecular mass by Landsberger's method

It should be noted that the solvent is heated by its own vapour to reduce superheating to a minimum. Further, the Beckmann thermometer measures differences in temperature over a scale of 5 °C, it does not measure actual temperatures.

Example   Calculate the relative molecular mass of a non-volatile solute if 1.25 g of it dissolved in 50 g benzene causes the boiling point to be elevated by 0.20 °C. The ebullioscopic constant for benzene is 2.7 °C kg^{-1}.

$$\Delta T = \frac{KW}{M} \quad \text{where} \quad \Delta T = 0.20 \text{ °C}, \quad K = 2.7 \text{ °C kg}^{-1}$$

W = mass of solute in 1000 g benzene = $\dfrac{1.25}{50} \times 1000 = 25$ g   and M = relative molecular mass of the solute.

$\therefore$
$$M = \frac{KW}{\Delta T} = \frac{2.7 \times 25}{0.2}$$

$$= 337.5$$

c. Depression of the freezing point

The vapour pressure of a solid, like that of a liquid, varies with temperature. If the vapour pressure curves of the solid and liquid states of the same substance are drawn on the same graph, it is found that they intersect (Figure 5.6). The point of intersection is the melting or freezing point of the substance, i.e. the point where solid and liquid can co-exist indefinitely. If a solute is dissolved in the liquid, the vapour pressure falls as explained above and so the vapour pressure curve of the solution cuts that of the solid solvent at a lower point, i.e. the freezing point is depressed. The depression of the freezing point is proportional to the lowering of the vapour pressure providing that dilute solutions are considered and so the same equation applies as for elevation of the boiling point (Equation (5.6)). In this case K is known as the *cryoscopic constant, the molecular depression constant*, or the *freezing point constant* and it is the depression of the freezing point which would theoretically occur if one mole of any solute were dissolved in 1000 g of the solvent. If the depression of the freezing point caused by a known mass of solute of known relative molecular mass is found and the relevant figures are substituted in equation (5.6), then K for the solvent may be calculated. The relative molecular mass of other solutes in the same solvent can then be calculated when the freezing point depressions, of known masses of them, have been determined.

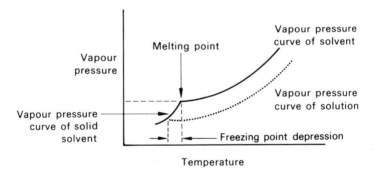

Figure 5.6   Vapour pressure curves of the solid and liquid forms of a solvent and of a solution

The freezing point depression may be determined by the Beckmann method using the apparatus shown in Figure 5.7.

In this method, it is essential that supercooling is reduced to an absolute minimum. This is achieved by very slow cooling via an air jacket and by adjusting the cooling mixture so that its temperature is not more than 5 °C below the freezing point of the solvent.

A known mass of the solvent is put in the centre tube and stirred vigorously. The temperature is read every quarter of a minute and a cooling curve is plotted. It will take the form shown in Figure 5.8, slight supercooling occurs but the temperature quickly rises to the true freezing point owing to the release of the enthalpy (latent heat) of fusion. The solvent is then warmed slightly to melt it and, after dissolving a weighed pellet of the solute in it, the above process is repeated to find the freezing point of the solution. The determination must be repeated if the supercooling exceeds 0.5 °C since undue deposition of solid solvent will considerably increase the concentration of the solution and cause an appreciable error.

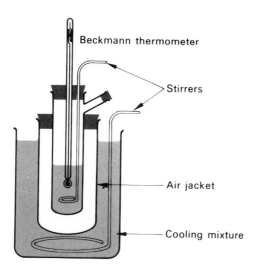

Figure 5.7 Beckmann apparatus for determining freezing point depression

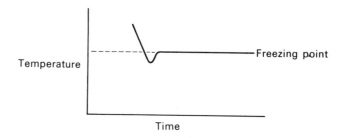

Figure 5.8 Cooling curve for the solvent in Beckmann's method of relative molecular mass determination

Knowing ΔT, K, and W, the relative molecular mass of the solute may be calculated.

Example Calculate the relative molecular mass of anthracene given that 0.813 g of it dissolved in 45 g benzene depress the freezing point by 0.51 °C. The cryoscopic constant for benzene is 4.99 °C kg^{-1}.

$$\Delta T = \frac{KW}{M}$$

where $K = 4.99$ °C kg^{-1}, $\Delta T = 0.51$ °C, and $W = $ mass anthracene in a kg of benzene $= \dfrac{0.813 \times 1000}{45}$

$\therefore$
$$M = \frac{KW}{\Delta T} = \frac{4.99 \times 0.813 \times 1000}{0.51 \times 45} = 176.8$$

i.e. relative molecular mass of anthracene $= 177$ ($C_{14}H_{10} = 178$).

d. Osmosis

If two solutions of unequal concentration are separated by a *semi-permeable membrane* (a membrane which will allow solvent but not solute molecules to pass through it), solvent will pass through the membrane until the two solutions have the same concentration. Similarly, if a solution is separated from pure solvent by a semi-permeable membrane there will be a net flow of solvent through the membrane. This process is known as *osmosis*.

Animal bladders have been used as semi-permeable membranes but they are not very strong or effective. They have been replaced by semi-permeable membranes which consist of copper(II) hexacyanoferrate(II) ($Cu_2[Fe(CN)_6]$) precipitated in porous pot.

Osmosis may be demonstrated with the apparatus shown in Figure 5.9. Initially the level of the solution in the column is the same as the water level but on standing the level in the column rises. Eventually, the pressure of the column of solution will be sufficient to balance the pressure of water trying to pass through the membrane, i.e. water will pass through the membrane in both directions at the same rate. The pressure at this point is known as the *osmotic pressure*. The osmotic pressure of a solution is the pressure which must be applied to it to just prevent osmosis occurring.

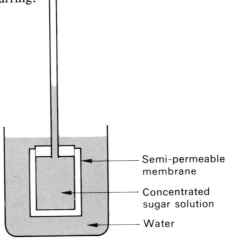

Semi-permeable membrane

Concentrated sugar solution

Water

Figure 5.9   Demonstration of osmosis

Solutions which have the same osmotic pressure are said to be *isotonic*.

The initial work on osmosis was done by Pfeffer and van't Hoff and this led to two laws.

(a)   The osmotic pressure of a dilute solution at constant temperature is directly proportional to the concentration of the solute.

(b)   The osmotic pressure of a given solution is directly proportional to its kelvin temperature.

According to (a),

$$\pi \propto C$$

where π = osmotic pressure   and C = concentration

but $C = \dfrac{n}{V}$ mol litre^{-1}, and if 1 mole is considered (i.e. $n = 1$) then $C = \dfrac{1}{V}$

$\therefore$ $\pi \propto \dfrac{1}{V}$ or $\pi V = $ constant

This is analogous to Boyle's law.

According to (b),

$$\pi \propto T \quad \text{where} \quad T = \text{kelvin temperature}$$

or

$$\frac{\pi}{T} = \text{constant}$$

Now since

$$\pi \propto \frac{1}{V} \quad \text{and} \quad \pi \propto T$$

then

$$\pi \propto \frac{T}{V}$$

or

$$\pi V = kT \quad \text{where} \quad k = \text{a constant} \tag{5.7}$$

This is analogous to the gas equation, i.e. $PV = RT$. If π is measured in $N\,m^{-2}$ and V in m^3 then

$$k = \frac{N\,m^{-2} \times m^3}{T} = N\,m\,mol^{-1}K^{-1}$$

but the force in newtons multiplied by length in metres equals work in joules and so $k = R$. The general equation is therefore

$$\pi V = nRT$$

where n is the number of moles of the solute.

It can be seen from the above equation that 1 mole of solute in 22.4 litres of solution has an osmotic pressure of 1 atmosphere.

$$\left(\pi V = RT \quad V = \frac{RT}{\pi} = \frac{8.31 \times 273}{101\,325} = 0.0224 \text{ m}^3 = 22.4 \text{ litres} \right)$$

The use of osmotic pressure measurements in relative molecular mass determinations is illustrated in the example below.

Example The osmotic pressure of a solution of 1.37 g mannitol in 100 g water is 1.779 atmospheres at 17 °C. Calculate the relative molecular mass of mannitol. 1 atm $= 1.013 \times 10^5\,N\,m^{-2}$, $R = 8.31\,J\,K^{-1}mol^{-1}$

$$\pi V = nRT \quad \text{where} \quad \pi = 1.779 \times 1.013 \times 10^5\,N\,m^{-2}$$
$$V = \text{volume of water in m}^3$$
$$= \frac{100}{10^6}\,m^3$$
$$n = \text{moles of solute} = \frac{1.37}{M}$$
$$R = 8.31\,J\,K^{-1}mol^{-1} \quad \text{and} \quad T = 290 \text{ K}$$

$$1.779 \times 1.013 \times 10^5 \times \frac{100}{10^6} = \frac{1.37}{M} \times 8.31 \times 290$$

$$\therefore \quad M = \frac{8.31 \times 290 \times 1.37}{1.779 \times 1.013 \times 10}$$

$$= 183.2$$

An alternative method of calculating the relative molecular mass is given below.

Since $\pi \propto C$ and $\pi \propto T$, then $\pi \propto CT$ or $\pi = kCT$ where π = osmotic pressure, k = constant, C = concentration and T = kelvin temperature. Rearranging gives

$$k = \frac{\pi}{CT}$$

At concentration C_1 and kelvin temperature T_1 for a given solution, the osmotic pressure is π_1 and so

$$k = \frac{\pi_1}{C_1 T_1}$$

At concentration C_2 and kelvin temperature T_2 for the same solution, the osmotic pressure is π_2 and so

$$k = \frac{\pi_1}{C_2 T_2}$$

$$\therefore \quad \frac{\pi_1}{C_1 T_1} = \frac{\pi_2}{C_2 T_2}$$

N.B. It is not necessary to use SI units in this method because similar terms occur on both sides of the equation.

Since a solution containing 1 mole of solute in 22 400 cm^3 solvent has an osmotic pressure of 1 atm

$$\frac{1.779}{\dfrac{1.37}{100} \times 290} = \frac{1}{\dfrac{M}{22\,400} \times 273}$$

where M = relative molecular mass, or

$$M = \frac{1.37 \times 290 \times 22\,400}{100 \times 273 \times 1.779}$$

$$= 183.2$$

Experimental determination of osmotic pressure

The method of Berkeley and Hartley, using the apparatus illustrated in Figure 5.10, is one of the most accurate ways of measuring osmotic pressure. A known pressure is applied to the solution such that the level gauge rises at a slow measured rate, i.e. solution solvent is just passing into the pure solvent. The pressure is then adjusted so that the level gauge falls at a slow measured rate, i.e. solvent is just passing into the solution. The osmotic pressure is taken as the mean of these two applied pressures.

The relative molecular masses of high relative molecular mass compounds are best determined from osmotic pressure measurements since they give only a very small freezing point depression.

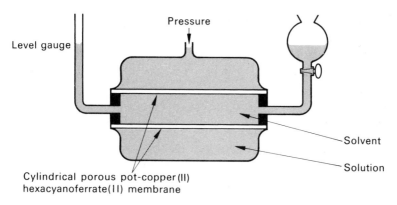

Figure 5.10   Berkeley and Hartley apparatus for osmotic pressure determination

Common examples of osmosis

When crystals of salts of copper, nickel, cobalt, chromium, iron, and magnesium, etc. are added to a 30% solution of sodium silicate(IV) (water glass), reaction occurs at their surface to give a layer of metallic silicate which acts as a semi-permeable membrane. Water from the sodium silicate(IV) solution flows to the concentrated metal salt solution until the membrane bursts. A jet of concentrated metallic salt solution shoots out, forms another membrane and the process is repeated. In this way a *chemical garden* is formed.

The skin round the white and yolk of an egg behaves as a semi-permeable membrane and it may be exposed intact by dissolving away the shell with dilute hydrochloric acid. If the egg is then placed in a concentrated solution of brine, the egg shrinks, whereas if it is placed in water, it swells. Red blood corpuscles similarly swell in water and shrink in brine.

Relative molecular masses of associated compounds

Relative molecular mass determinations from the colligative properties of many organic acids give higher results than expected. For example, when ethanoic acid (acetic acid) is dissolved in non-polar solvents its relative molecular mass is found to be 120 instead of the expected 60. This implies that there are only half the expected number of molecules present and from this and other evidence it is thought that association or hydrogen bonding (page 37) occurs:

In acids where this association is not complete, the relative molecular mass will lie somewhere between the expected value and twice the expected value.

Relative molecular masses of electrolytes

If the solute partially or completely dissociates (i.e. if it is an electrolyte), the observed relative molecular mass will be lower than the true value due to the extra particles present.

Van't Hoff tried to relate the observed colligative property to the calculated value when no dissociation occurs by use of a factor, i, called the *van't Hoff factor* where

$$i = \frac{\text{observed colligative property}}{\text{calculated colligative property}}$$

This factor is applicable to all the colligative properties and so for osmosis of electrolytes, for example, he used the formula

$$\pi V = iRT \qquad \text{(for 1 mole)}$$

Consider one mole of any substance, A, dissociating into n ions, i.e.

$$A \; \rightleftharpoons \; n \; \text{ions}$$

If undissociated	1 mol	0	
At equilibrium	$1 - \alpha$	$n\alpha$	where α = degree of dissociation.

Now

$$i = \frac{\text{total number of particles in solution}}{\text{number of particles if no dissociation occurs}}$$

$$= \frac{1 - \alpha + n\alpha}{1}$$

$\therefore$

$$i - 1 = \alpha(-1 + n)$$

or

$$\alpha = \frac{i - 1}{n - 1} \qquad (5.8)$$

The degree of dissociation may therefore be calculated.

Example Dissolving 0.12 mole of a monobasic organic acid in 1 kg water depresses the freezing point by 0.268 °C.

Calculate the degree of dissociation of the acid given that the cryoscopic constant for water is 1.86 °C kg^{-1}.

If no dissociation occurs, 1 mole of acid in 1 kg of water depresses the freezing point by 1.86 °C and so 0.12 mole acid would depress it by $1.86 \times 0.12 = 0.223$ °C.

Now

$$i = \frac{\text{observed freezing point depression}}{\text{calculated freezing point depression}} = \frac{0.268}{0.223}$$

and

$$\alpha = \frac{i - 1}{n - 1} = \frac{\dfrac{0.268}{0.223} - 1}{2 - 1} \qquad \begin{array}{l} (n = 2 \text{ since two ions are formed} \\ RCOOH \rightleftharpoons RCOO^- + H^+) \end{array}$$

$$= 1.2 - 1$$

$$= 0.2$$

i.e. the acid is 20% dissociated in this solution.

Distribution of a solute between two immiscible solvents

Ethers, trichloromethane, tetrachloromethane, and benzene all dissolve in water to a slight extent but, for most practical purposes, they may be considered to be immiscible.

If some solute is added to a pair of immiscible solvents and the mixture is shaken until equilibrium is attained, it is found that the ratio of its concentration in the two layers is constant at constant temperature, providing that the solute is in the same state in both solvents. This is known as the *distribution* or *partition law* and it may be expressed mathematically as

$$\frac{\text{Concentration of solute in solvent } X}{\text{Concentration of solute in solvent } Y} = K \tag{5.9}$$

where K is known as the *partition* or *distribution coefficient*. The law does not apply if the solubility of the solute is exceeded in the solvents.

The law needs modification if the solute is not in the same state in the two solvents.

a. Association in one solvent

An example of this occurs when ethanoic acid (acetic acid) or benzenecarboxylic acid (benzoic acid) is added to a pair of immiscible solvents such as water and benzene; association occurs in the latter solvent. The slight dissociation of the acids in water is ignored here.

Consider any solute, S, having its normal relative molecular mass in one solvent in which its concentration is C_1 whilst in a second solvent it exists largely as the associated molecule Sn, the total concentration being C_2. In the second solvent, the following equilibrium exists:

$$\text{Sn} \rightleftharpoons n\text{S}$$

According to the equilibrium law (page 111)

$$K = \frac{[\text{S}]^n}{[\text{Sn}]} \quad \text{where } K \text{ is the equilibrium constant}$$
$$\text{and [ ] represents concentration}$$

and so

$$[\text{S}] = \text{constant} \times \sqrt[n]{[\text{Sn}]} \tag{5.10}$$

where the constant $= \sqrt[n]{K}$

Since the solute is mainly in the associated form (except when the solution is extremely dilute), [Sn] is approximately equal to C_2, the total concentration in the second layer, and so substituting this in equation (5.10) gives

$$[\text{S}] \approx \text{constant} \times \sqrt[n]{C_2} \tag{5.11}$$

Now the distribution law applies only to the unassociated molecules in each solvent (i.e. the common state in each solvent) and so

$$\frac{C_1}{[\text{S}]} = \text{constant} \tag{5.12}$$

Substituting the value of [S] from equation (5.11) in equation (5.12) gives

$$\frac{C_1}{\sqrt[n]{C_2}} = \text{a constant} \qquad (5.13)$$

Equation (5.13) is confirmed by experimental results.

b. Dissociation in one solvent

When the solute is slightly dissociated in one solvent but not in the other, it is the undissociated molecules in each solvent which obey the law.

Consider a solute, S, which is undissociated in solvent X and slightly dissociated in solvent Y. If C_1 is the concentration of S in X and C_2 is the total concentration of S in Y, the undissociated solute in Y has a concentration of $C_2(1 - \alpha)$ where α is the degree of dissociation. In this case,

$$\frac{C_1}{C_2(1 - \alpha)} = \text{constant} \qquad (5.14)$$

The use of the distribution law and its various modifications is illustrated below.

Example   The partition coefficient of a solid, S, between two immiscible liquids A and B is 0.15. If 6 g of S were dissolved in 100 cm³ of A and extracted with 100 cm³ of B, what mass of S would remain in A?

Let the mass of S in B be x g, then the mass remaining in A is $6-x$ g.

$$C_A = \text{concentration of S in solvent A} = (6-x)/100$$
$$C_B = \text{concentration of S in solvent B} = x/100$$

According to equation (5.9)

$$\frac{C_A}{C_B} = 0.15$$

i.e.

$$\frac{(6-x)/100}{x/100} = 0.15$$

and so

$$1.15\,x = 6$$

or

$$x = 5.218$$

Therefore, the mass of S remaining in   A $= 6 - 5.218 = 0.782$ g.

Example   The following concentrations, in g l⁻¹, were found for the distribution of ethanoic acid (acetic acid) between water and tetrachloromethane:

in water	253	337	467	631
in tetrachloromethane	13.2	23.5	44.7	82.2

Given that the acid has its normal relative molecular mass in aqueous solution, show either numerically or graphically, that it exists as the dimer (two molecules paired up) in tetrachloromethane.

Numerical solution: Since association is occurring,

$$\frac{C_1}{\sqrt[n]{C_2}} = \text{constant},$$

i.e. equation (5.13) will be the equation involved. If the acid is present as a dimer in tetrachloromethane then

$$\frac{C_1}{\sqrt{C_2}} = \text{constant}$$

where C_1 = concentration of acid in water C_2 = concentration of acid in tetrachloromethane.

Therefore, in the four cases

$$K = \frac{253}{\sqrt{13.2}}; \quad K = \frac{337}{\sqrt{23.5}}; \quad K = \frac{467}{\sqrt{44.7}}; \quad K = \frac{631}{\sqrt{82.2}}$$

or $\qquad K = 69.63; \quad K = 69.51; \quad K = 69.85; \quad K = 69.60$

Since K is the same in each case, $n = 2$.

Graphical solution:

$$\frac{\text{Concentration of acid in water}}{\sqrt[n]{\begin{array}{c}\text{Concentration of acid in}\\ \text{tetrachloromethane}\end{array}}} = \text{constant}$$

Therefore

$$\text{Concentration of acid in water} = \sqrt[n]{\begin{array}{c}\text{Concentration of acid}\\ \text{in tetrachloromethane}\end{array}} \times \text{constant}$$

or $\quad$ lg conc. in water $= \dfrac{1}{n}$ lg conc. in tetrachloromethane + lg constant

where $n = 2$ if the acid is present as a dimer in tetrachloromethane.

Since this equation is of the form $a = bx + c$, a plot of lg conc. in water against lg conc. in tetrachloromethane (Figure 5.11) will give a straight line graph with gradient equal to $1/n$ and intercept equal to lg constant. The gradient n is seen to be equal to $\frac{1}{2}$ and so $n = 2$ as predicted.

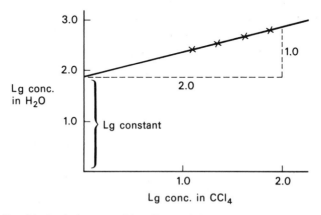

Figure 5.11 Graphical solution to problem illustrating association of solute in one of a pair of immiscible solvents

Example When 3.35 g of an aromatic hydroxy compound was shaken with 50 cm³ of tetrachloromethane and 100 cm³ of water it was found that 0.65 g of it were present in the organic layer. Calculate the degree of dissociation in the aqueous layer given that the distribution coefficient for the compound between tetrachloromethane and water is 0.547.

$$C_1 = \text{concentration of solute in } CCl_4 = 0.65/50 \, \text{g cm}^{-3}$$
$$C_2 = \text{concentration of solute in } H_2O = 2.7/100 \, \text{g cm}^{-3}$$

From equation (5.14) it follows that

$$1-\alpha = \frac{C_1}{\text{constant} \times C_2}$$

i.e.

$$1-\alpha = \frac{0.65/50}{0.547 \times 2.7/100}$$

and so

$$1 - \alpha = 0.880$$

or

$$\alpha = 0.12$$

The compound is therefore 12% dissociated in the aqueous solution.

3. Solutions of liquids in liquids

Consider two completely miscible ideal liquids, i.e. two liquids which show no heat or volume change when mixed and obey Raoult's law over the whole range of composition. The vapour pressure exerted by each liquid is proportional to its mole fraction, and a plot of partial pressure of each liquid against its mole fraction in the mixture will be a straight line passing through the origin (Figure 5.12). The total pressure, which is the sum of the partial pressures, is also a straight line. Several mixtures are known which obey Raoult's law fairly closely, e.g. benzene and methylbenzene or hexane and heptane. Figure 5.13 shows the type of vapour pressure and boiling point curves which are given by mixtures in which deviation from ideal behaviour is slight. The dotted lines in Figure 5.13(a) represent ideal behaviour.

In the temperature–composition diagram (Figure 5.13(b)) the curve marked liquid represents the variation of boiling point with the composition of the liquid mixture. The vapour curve shows the composition of the vapour in equilibrium with a liquid mixture at its boiling point. It can be seen from this that the vapour, which is in equilibrium with the liquid mixture at any

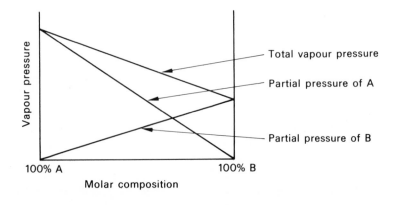

Figure 5.12   Vapour pressure–composition diagram for an ideal liquid mixture

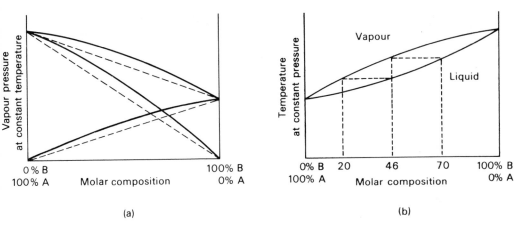

(a)

(b)

Figure 5.13   Vapour pressure and boiling point diagrams for near ideal liquid mixtures

temperature, always contains a greater proportion of the more volatile component. For example, if a mixture containing 70% B is distilled, the initial distillate contains only 46% B and if this is distilled the new distillate contains only 20% B. Instead of separating two liquids by several distillations in this manner, the distillation processes are combined in one operation by using a *fractionating column* between the distillation flask and the take-off head as illustrated in Figure 5.14. As the vapour ascends the column the higher boiling component condenses and runs back down. The returning high boiling liquid washes the rising vapour and helps to condense the high boiling component, and so, as the vapour ascends the column, it becomes richer in the lower boiling component. The greater the contact between the rising vapour and the returning liquid the greater the purification and this accounts for most columns being packed with glass helices, etc.

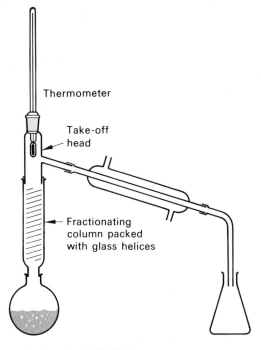

Figure 5.14   Apparatus for fractional distillation

The enrichment produced by boiling a liquid mixture and condensing the vapour back to liquid is known as *theoretical plate*. Fractionating columns are compared by the number of theoretical plates they achieve. The *height equivalent per theoretical plate* (H.E.T.P.) is often quoted for packed columns, for example, if a column with a packed length of 90 cm produces an enrichment corresponding to six theoretical plates, its H.E.T.P. value is 15 cm per plate.

Deviations from Raoult's law

If trichloromethane and ethoxyethane (diethyl ether) are mixed, it is found that the temperature rises thus indicating that intermolecular attractions are occurring. These attractions occur because the strongly electronegative chlorine atoms in trichloromethane attract electrons from the carbon which in turn takes more than its fair share of the electrons in the C—H bond. The hydrogen therefore becomes slightly positively charged and it attracts a lone pair of electrons on the oxygen in the ethoxyethane. (This is an example of hydrogen bonding — see page 37.)

$$\underset{\delta^-\mathrm{Cl}}{\overset{\mathrm{Cl}^{\delta-}}{\mathrm{Cl}}}{-}\overset{}{\underset{}{\mathrm{C}}}{-}\overset{\delta+}{\mathrm{H}} \text{-------------} :\overset{\mathrm{C_2H_5}}{\underset{\mathrm{C_2H_5}}{\mathrm{O}}}:$$

In contrast, when tetrachloromethane is mixed with ethanol the temperature falls and the inference here is that intermolecular attractions are being disrupted. In fact, the tetrachloromethane hinders the hydrogen bonding between the alcohol molecules.

It is apparent from the above examples that, if two liquids of different character are mixed, the tendency of their molecules to escape as vapour will be affected. If the molecules of the two liquids strongly attract one another the vapour pressure of each will be less than expected and *negative deviation* from Raoult's law is said to occur. In such cases, the vapour pressure curve may have a minimum value for a particular composition and the boiling point curve has a maximum (Figure 5.15). If a liquid mixture of this type is fractionally distilled, the initial distillate will be pure A or pure B depending on which component is present in excess of the mixture with composition, X (Figure 5.15). The boiling

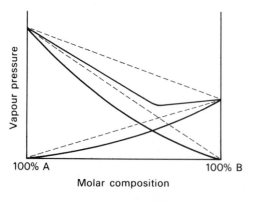

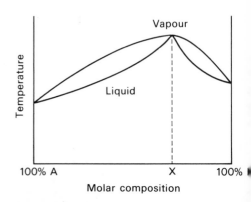

Figure 5.15   Vapour pressure and boiling point diagrams for a mixture with a maximum boiling point

point of the residue will slowly rise until the conditions at X prevail and the mixture with composition X will then distil over unaltered at a constant temperature. Liquid mixtures which distil at a definite temperature are known as *azeotropic* or *constant boiling point mixtures*. Aqueous solutions of nitric(V), sulphuric(VI), and hydrochloric acids behave in the above manner. The boiling point and composition of the azeotrope will naturally alter with changes in the external pressure.

If the attraction between the molecules of one liquid in a mixture are much greater than those between the molecules of the other liquid, the escaping tendency of the latter molecules will be increased. The partial pressure of this liquid will therefore be greater than expected and *positive deviation* from Raoult's law is said to occur. Liquid mixtures behaving like this include ethanol and water, ethanol and tetrachloromethane, and ethanol and ethyl ethanoate (ethyl acetate). The vapour pressure and boiling point curves for this type of mixture are shown by Figure 5.16. Fractional distillation will give the constant boiling point mixture as the distillate until one component is completely used up and then the remainder of the other component will distil over.

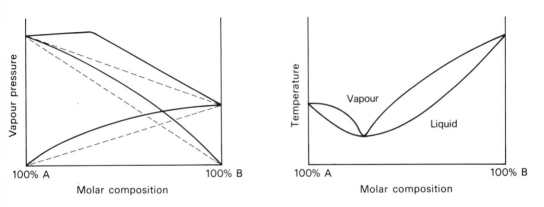

Figure 5.16   Vapour pressure and boiling point diagrams for a mixture with a minimum boiling point

Immiscible liquids

Pairs of liquids which are markedly different in polarity have very low mutual solubilities. For example, hexane (non-polar) and water (highly polar) are, for all practical purposes, immiscible. They act independently of each other and so there are three phases present, each liquid being in equilibrium with its own vapour. At any given temperature, the total pressure above the mixture is equal to the sum of the vapour pressures of the individual liquids and is independent of the amount of each liquid present. Equilibrium is attained faster if the mixture is agitated. The mixture will boil when the total pressure equals atmospheric pressure and consequently this temperature will be below the boiling point of either of the pure liquids. For example, it can be seen from Figure 5.17 that, whilst benzene and water boil at 80 and 100 °C respectively, a mixture of the two boils at 69.6 °C.

Advantage is taken of the above type of situation in the process of steam distillation described on page 272. Steam distillation is used in the purification of some organic compounds. It is based on the fact that a liquid which is insoluble in

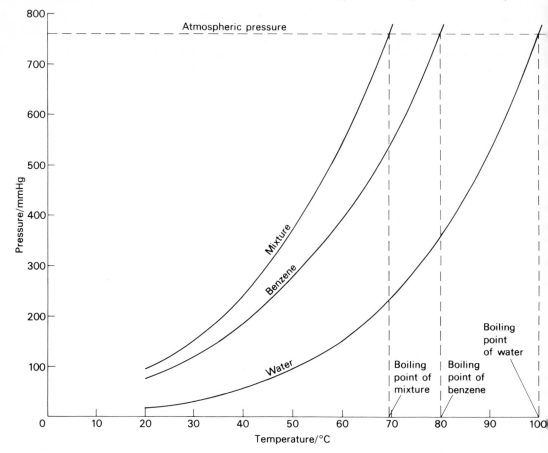

Figure 5.17   Vapour pressure curves of immiscible liquids

water distils in a current of steam. The composition of the distillate may be calculated from the relationship:

$$\frac{\text{moles of X}}{\text{moles of water}} = \frac{\text{vapour pressure of X}}{\text{vapour pressure of water}}$$

Example   Nitrobenzene, $C_6H_5NO_2$, steam distilled at a temperature of 98.7 °C when the atmospheric pressure was 747 mm. Calculate the percentage composition by mass of the distillate given that the vapour pressures of water and nitrobenzene at 98.7 °C are 725 mm and 22 mm respectively.

Consider 100 g of distillate. If it contains x g of nitrobenzene, it will contain $100 - x$ g of water. Now,

$$\frac{\text{moles of } C_6H_5NO_2}{\text{moles of } H_2O} = \frac{P_{C_6H_5NO_2}}{P_{H_2O}}$$

$$\frac{x/123}{(100-x)/18} = \frac{22}{725}$$

$$15756\,x = 270600$$

$$x = 17.2$$

Hence the distillate contains 17.2% nitrobenzene and 82.8% water.

It is apparent that, in order to obtain a reasonable amount of the organic compound in the distillate, it must have an appreciable vapour pressure near the boiling point of water. Like distillation under reduced pressure, the method is particularly useful when the compound decomposes at or near its boiling point.

Questions

5.1   State Henry's law. Outline how the solubility in water of (a) a sparingly soluble gas and (b) a very soluble gas at room temperature could be determined.
        The absorption coefficient at 0 °C and 760 mm pressure for sulphur dioxide is 79.8. Explain what this means. Discuss whether Henry's law applies in the case of this gas.   [University of London]

5.2   The vapour pressure P of a liquid is related to the thermodynamic (kelvin) temperature, T, and to the enthalpy of vaporisation ΔH_{vap} by the equation

$$2.303 \log_{10}P = - \frac{\Delta H_{vap}}{RT} + \text{constant.}$$

The table below gives values for the vapour pressures of water (expressed in mm. Hg) at various temperatures (expressed in kelvins).

Temperature	Reciprocal Temperature/K^{-1}	P/mm Hg	$\log_{10}$ (P/mm Hg)
343	0.002915	233.7	2.369
353	0.002833	355.1	2.550
363	0.002755	525.8	2.721
373	0.002681	760.0	2.881

(a) Plot $\log_{10}P$ against reciprocal temperature using the data in the table.

(b) Calculate ΔH_{vap} for water from your graph.

(c) ΔH_{vap} for hydrogen sulphide is 18.8 kJ mol^{-1}. Compare this value with your answer in (b) and explain the difference qualitatively in terms of the bonding in water and in hydrogen sulphide.

(d) What is the boiling point of water when the external pressure is 233.7 mm Hg?

(e) Give reasons for your answer in (d).   [J.M.B.]

5.3   State Raoult's law of vapour pressure lowering and explain qualitatively why a solution of a non-volatile solute in a volatile solvent has a higher boiling point than the solvent.
        At 25 °C, a saturated aqueous solution of calcium hydroxide, $Ca(OH)_2$, possesses a vapour pressure of 23.765 mm Hg. At the same temperature, the vapour pressure of water alone is 23.790 mm Hg. Assuming the calcium hydroxide to be completely ionised in solution, what is its solubility in moles per litre at 25 °C?   [University of London]

5.4   State Raoult's law and explain concisely its application in the determination of relative molecular masses by the boiling point method. Outline the experimental procedure which you would adopt for the determination of the relative molecular mass of a non-electrolyte which is soluble in water. Draw a sketch of the apparatus that you would employ.
        In an experiment, a 5% solution of glucose ($C_6H_{12}O_6$) in water was found to give the same boiling point elevation as a 3.3% aqueous solution of the simple carbohydrate erythrose. If the composition of simple carbohydrates may be represented by the general formula $C_nH_{2n}O_n$, what is the molecular formula of erythrose?   [University of London]

5.5   Define the boiling point and freezing point of water in relation to its vapour pressure.

By means of sketch graphs (not on graph paper) show qualitatively the relationship between the vapour pressure-temperature curves for water and for dilute aqueous solutions of (a) a non-electrolyte, (b) potassium chloride of the same molarity. From these graphs deduce qualitatively the effect of the solute on (i) the freezing point, (ii) the boiling point of the solvent.

Calculate the vapour pressure at 20 °C of an aqueous solution containing 20 g of sodium hydroxide in 110 g of solution, assuming that the base is completely dissociated and that the vapour pressure of water at 20 °C is 17.4 mm of mercury.   [J.M.B.]

5.6   Explain what is meant by the boiling point elevation constant of a liquid. How would you measure the relative molecular mass of a compound by the method of the elevation of boiling point of a liquid? Give concise experimental details. State, with reasons, why this method could be used with aqueous solutions to obtain the relative molecular mass of urea, but not that of ethanoic (acetic) acid.

A solution of 2.8 g of cadmium iodide (CdI_2) in 20 g of water boiled at 100.2 °C at normal pressure. Calculate the relative molecular mass of the solute and comment on the result. (The boiling point elevation constant for water is 0.52 °C per 1000 g.)   [J.M.B.]

5.7   Describe how you would determine the relative molecular mass of a non-volatile substance by freezing point measurements. Briefly indicate what precautions you would take to obtain an accurate result and what are the likely sources of error in the method you describe.

A solution of 2.0 g of glucose ($C_6H_{12}O_6$) in 100 g of water freezes at –0.206 °C whereas a 0.02 M iron(III) chloride solution freezes at –0.137 °C. What information can you obtain from these results?

[University of London]

5.8   Show how the lowering of vapour pressure of a solvent by a non-volatile solute is related to the depression of the freezing point of the solvent by the solute.

A solution of 3.136 g of sulphur in 100 g of naphthalene (m.p. 80.1 °C) showed a lowering of freezing point of 0.830 °C, while a solution of 3.123 g of iodine in 100 g of naphthalene gave a depression of 0.848 °C. If the molecular formula of iodine in naphthalene solution is I_2, what is that of sulphur?

[University of London]

5.9   Explain what is understood by the term 'colligative property'. Name three such properties and describe the practical determination of one such property. Explain how such a measurement can be used to determine relative molecular mass. Give two instances when the method selected would give unexpected results. A 0.10 M solution of $HgCl_2$ in water freezes at –0.186 °C whereas a 0.10 M solution of $Hg(NO_3)_2$ freezes at –0.558 °C. Deduce the structural states of these two compounds. Given that the melting point of mercury(II) chloride is 280 °C, what is the predominant type of bonding in this compound?

(The freezing point depression constant for water is 1.86 K for 1 mole of solute in 1 litre of solution.)   [J.M.B.]

5.10   Describe, giving all essential practical details, how the relative molecular mass of a compound may be found by the method of the depression of freezing point.

0.148 g of a weak acid, titrated with M/10 sodium hydroxide, required 20.0 cm³ of the alkali for neutralisation. 1.00 g of the acid, dissolved in 100 g of water, depressed the freezing point by 0.25 °C. 1.00 g of the acid, dissolved in 100 g of benzene, depressed the freezing point by 0.35 °C. (The freezing point constants of water and benzene, per 1000 g, are 1.85 °C and 5.12 °C respectively.)

What deductions can be made from the above information?

[J.M.B.]

5.11   Explain how the relative molecular mass of a solute, dissolved in a solvent, may be determined by boiling point or freezing point measurement. In your explanations sketch curves of vapour pressure against temperature. Outline an experimental technique for one of these measurements.

What method based on colligative properties would you use to measure a very high relative molecular mass?

What volume of ethane-1,2-diol (ethylene glycol), $C_2H_6O_2$ (density = $1.116\,g\,cm^{-3}$) must be added to 8 litres of water to produce a solution which will freeze at $-2.0\,°C$?

The depression of freezing point for 1 mole of solute dissolved in 1 litre of water is $1.86\,°C$. [J.M.B.]

5.12 Describe the phenomenon of osmosis, illustrating your answer with one appropriate example.

Define the term 'osmotic pressure' and describe, in outline only, a method for the determination of the osmotic pressure of a solution of sugar. Calculate the osmotic pressure at $25\,°C$ of an aqueous solution of cane sugar ($C_{12}H_{22}O_{11}$) containing 3.42 grams per litre. [University of London]

5.13 An aqueous solution A contains 1.71 g of cane sugar ($C_{12}H_{22}O_{11}$) in $112\,cm^3$ of solution. An aqueous solution B contains 0.15 g of urea $CO(NH_2)_2$ in $44.8\,cm^3$ of solution.

(a) Calculate the molarity of each of the solutions A and B.

(b) Calculate the osmotic pressure of each solution, in atmospheres, at $25\,°C$.

(c) (i) If the two solutions at $25\,°C$, exposed to the atmosphere, are separated by a semi-permeable membrane, show how you would deduce theoretically that some flow will take place. State what will flow and in which direction.

(ii) If you wished to prevent any flow in the system described in (c) (i), what pressure (in atm) would have to be applied and where?

(iii) What would be the consequences of applying a pressure of 0.5 atmospheres to solution B?

(d) If the solutions A and B were cooled, which would require the lower temperature for commencement of freezing? [J.M.B.]

5.14 Define the term osmotic pressure.

State the laws governing the osmotic pressure of a dilute solution and show how they are analogous to the gas laws.

A sample of poly(phenylethene) (polystyrene) has a relative molecular mass of 1.5×10^5; calculate the concentration in $g\,litre^{-1}$ of the solution of it in methylbenzene which would have an osmotic pressure of 3.05×10^{-4} atm at $25\,°C$.

5.15 State the law governing the distribution of a solute between two immiscible solvents.

Describe an experiment to determine the partition coefficient of butanedioic acid (succinic acid) between ethoxyethane (ether) and water.

In experiments on the distribution of a compound X between (a) the immiscible solvents S_1 and S_2, and (b) the immiscible solvents S_1 and S_3 the following results were obtained:

(a) Distribution between S_1 and S_2

Conc. of X in S_1/gl^{-1}	12.1	7.0	2.4
Conc. of X in S_2/gl^{-1}	2.2	1.3	0.44

(b) Distribution between S_1 and S_3

Conc. of X in S_1/gl^{-1}	1.5	1.95	2.9
Conc. of X in S_3/gl^{-1}	24.2	41.2	97.0

In S_1 the relative molecular mass of X is M. Use the above data to determine the relative molecular mass of X: (a) in S_2, (b) in S_3. [J.M.B.]

5.16 State the law governing the distribution of a solute between immiscible solvents. 0.008 moles of a compound Q in $10\,cm^3$ benzene solution can be in equilibrium with 0.0005 moles in $50\,cm^3$ aqueous solution. Calculate the distribution (partition) coefficient of Q between benzene and water assuming that Q has the same relative molecular mass in both solvents.

Another compound P reacts with Q in aqueous solution thus:

$$P + Q \rightleftharpoons PQ$$

All three compounds are soluble in water but only Q is soluble in benzene. An aqueous solution of P having a concentration of 0.102 moles per litre is shaken with a benzene solution of Q until equilibrium is reached. The benzene layer then contains 0.080 moles Q/litre and the aqueous layer 0.003 moles Q/litre.

Use the distribution coefficient to determine the concentration of free Q in the aqueous layer.

Then find (a) the concentration of combined Q, and (b) that of free P.

[J.M.B.]

5.17   Draw clear diagrams to show the variation in saturated vapour pressure with composition for pairs of liquids which (i) obey Raoult's law, (ii) show positive deviations from that law. Your diagrams should show both liquid and vapour compositions.
Discuss

(a)   what happens when a mixture as in (i) is distilled, and

(b)   the fractional distillation of a pair of liquids which form a mixture of maximum boiling point, considering both the actual maximum boiling point mixture and mixtures of other compositions.

[University of London]

5.18 (a)   Discuss the factors affecting the distribution of solutes between pairs of non-miscible solvents.

(b)   Explain how you would use the phenomenon of distribution to determine the equilibrium constant of the system

$$I_2 + I^- \rightleftharpoons I_3^-$$

or any other system of your choice.

(c)   An organic liquid X and water are non-miscible. A mixture of X and water boils at 95 °C and 760 mm Hg. At this temperature water has a vapour pressure of 640 mm Hg. What is the vapour pressure of X at 95 °C? Explain your calculation. The distillate contains $X : H_2O$ in the ratio 1.65 : 1 by mass. What is the relative molecular mass of X?   [J.M.B.]

6 Chemical Equilibria

Theoretically, all reactions can be considered to be reversible to some extent, some more so than others. For example, ethanoic acid (acetic acid) and ethanol react together to give ethyl ethanoate (ethyl acetate) and water, and these products react together to give ethanoic acid and ethanol:

$$CH_3 \cdot COOH + C_2H_5OH \rightleftharpoons CH_3 \cdot COOC_2H_5 + H_2O$$

As will be seen later, the rate of a reaction is proportional to the concentration of the reactants. Hence, if ethanoic acid and ethanol are mixed, the ester and water will initially be formed at a fast rate, but this will slow up as the concentrations of the reactants fall. Conversely, the ester and water will have gradually increasing concentrations, and so the rate of their reaction will increase. After some time, the rates of the two reactions will become equal and a *dynamic equilibrium* will be reached.

THE EQUILIBRIUM LAW

The equilibrium law may be derived by use of the *law of mass action* (Guldberg and Waage, 1864) which states that at constant temperature, the rate of a chemical reaction is directly proportional to the active mass of the reacting substances. For most practical purposes the *active mass* may be taken as the concentration expressed in $mol\,litre^{-1}$. (Active mass takes account of intermolecular attractions.)

Consider any reaction

$$A + B \rightleftharpoons C + D$$

If initially there are a moles of A and b moles of B, and at equilibrium x moles of A and B have been used up, then:

	A	+	B	$\rightleftharpoons$	C	+	D
Initially	a		b		0		0
At equilibrium	$(a-x)$		$(b-x)$		x		x

If V is the volume at equilibrium, then according to the law of mass action (and assuming, for the sake of simplicity, that the reactions are first order with respect to each reactant):

$$\text{rate of reaction } A + B \rightarrow C + D \propto \frac{(a-x)}{V} \times \frac{(b-x)}{V}$$

$$= k_1 \frac{(a-x)\,(b-x)}{V^2}$$

$$\text{rate of reaction } C + D \rightarrow A + B \propto \frac{x}{V} \times \frac{x}{V} = k_2 \frac{x^2}{V^2}$$

where k_1 and k_2 are *velocity constants*.

At equilibrium these rates of reaction are equal, i.e. $k_1 \dfrac{(a-x)\,(b-x)}{V^2} = k_2 \dfrac{x^2}{V^2}$

or

$$\frac{x^2}{(a-x)(b-x)} = \frac{k_1}{k_2} = K \qquad (6.1)$$

where K is known as the *equilibrium constant*. It should be noted that, by convention, K is always $\frac{k_1}{k_2}$ and not $\frac{k_2}{k_1}$.

Equation (6.1) may also be written as

$$K = \frac{[C][D]}{[A][B]} \qquad (6.2)$$

where [ ] stands for equilibrium concentration. If the reaction involves solids, liquids, or solutions the equilibrium constant is denoted by K_c, the suffix c indicating concentration in moles or moles litres^{-1} but, for gaseous reactions, K_p is used, the suffix p indicating that the concentrations are in terms of partial pressures. In the general reaction

$$aA + bB \rightleftharpoons cC + dD$$

where the small letters represent the number of molecules of each substance in the equation, equation (6.2) becomes

$$K = \frac{[C]^c[D]^d}{[A]^a[B]^b} \qquad (6.3)$$

Equation (6.3) represents the *equilibrium law*, i.e. if the concentration of all the substances at equilibrium are raised to the power of the number of their molecules in the equation, then the product of the concentrations of the products, divided by the product of the concentrations of the reactants, is a constant, provided that the temperature is constant. If the temperature is raised, a new equilibrium is set up and K assumes a different value.

The equilibrium constant for a reaction at a given temperature may be calculated once the composition of the equilibrium mixture, obtained from a known amount of the reactants, has been determined. Consider, for example, the reaction between ethanoic acid (acetic acid) and ethanol. When one mole of each are heated together at a constant temperature of 25 °C until equilibrium has been reached, titration of the reaction mixture with standard alkali shows that two-thirds of the acid have been used up

$$CH_3 \cdot COOH + C_2H_5OH \rightleftharpoons CH_3 \cdot COOC_2H_5 + H_2O$$

Initially	1	1	0	0
At equilibrium	$\frac{1}{3}$	$\frac{1}{3}$	$\frac{2}{3}$	$\frac{2}{3}$

Now, according to equation (6.3)

$$K_c = \frac{[CH_3 \cdot COOC_2H_5][H_2O]}{[CH_3 \cdot COOH][C_2H_5OH]}$$

$$= \frac{\frac{2}{3} \times \frac{2}{3}}{\frac{1}{3} \times \frac{1}{3}}$$

$$= 4$$

According to the equilibrium law, K_c should remain constant no matter how the concentrations are varied and this is confirmed experimentally provided that all the experiments are performed at the same temperature.

Once the value of the equilibrium constant has been established, it is possible to calculate the composition of the equilibrium mixture obtained from any given quantity of the reactants.

The use of the equilibrium law is illustrated by the examples below.

Example Calculate the composition of the equilibrium mixture when 240 g ethanoic acid (acetic acid) are warmed with 46 g ethanol at 25 °C, the equilibrium constant at this temperature being 4.

$$CH_3 \cdot COOH + C_2H_5OH \rightleftharpoons CH_3 \cdot COOC_2H_5 + H_2O$$

Initially	$\dfrac{240}{60}$ mole	$\dfrac{46}{46}$ mole	0	0
At equilibrium	$4-x$	$1-x$	x	x

where $x =$ moles of ethanoic acid and ethanol that have reacted.

$$K_c = \frac{[CH_3 \cdot COOC_2H_5]\,[H_2O]}{[CH_3 \cdot COOH]\,[C_2H_5OH]}$$

hence

$$4 = \frac{x^2}{(4-x)(1-x)}$$

This gives a quadratic equation $3x^2 - 20x + 16 = 0$ which is solved by using the formula

$$x = \frac{-b \pm \sqrt{b^2 - 4ac}}{2a}$$

i.e

$$x = \frac{20 \pm \sqrt{20^2 - 4 \times 3 \times 16}}{2 \times 3} = 5.74 \quad \text{or} \quad 0.93$$

The former value is impossible and so the equilibrium mixture has the following composition.

Ethanoic acid	$4 - 0.93 = 3.07$ mole
Ethanol	$1 - 0.93 = 0.07$ mole
Ethyl ethanoate	$= 0.93$ mole
Water	$= 0.93$ mole

Example Nitrogen and oxygen combine endothermically at elevated temperatures according to the equation

$$N_2 + O_2 \rightleftharpoons 2NO$$

If the equilibrium constant for the reaction is 4.3×10^{-3} at 3000 °C and 1 atmosphere total pressure, calculate the composition of the equilibrium mixture when equal volumes of nitrogen and oxygen are heated under these conditions.

Explain the effect on the position of equilibrium when (a) the temperature is raised, (b) the pressure is increased, and (c) a catalyst is added.

$$N_2 + O_2 \rightleftharpoons 2NO$$

Initially	1	1	0
At equilibrium	$1-x$	$1-x$	$2x$

$$K_c = \frac{[NO]^2}{[N_2][O_2]}$$

and so

$$4.3 \times 10^{-3} = \frac{(2x)^2}{(1-x)(1-x)}$$

or

$$3.9957x^2 + 0.0086x - 0.0043 = 0$$

Solving this equation gives $x = 0.032$ and so the equilibrium mixture contains 0.968 mole of nitrogen, 0.968 mole of oxygen, and 0.064 mole of nitrogen oxide.

(a) Since the reaction is endothermic, increasing the temperature will be counteracted by the equilibrium moving to the right, i.e. more nitrogen oxide will be formed in an attempt to lower the temperature.

(b) Increasing the pressure will not affect the position of equilibrium since there is no volume change in the reaction.

(c) There will be no change in the position of equilibrium since catalysts affect only the rate of a reaction.

(The answer to parts (a) and (b) require a knowledge of Le Chatelier's principle— page 116.)

Example   When one mole of nitrogen and three moles of hydrogen were heated under a pressure of 80 atmospheres, the equilibrium mixture was found to contain 0.7 mole of ammonia. Calculate the value of K_p under these conditions.

$$N_2 + 3H_2 \rightleftharpoons 2NH_3$$

Initially	1	3	0
At equilibrium	$1-x$	$3-3x$	$2x$

At equilibrium there are $(1-x) + (3-3x) + 2x = 4-2x$ moles present and since the partial pressure of each gas is equal to its mole fraction times the total pressure,

$$P_{N_2} = \frac{1-x}{4-2x} \times 80, \qquad P_{H_2} = \frac{3-3x}{4-2x} \times 80$$

and

$$P_{NH_3} = \frac{2x}{4-2x} \times 80$$

But it is known that $2x = 0.7$, and so $x = 0.35$.

$$\therefore \qquad P_{N_2} = \frac{0.65 \times 80}{3.3} \qquad P_{H_2} = \frac{1.95 \times 80}{3.3}$$

and

$$P_{NH_3} = \frac{0.7 \times 80}{3.3}$$

Now

$$Kp = \frac{(P_{NH_3})^2}{(P_{N_2}) \times (P_{H_2})^3}$$

$$= \frac{\left(\dfrac{0.7 \times 80}{3.3}\right)^2}{\left(\dfrac{0.65 \times 80}{3.3}\right)\left(\dfrac{1.95 \times 80}{3.3}\right)^3}$$

$$= 1.73 \times 10^{-4} \text{ atm}^{-2}$$

Note that in this case K_p has units, i.e.

$$\frac{\text{atm}^2}{\text{atm} \times \text{atm}^3} = \text{atm}^{-2}$$

The equilibrium constant could also have been calculated in terms of moles.

Calculation of equilibrium constants for gas–solid equilibria

Strictly speaking, the equilibrium law can be applied only to homogeneous systems (systems in which the reactants and products are in a single phase) and so in gas–solid equilibria it is just the vapour phase which is considered.

In a solid's crystal lattice, the surface molecules are subject to uneven attractive forces and so there is a tendency to escape as vapour since they are less firmly held than the rest. Hence all solids exert a vapour pressure and at any temperature it is independent of the amount of solid present. The concentration of a solid in the vapour phase is therefore taken as its partial pressure.

Consider the thermal decomposition of calcium carbonate in an evacuated sealed tube:

$$CaCO_3(s) \rightleftharpoons CaO(s) + CO_2(g)$$

For the vapour phase
$$\frac{(P_{CaO}) \times (P_{CO_2})}{(P_{CaCO_3})} = \text{constant}$$

But at any given temperature, P_{CaO} and P_{CaCO_3} are constant and so

$$\text{constant} = \frac{\text{constant} \times (P_{CO_2})}{\text{constant}}$$

or
$$K_p = P_{CO_2}$$

This means that, at any given temperature, the partial pressure of carbon dioxide is constant and independent of the amounts of calcium carbonate and calcium oxide present.

The reversible reaction between iron and steam may be treated similarly.

$$3Fe(s) + 4H_2O(g) \rightleftharpoons Fe_3O_4(s) + 4H_2(g)$$

Applying the equilibrium law to the gaseous phase

$$\frac{(P_{Fe_3O_4}) \times (P_{H_2})^4}{(P_{Fe})^3 \times (P_{H_2O})^4} = \text{constant}$$

But at any given temperature, $P_{Fe_3O_4}$ and P_{Fe} are constant and so

$$\text{constant} = \frac{(P_{H_2})^4}{(P_{H_2O})^4}$$

or
$$K_p = \frac{P_{H_2}}{P_{H_2O}}$$

This means that, at any temperature, there is a constant ratio between the amounts of hydrogen and water vapour present, provided that the system is at equilibrium.

It should be noted that, if the hydrogen is allowed to escape, all the water will be converted to hydrogen and the iron to iron(II) diiron(III) oxide, Fe_3O_4. A considerable amount of hydrogen is, in fact, made industrially by this process. The reverse process will similarly proceed to completion if the water is allowed to escape.

Le Chatelier's principle

The way in which the position of equilibrium is affected by altering the conditions is summarised by *Le Chatelier's principle:* if the conditions of a system at equilibrium are changed, the system moves in such a way as to oppose the effects of the change. Thus, increasing the concentration of the reactants in a reaction results in the formation of more of the products so that the equilibrium is maintained.

Increasing the temperature of a reversible reaction results in an increase in the rate of the forward and reverse reactions, but to different extents. The equilibrium moves so as to oppose the increase in temperature. Therefore, when the temperature is increased for an exothermic reaction, the system counteracts this by an absorption of heat and the yield of product is decreased. On the other hand, if the temperature is lowered, the system responds by proceeding further and giving out more heat. However, lowering the temperature lowers the reaction rate (Chapter 8) and so a catalyst is normally used.

Decreasing the temperature of an endothermic reaction decreases its rate whilst an increase in temperature increases its rate. Since increasing the temperature increases the rate, a catalyst is not normally used.

If in a reaction the number of moles of gaseous products exceeds the number of moles of gaseous reactants, then increasing the total pressure will result in the reaction moving to the left in order to reduce the pressure. However, if the number of moles of gaseous products is less than those of the gaseous reactants, the system will counteract an increase in the total pressure by moving to the right. When the number of moles of gaseous reactants and products are the same, the position of equilibrium will be unaffected by a change in the total pressure.

APPLICATIONS OF LE CHATELIER'S PRINCIPLE

1. The Haber process

At room temperature nitrogen and hydrogen are in equilibrium

$$N_2(g) + 3H_2(g) \rightleftharpoons 2NH_3(g); \quad \Delta H^{\ominus}_{298} = -92 \text{ kJ}$$

The position of equilibrium will move over to the right as the temperature is lowered, but at these temperatures the rate of attaining equilibrium is so slow that the reaction is not practicable. A catalyst is therefore required so that the rate will be increased without altering the yield of ammonia.

Since the volume decreases as ammonia is formed (four volumes of reactants give two volumes of product), an increase in pressure will result in an increase in the yield of ammonia.

In this process a compromise has to be found between the opposing factors of reaction rate and yield. The catalyst is more effective at high temperatures and so it is used at the lowest temperature consistent with a satisfactory rate of reaction, i.e. 550 °C. Pressures of up to 1000 atmospheres are used and the catalyst consists of finely divided iron with molybdenum as promoter.

2. The Contact process

$$2SO_2(g) + O_2(g) \rightleftharpoons 2SO_3(g); \quad \Delta H^{\ominus}_{298} = -197 \text{ kJ}$$

At room temperature, good yields of sulphur(VI) oxide will be obtained but the rate of reaction is slow. Again it is necessary to compromise between the reaction rate and yield and so the platinised asbestos or vanadium(V) oxide catalyst is used at temperatures of up to 450 °C. High pressures would increase the yield (three volumes of reactants give two volumes of product), but this is satisfactory at about two atmospheres.

3. The Birkeland-Eyde process (now obsolete)

$$N_2(g) + O_2(g) \rightleftharpoons 2NO(g); \quad \Delta H^{\ominus}_{298} = +180 \text{ kJ}$$

The position of equilibrium will be unaffected by pressure since the volume of the reactants and product are the same. The yield (and rate) will be increased by increasing the temperature; this is illustrated by the yields at 1500 °C, 2000 °C, and 3000 °C being 0.1, 2, and 5% by volume respectively.

4. Salt solubility

Most solids dissolve with absorption of heat, and so, if a saturated solution is in contact with solid, and the temperature is raised, more solid dissolves. In this way the applied heat is counteracted by an absorption of heat.

Questions

6.1   State Le Chatelier's principle and the law of mass action. Write expressions for the equilibrium constants of the following reactions:

(a) $N_2 + 3H_2 \rightleftharpoons 2NH_3$

(b) $CH_3 \cdot COOH + C_2H_5OH \rightleftharpoons CH_3 \cdot COOC_2H_5 + H_2O$

(c) $3Fe + 4H_2O \rightleftharpoons Fe_3O_4 + 4H_2$

(d) $H_2 + I_2 \rightleftharpoons 2HI$

The reaction $2SO_2 + O_2 \rightleftharpoons 2SO_3$ is exothermic; explain the effect on the rate of reaction and the position of equilibrium of (i) altering the pressure, (ii) changing the temperature, and (iii) using a vanadium(V) oxide catalyst.

6.2   Give an account of the influences of pressure and temperature on (a) the rate and (b) the position of equilibrium in a reversible reaction. Illustrate your answer by reference to the systems:

$$2SO_2 + O_2 \rightleftharpoons 2SO_3 \qquad \text{exothermic}$$
$$CO_2 + H_2 \rightleftharpoons CO + H_2O \qquad \text{endothermic}$$

Comment on the feasibility of using a catalyst in the reactions.

6.3   Explain what is meant by the terms 'velocity constant' and 'equilibrium constant'. Nitrogen and oxygen combine at high temperatures, with absorption of heat, according to the equation:

$$N_2 + O_2 \rightleftharpoons 2NO$$

The equilibrium constant for this reaction, at 2680 K and 1 atm total pressure, is 3.6×10^{-3}. Equal volumes of nitrogen and oxygen are mixed, at 2680 K and 1 atm total pressure, and allowed to react until equilibrium is reached. Calculate the fraction of the original nitrogen which is used in the reaction, and the fraction (by volume) of nitrogen oxide in the equilibrium mixture. Is the yield increased, decreased, or unchanged when (a) the pressure is increased to 10 atm, (b) the temperature is raised to 2780 K, (c) a catalyst is added? Give reasons.   [J.M.B.]

6.4   Explain the meaning of the terms 'equilibrium constant' and 'velocity constant'. Write an expression for the equilibrium constant for the reaction $H_2 + I_2 \rightleftharpoons 2HI$. Show how the equilibrium constant is related to the velocity constants of the forward and backward reactions. At 337 °C, 0.01 mole hydrogen and 0.01 mole iodine vapour are mixed together in a total volume of 1 litre with a total pressure of 1 atmosphere. The initial rate of reaction is 9×10^{-8} mol s^{-1}.

For this reaction, at a constant temperature of 337 °C:

(a) Why would you expect the reaction rate to decrease with time?

(b) What would be the initial rate of reaction if the pressure of the mixture were doubled?

(c) Calculate the velocity constant for the reaction between hydrogen and iodine.

(d) If the velocity constant for the reverse reaction (decomposition of hydrogen iodide) equals

$$6.0 \times 10^{-6} \left(\frac{mol \times s}{litre} \right)^{-1}$$

calculate the equilibrium constant.

(e) What would be the effect of doubling the pressure on the equilibrium mixture from this reaction?   [J.M.B.]

6.5   State the equilibrium law.

Nitrogen and hydrogen, in the ratio of $1:3$, were allowed to react at 450 °C under a pressure of 45 atm. When equilibrium has been attained, the mixture was found to contain 8% by volume of ammonia. Calculate K_p for the reaction.

6.6 When the forward reaction leading to the high temperature equilibrium

$$CaCO_3 \rightleftharpoons CaO + CO_2$$

takes place, heat is absorbed. Show what expression is obtained for the equilibrium constant by application of the law of mass action to the equilibrium and explain what this expression implies with regard to the pressure of the system. Why is this expression of a different form from that derived for a homogeneous gaseous equilibrium?

If the reaction has come to an equilibrium at a high temperature, deduce what would be the effect of

(a) an increase in temperature;

(b) an increase of pressure at constant temperature;

(c) the replacement of half the carbon dioxide in the gas space by an equal volume of oxygen;

(d) the addition of more calcium carbonate. [J.M.B.]

6.7 State the law of mass action. How does its application to a chemical equilibrium in an entirely gaseous system differ from its application to a chemical equilibrium involving both solid and gaseous substances?

The equation for the action of steam on iron at temperatures of about 700–800 °C is

$$3Fe + 4H_2O \rightleftharpoons Fe_3O_4 + 4H_2$$

What relationship showing the behaviour of this reaction at equilibrium is obtained by application of the Law of Mass Action?

If the reaction has reached equilibrium at 750 °C, state and explain what you would expect to happen when

(a) the total pressure on the system is increased;

(b) more iron is added;

(c) more water vapour is added, but the total volume is kept constant;

(d) more water vapour is added, but the total volume is increased so that the partial pressure of the water vapour remains the same.

[J.M.B.]

7 Ionic Equilibria

Solubility products

The equilibrium law may be applied to the equilibrium that exists between an almost insoluble solid and its ions in solution and this gives the conditions for precipitation of that solid from solution. Suppose solid silver chloride is stirred with water, at constant temperature, until the solution is saturated. When this state is attained, the rate at which Ag^+ and Cl^- ions go into solution is the same as the rate at which they come out of solution back into the crystal lattice, i.e. the system is at equilibrium:

$$AgCl(s) \rightleftharpoons Ag^+(aq) + Cl^-(aq)$$

Applying the equilibrium law gives

$$\frac{[Ag^+][Cl^-]}{[AgCl]} = \text{constant}$$

Now, in heterogeneous systems, the active mass of any solid is taken as being constant at constant temperature and so:

$$[Ag^+][Cl^-] = K_s \qquad (7.1)$$

where K_s is known as the *solubility product*. Equation (7.1) shows the maximum value of the ionic product of Ag^+ and Cl^- ions that can exist in solution without precipitation taking place. If the conditions are altered so that $[Ag^+][Cl^-]$ would exceed the value of K_s at that temperature, as for example, if Cl^- (HCl) is added, silver chloride is precipitated until the value of K_s is restored. This precipitation from a solution of a sparingly soluble salt, by adding another substance with the same anion or cation, is known as the *common ion effect*.

It should be noted that solubility product is applicable only to sparingly soluble salts. With more soluble salts the concentration of ions in solution is such that interionic attractions occur to an appreciable extent and so active mass and concentration are not the same. Solubility product may be defined as the maximum product of the concentration of the ions of a sparingly soluble electrolyte that can exist in solution at a given temperature.

Solubility product has many applications, as for example in qualitative inorganic analysis. Thus, hydrogen sulphide dissociates as follows:

$$H_2S \rightleftharpoons 2H^+ + S^{2-}$$

Now if hydrochloric acid is added, $[H^+]$ is greatly increased and so $[S^{2-}]$ is decreased in order to maintain the value of K. Therefore, if hydrogen sulphide is passed through an acidified solution of Bi^{3+}, Hg^{2+}, Cu^{2+}, Cd^{2+}, Sn^{4+}, Ni^{2+}, Zn^{2+}, Co^{2+}, and Mn^{2+} ions, only the sulphides with very low solubility products (Bi, Hg, Cu, Cd, and Sn) will be precipitated. The solubility products of the remaining sulphides are not exceeded and so they stay in solution. In this way the ions are separated into two groups.

Examples of calculations involving solubility products are given opposite.

Example Calculate, in g litre^{-1}, the solubility of silver chloride at 25 °C if its solubility product is $2 \times 10^{-10} \, mol^2 litre^{-2}$ at that temperature. Ag = 108; Cl = 35.5

$$[Ag^+] [Cl^-] = 2 \times 10^{-10}$$

$$[Ag^+] = [Cl^-] = \sqrt{2 \times 10^{-10}}$$

$$= 1.414 \times 10^{-5} \, mol \, litre^{-1}$$

But, the silver chloride concentration will be the same as that of its ions, i.e.

$$[AgCl] = 1.414 \times 10^{-5} \, mol \, litre^{-1}$$

$$= 1.414 \times 10^{-5} \times 143.5 \, g \, litre^{-1}$$

$$= 2.029 \times 10^{-3} \, g \, litre^{-1}$$

Example Calculate the solubility product of calcium fluoride at 25 °C given that its solubility at this temperature is $2.31 \times 10^{-4} \, mol \, litre^{-1}$. What mass of calcium fluoride will dissolve in 1 litre of 0.1 M hydrogen fluoride?

$$K_s = [Ca^{2+}] [F^-]^2$$

$$= (2.31 \times 10^{-4}) (2 \times 2.31 \times 10^{-4})^2 \qquad \text{(Note } [F^-] = 2 \times [Ca^{2+}])$$

$$= 4.93 \times 10^{-11} \, mol^3 litre^{-3}$$

$$K_s = [Ca^{2+}] [F^-]^2$$

$$4.93 \times 10^{-11} = [Ca^{2+}] \times (0.1)^2 \qquad ([F^-] \text{ from } CaF_2 \ll [F^-]$$
$$\text{from HF)}$$

$$[Ca^{2+}] = \frac{4.93 \times 10^{-11}}{0.01}$$

$$= 4.93 \times 10^{-9} \, mol \, litre^{-1}$$

Hence 1 litre of 0.1 M HF dissolves $4.93 \times 10^{-9} \, mol \, CaF_2$
$$= 4.93 \times 10^{-9} \times 78 \, g \, CaF_2$$
$$= 3.845 \times 10^{-7} \, g \, CaF_2$$

ACIDS AND BASES

According to Brønsted and Lowry (1923), acids may be defined as proton donors and bases as proton acceptors. Acids and bases are therefore related in the following way:

$$\underset{\text{acid}}{HA} \rightleftharpoons \underset{\text{proton}}{H^+} + \underset{\text{base}}{A^-}$$

A^- is said to be the conjugate base of the acid HA. Every acid has a conjugate base and every base has a conjugate acid.

For an acid to behave as an acid, it is necessary to have a base of some kind present to accept the proton. Hydrogen chloride, for example, does not behave as an acid when it is dissolved in benzene because there are no molecules present that can accept protons. However, if a base is added, the proton can be transferred to it. Water can behave as a base:

$$HCl + H_2\ddot{O} \rightleftharpoons H_3O^+ + Cl^-$$

$$\text{acid} \qquad \text{base} \qquad \text{acid} \qquad \text{base}$$

Here, Cl^- is the conjugate base of HCl and HCl is the conjugate acid of the base Cl^-.

An equilibrium is set up when any acid is added to water:

$$HA + H_2\ddot{O} \rightleftharpoons H_3O^+ + A^-$$

the position of equilibrium depending on the strength of the acid. For strong acids the equilibrium is well over to the right and it follows that in such cases the conjugate base A^- must be a very weak base (examples are Cl^-, NO_3^-, etc.). On the other hand, with weak acids such as ethanoic acid (acetic acid) the equilibrium is well over to the left and so A^- (in this case $CH_3 \cdot COO^-$) must be a fairly strong base.

Similar considerations apply in the case of bases, i.e. for a base to act as a base it is necessary to have an acid present to supply protons. Water can behave as an acid as well as a base (such compounds are said to be *amphiprotic*) and so ammonia acts as a base when dissolved in water:

$$\ddot{N}H_3 + H_2\ddot{O} \rightleftharpoons NH_4^+ + HO^-$$

$$\text{base} \qquad \text{acid} \qquad \text{acid} \qquad \text{base}$$

It is, however, a very weak base because the equilibrium is well over to the left.

The strong bases, such as the hydroxides of sodium, potassium, and calcium, are fully ionised even in the solid state and so interaction with the solvent is not essential since it is the HO^- which acts as the base. This is in contrast to the strong acids such as sulphuric(VI) acid, chloric(VII) acid (perchloric acid), hydrochloric acid, etc. which are covalent in the pure state.

Since the dissociations of weak acids and bases are equilibrium processes, the equilibrium law may be applied to them. Consider the dissociation of one mole of any weak acid, HA in V litres of solution. If α is the degree of dissociation then:

$$HA \rightleftharpoons H^+ + A^-$$

Initially   1 mole   0   0

At equilibrium $1 - \alpha$ α α

Applying the equilibrium law

$$\frac{[H^+][A^-]}{[HA]} = \text{constant at constant temperature,}$$

i.e.
$$\frac{\dfrac{\alpha}{V} \times \dfrac{\alpha}{V}}{\dfrac{1-\alpha}{V}} = \text{constant}$$

or
$$\frac{\alpha^2}{V(1-\alpha)} = K \qquad\qquad (7.2)$$

where K is known as the *dissociation constant*. The expression (7.2) is known as

Ostwald's dilution law. For very weak electrolytes, $(1 - \alpha)$ approximates to 1 and so the equation simplifies to

$$\frac{\alpha^2}{V} = K$$

or
$$\alpha = \sqrt{KV} \qquad\qquad (7.3)$$

i.e. the degree of dissociation is directly proportional to the square root of the volume containing one mole of the weak electrolyte.

The degree of dissociation of a weak electrolyte varies with concentration, but the dissociation constant remains unaltered provided that the temperature is not changed. The relative strengths of weak acids and bases are therefore indicated by their dissociation constants, the larger the value the stronger the acid or base.

The degree of dissociation of a weak electrolyte at a given dilution may be calculated from its dissociation constant.

Example Calculate (a) the approximate degree of dissociation and (b) the hydrogen ion concentration for a 0.04 M solution of propanoic acid (propionic acid) at 25 °C given that its dissociation constant is 1.3×10^{-5} mol^{-1} under these conditions.

$$CH_3 \cdot CH_2 \cdot COOH \rightleftharpoons CH_3 \cdot CH_2 \cdot COO^- + H^+$$

At equilibrium

$$(1 - \alpha) \text{ mole} \qquad \alpha \text{ mole} \qquad \alpha \text{ mole}$$

According to Ostwald's dilution law (equation 7.3), $\alpha = \sqrt{KV}$ where V is the volume in litres containing 1 mole = $1/0.04 = 25$.

Therefore
$$\alpha = \sqrt{1.3 \times 10^{-5} \times 25}$$

$$= 0.018$$

Thus, the degree of dissociation in the solution is 0.018 or 1.8%.
Now

$$[H^+] = \text{molarity of acid} \times \text{degree of dissociation}$$

$$= 0.04 \times 0.018$$

$$= 7.2 \times 10^{-4} \text{ mol litre}^{-1}$$

The ionic product of water

If water is purified by repeated distillation under vacuum, its electrical conductivity falls to a constant figure which is entirely due to ionisation of the water:

$$H_2O \rightleftharpoons H^+ + HO^-$$

It should be noted that the ions are in fact hydrated.
Applying the equilibrium law

$$K_c = \frac{[H^+][HO^-]}{[H_2O]}$$

Since such a small fraction of the water is dissociated, $[H_2O]$ may be regarded as being constant under all conditions and so

$$[H^+][HO^-] = K_c \times \text{constant} = K_w \qquad (7.4)$$

where K_w is known as the *ionic product of water*.

Conductivity experiments have shown that the value of K_w at 25 °C is $1 \times 10^{-14} \text{ mol}^2 \text{litre}^{-2}$. Now one molecule of water gives one H^+ ion and one HO^- ion and so $[H^+]$ and $[HO^-]$ must be the same, i.e. since

$$[H^+][HO^-] = 1 \times 10^{-14}$$

$$[H^+] \text{ and } [HO^-] = 1 \times 10^{-7} \text{mol litre}^{-1} \text{ at 25 °C.}$$

Aqueous solutions which contain $1 \times 10^{-7} \text{ mol litre}^{-1}$ H^+ (or of HO^-) are said to be neutral, but it should be remembered that this value will increase if the temperature is raised. Neutralisation may be described as the process of adding acid or alkali to a solution until $[H^+]$ and $[HO^-]$ become equal.

If acid is added to water, the concentration of H^+ ions is increased, and since K_w is unaltered, the concentration of HO^- ions must fall, i.e. H^+ ions combine with HO^- ions so that K_w is maintained.

Example Calculate the hydroxide ion concentration in 0.1 M hydrochloric acid.

$$[H^+][HO^-] = 1 \times 10^{-14}$$

But 0.1 M   HCl contains $1 \times 10^{-1} \text{mol}$ H^+ litre^{-1}.

Therefore $1 \times 10^{-1} \times [HO^-] = 1 \times 10^{-14}$

or $$[HO^-] = \frac{1 \times 10^{-14}}{1 \times 10^{-1}} = 1 \times 10^{-13} \text{ mol litre}^{-1}$$

Similarly, if alkali is added to water the HO^- ion concentration increases and so the H^+ ion concentration decreases.

Example Calculate the hydrogen ion concentration in 0.01 M sodium hydroxide solution.

$$[H^+][HO^-] = 1 \times 10^{-14}$$

But 0.01 M   NaOH contains $1 \times 10^{-2} \text{ mol of } HO^-$ per litre.

Therefore $[H^+] \times 1 \times 10^{-2} = 1 \times 10^{-14}$

or $$[H^+] = \frac{1 \times 10^{-14}}{1 \times 10^{-2}} = 1 \times 10^{-12} \text{ mol litre}^{-1}$$

pH OR HYDROGEN ION CONCENTRATION

It has been seen above that the hydrogen ion (or hydroxide ion) concentration determines the acidity or alkalinity of a solution, and that compounds which increase it are acids, whilst those which decrease it are bases. A knowledge of the acidity or alkalinity is important in many processes, but it is inconvenient to work in values such as $1.67 \times 10^{-3} \text{ mol } H^+ \text{ litre}^{-1}$. Instead, pH is used where

$$pH = \lg \frac{1}{[H^+]}$$

or

$$pH = -\lg [H^+] \qquad (7.5)$$

Example Calculate the pH of a solution whose hydrogen ion concentration is $0.001 \, \text{mol litre}^{-1}$.

$$\lg \text{ of } 0.001 = -3$$

but

$$pH = -\lg [H^+] = -(-3) = 3$$

Example Calculate the pH of a solution with a hydrogen ion concentration of $6.28 \times 10^{-6} \, \text{mol litre}^{-1}$.

$$
\begin{aligned}
pH &= -\lg (6.28 \times 10^{-6}) \\
&= -(0.798 + (-6)) \\
&= -0.798 + 6 \\
&= 5.202
\end{aligned}
$$

It is also possible to describe alkalinity by pH because:

$$[H^+] [HO^-] = 1 \times 10^{-14}$$

and taking negative logarithms gives

$$pH + pOH = 14$$

or

$$pH = 14 - pOH$$

Example Calculate the pH of a 0.001 M solution of sodium hydroxide.

$$pOH = -\lg [HO^-] = -\lg 1 \times 10^{-3} = 3$$

but

$$
\begin{aligned}
pH &= 14 - pOH \\
&= 14 - 3 = 11.
\end{aligned}
$$

The pH scale normally covers the range 0–14, i.e. molar solutions of strong acids, HA at one extreme to molar solutions of strong bases, BOH at the other.

The pH of a molar solution of a strong acid, $\quad HA = -\lg 1 = 0$
The pH of a neutral solution $= -\lg 10^{-7} = 7$
The pH of a molar solution of a strong base, $\quad BOH = 14 - \lg 1 = 14$

Stronger acid and alkali solutions are possible, but at these concentrations active mass (page 111) and concentration are not the same and so errors creep in.

It should be noted that a ten-fold dilution of a solution of a strong acid will cause the pH to fall by one unit. However, similar dilution of a solution of a weak acid will have less effect than this since the acid is only partially ionised.

It is sometimes necessary to calculate the hydrogen ion concentration in a solution of known pH; this type of calculation is illustrated in the example below.

Example Calculate the hydrogen ion concentration in a solution with a pH of 8.54.

$$-\lg [H^+] = 8.54$$

and so
$$\lg[H^+] = -8.54$$
$$= \overline{9}.46$$

(The mantissa — the part of the logarithm behind the decimal point — is always positive. $\overline{9}.46$ is shorthand for $-9 + 0.46$.)

or
$$[H^+] = 2.88 \times 10^{-9} \, \text{mol litre}^{-1}.$$

Comparing the strengths of weak acids and bases

The equilibrium resulting from dissociation of a weak acid HA in aqueous solution is represented by:

$$HA + H_2\ddot{O} \rightleftharpoons H_3\ddot{O}^+ + A^-$$

Applying the equilibrium law,

$$\frac{[H_3O^+]\,[A^-]}{[HA]\,[H_2O]} = \text{constant}$$

But $[H_2O]$ is so large that it effectively remains constant and so

$$\frac{[H_3O^+]\,[A^-]}{[HA]} = \text{constant} \times \text{constant} = K_a \tag{7.6}$$

where K_a is the dissociation constant of the acid. For weak acids K_a will obviously be very small and so it is more convenient to use pK_a where $pK_a = -\lg K_a$. Weaker acids will have higher pK_a values than stronger ones as illustrated by the following: $CH_3 \cdot COOH$ $pK_a = 4.76$ $CH_2Cl \cdot COOH$ $pK_a = 2.86$, and $CH_2F \cdot COOH$ $pK_a = 2.66$. (The reasons for this order are discussed in Chapter 20.)

The pH of an acid solution at a given dilution may be calculated if its K_a value is known.

Example   Calculate the pH of a 0.1 M solution of ethanoic (acetic) acid if $K_a = 1.7 \times 10^{-5} \, \text{mol litre}^{-1}$. According to equation (7.6)

$$K_a = \frac{[CH_3 \cdot COO^-]\,[H_3O^+]}{[CH_3 \cdot COOH]}$$

$[CH_3 \cdot COOH]$ will be very nearly the same as the initial concentration since it is a weak acid and little dissociation will have occurred.

Let
$$[CH_3 \cdot COO^-] = [H_3O^+] = x$$

Then
$$1.7 \times 10^{-5} = \frac{x^2}{0.1}$$

$$x = \sqrt{1.7 \times 10^{-6}}$$

Therefore
$$pH = -\lg(\sqrt{1.7 \times 10^{-6}})$$
$$= -(\overline{3}.1152)$$
$$= -(-3 + 0.1152)$$
$$= -(-2.8848)$$
$$= 2.88$$

In aqueous solutions of weak bases the equilibrium is:

$$B: + H_2O \rightleftharpoons B^+H + HO^-$$

The concentration of water is again large and, for all practical purposes, constant throughout. Hence, application of the equilibrium law gives

$$K_b = \frac{[B^+H][HO^-]}{[B:]}$$

where K_b is the equilibrium constant of the base. The K_b values are very small and so it is less cumbersome to use pK_b where $pK_b = -\lg K_b$. The smaller the numerical value of pK_b the stronger the base, e.g.

$$NH_3, \quad pK_b = 4.75 \quad \text{and} \quad CH_3 \cdot NH_2, \quad pK_b = 3.36$$

The factors affecting the strength of organic bases are discussed in Chapter 20.

The strength of bases can also be expressed in terms of pK_a so that a continuous scale for acids and bases results. K_a and so pK_a for a base is a measure of the ease with which B^+H will lose a proton:

$$B^+H + H_2\ddot{O} \rightleftharpoons B: + H_3\underset{\cdot\cdot}{O}^+$$

i.e.
$$K_a = \frac{[B:][H_3\underset{\cdot\cdot}{O}^+]}{[B^+H]}$$

pK_b can be converted into pK_a by use of the equation $pK_a + pK_b = 14$.

Determination of pK_a for weak acids

To determine the dissociation constant of a weak acid, it is necessary to find the pH and concentration of one of its solutions. The pH may be found by means of a pH meter (page 144) or by an indicator method (page 136). The concentration can be determined by a conventional titration with sodium hydroxide solution using an indicator or alternatively by a conductometric titration, the basis of which is outlined below.

The *conductance* (G) of a substance is by definition the reciprocal of its resistance. The resistance, R, of a substance is dependent upon its length, l, cross sectional area, a, across the conducting path and its *resistivity*, ρ, according to the equation

$$R = \rho \frac{l}{a}$$

(The resistivity of an electrolyte at a given concentration and temperature is defined as the resistance, in ohms, between the opposite faces of a metre cubed of the electrolyte under these conditions.)

Thus, the conductance,
$$G = \frac{1}{R}$$

$$= \frac{a}{\rho l} = \kappa \frac{a}{l}$$

where $\kappa = 1/\rho$ and is known as the *electroytic conductivity*, the units being $\Omega^{-1}m^{-1}$.

The conductance of a solution may therefore be determined from the measurement of its resistance in a calibrated cell. Alternating current has to be used in these measurements since d.c. would cause electrolysis.

The course of reactions may be followed by conductance measurements if ions are removed or formed as one solution is added to another. This is applicable to acid–base titrations, and they then come under the general name of *conductometric titrations*. It should be noted here that most ions in solution have similar velocities because, although their sizes and charges may be different, they are hydrated and their movement is hindered by this and by the oppositely charged ions present. However, H^+ and HO^- ions are particularly good at conducting a current because they operate in a rather different manner:

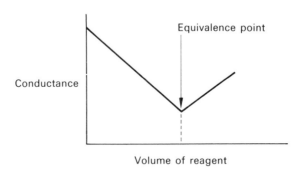

Thus, different ions are involved throughout as new bonds are formed and others are broken.

If a strong acid, e.g. hydrochloric acid, is titrated with a strong base, a plot of conductance against volume of reagent used takes the form shown in Figure 7.1.

Figure 7.1   Variation of conductance in a strong acid–strong base titration

The conductance is initially high, because of the high conductivity of the hydrogen ion. However, it gradually falls as the H^+ combines with HO^- to give covalent water, and is replaced by the slower moving Na^+ ions. The conductance rises rapidly after the equivalence point and excess HO^- is added. The equivalence point is therefore at the intersection of the two lines.

In a weak acid–strong base titration the graph is as shown in Figure 7.2, The conductance rises steadily from the start owing to acid anion and base cation replacing the H^+ present. However, it rises more sharply after the equivalence point as excess HO^- ion is added. Again the equivalence point is at the intersection of the two lines.

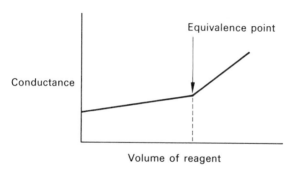

Figure 7.2   Variation of conductance in a weak acid–strong base titration

Calculation of K_a

The calculation of the dissociation constant from the concentration of the acid solution and its pH is illustrated by the example below.

Example   In a titration, 23.8 cm³ of a solution of ethanoic acid (acetic acid), $CH_3 \cdot COOH$ were neutralised by 25.0 cm³ of 0.1059 M sodium hydroxide solution while, in a separate experiment, the acid solution was found to have a pH of 2.86. Calculate K_a for ethanoic acid.

$$CH_3 \cdot COOH + NaOH \rightarrow CH_3 \cdot COONa + H_2O$$

Now $bV_A M_A = aV_B M_B$ where a, V_A, and M_A denote moles, volume, and molarity respectively of ethanoic acid, and b, V_B, and M_B represent the corresponding figures for sodium hydroxide.

∴ $$1 \times 23.8 \times M_A = 1 \times 25 \times 0.1059$$

or $$M_A = \frac{25 \times 0.1059}{23.8} = 0.1112$$

i.e., the acid solution contains 0.1112 mol litre⁻¹ ethanoic acid.

Since
$$pH = -\lg[H^+]$$
$$2.86 = -\lg[H^+]$$
∴
$$\lg[H^+] = -2.86 = \bar{3}.14$$
i.e.
$$[H^+] = 1.38 \times 10^{-3} \text{ mol litre}^{-1}$$

Substituting the relevant figures in $K_a = \dfrac{[H^+][CH_3 \cdot COO^-]}{[CH_3 \cdot COOH]}$ gives

$$K_a = \frac{1.38 \times 10^{-3} \times 1.38 \times 10^{-3}}{(0.1112 - 0.00138)} \qquad ([H^+] = [CH_3 \cdot COO^-])$$

$$= 1.73 \times 10^{-5} \text{ mol litre}^{-1}$$

Buffer solutions

A *buffer solution* is a solution which undergoes negligible change in pH on addition of moderate quantities of acid or alkali. Buffer solutions with a pH of less than 7 (acid buffers) are made by dissolving a weak acid and its sodium or potassium salt in water. A typical acid buffer contains ethanoic acid (acetic acid) and sodium ethanoate, the higher the proportion of acid the lower the pH of the buffer. The operation of the buffer is outlined below.

When ethanoic acid is added to water two equilibria are set up

$$CH_3 \cdot COOH \rightleftharpoons CH_3 \cdot COO^- + H^+$$

and

$$H_2O \rightleftharpoons HO^- + H^+$$

Ignoring the hydration of the ions.

On addition of sodium ethanoate the $CH_3 \cdot COO^-$ concentration is greatly increased and so it combines with H^+ until K_a is restored. The resultant solution contains a high concentration of undissociated ethanoic acid and the ethanoate ion, i.e. a high concentration of the acid and its conjugate base. If a strong acid is now added to the system, the additional H^+ combines with $CH_3 \cdot COO^-$ to give more undissociated acid and so K_w is restored with little change in pH. On the other hand, if a strong base is added, the additional HO^- combines with H^+ to restore K_w and then more ethanoic acid dissociates to restore K_a. The new equilibrium is therefore attained with only a small pH change.

Buffers on the alkaline side are made by dissolving a weak base and one of its soluble salts in water, for example, ammonia and ammonium chloride. When ammonia dissolves in water the equilibria present are:

$$NH_3 + H_2O \rightleftharpoons NH_4^+ + HO^-$$

and

$$H_2O \rightleftharpoons H^+ + HO^-$$

Ignoring the hydration of the various species.

When ammonium chloride is added, the concentration of NH_4^+ is greatly increased and so NH_4^+ combines with HO^- until K_b is restored. The solution then contains a high concentration of the base, NH_3 and its conjugate acid, NH_4^+. If a strong acid is added the additional H^+ combines with NH_3 to give NH_4^+, whilst if a strong base is added, the extra HO^- combines with NH_4^+ to give ammonia and water. In both cases K_b and K_w are restored with negligible change in pH.

It is seen, therefore, that acid buffers depend upon the fact that they contain high concentrations of an undissociated weak acid and its conjugate base, the former compound acting as a trap for HO^- and the latter as a trap for H^+. Alkaline buffers, on the other hand, contain high concentrations of a weak base, and its conjugate acid. The base mops up any additional H^+ and its conjugate acid takes care of any additional HO^-.

The pH of a buffer solution may be calculated as illustrated in the example below.

Example   Calculate the pH of a solution containing 0.1 mole of ethanoic acid and 0.01 mole of sodium ethanoate litre^{-1}. K_a for ethanoic acid is 1.7×10^{-5} mol litre^{-1}.

According to equation (7.6)

$$K_a = \frac{[H^+][CH_3 \cdot COO^-]}{[CH_3 \cdot COOH]}$$

where

$$K_a = 1.7 \times 10^{-5}$$

$[CH_3 \cdot COOH] = 10^{-1}$ mol litre^{-1} neglecting the small amount which is ionised and $[CH_3 \cdot COO^-] = 10^{-2}$ mol litre^{-1} neglecting the small amount from dissociation of the acid.

$\therefore$ $1.7 \times 10^{-5} = \dfrac{[H^+] \times 10^{-2}}{10^{-1}}$

or $[H^+] = 1.7 \times 10^{-4}$

but $pH = -\lg[H^+]$

$$= -\lg 1.7 \times 10^{-4}$$
$$= -(\bar{4}.2304)$$
$$= -(-3.7696)$$
$$= 3.77$$

Buffer solutions are used to calibrate indicators and pH meters and to control the pH of solutions in which chemical reactions are performed (page 324). They are extremely important in biochemistry and medicine. For example, the pH of blood has to be controlled within strict limits and so it contains various salts which behave as buffers. It is sometimes necessary to buffer injections into the blood stream because disturbing the pH could be fatal. Buffer solutions are also used in the preservatives industry to limit the acidity.

HYDROLYSIS OF SALTS

When a salt is dissolved in water, the solution is not necessarily neutral. This is because the salt to some extent undergoes decomposition or *hydrolysis* with the water and, although equivalent quantities of acid and base are produced, the one may be stronger than the other. If the acid and base are dissociated to different extents, the resultant solution will be acidic or alkaline, because the normal equilibrium between H^+ ions, HO^- ions, and water will be upset. There are four types of salt to consider.

a. Salts of weak acids and strong bases

In a solution of sodium ethanoate (sodium acetate), the situation may be represented as:

$$\left. \begin{array}{l} Na^+(aq) + CH_3 \cdot COO^-(aq) \\ \\ H_2O(l) \rightleftharpoons HO^-(aq) + H^+(aq) \end{array} \right\} \rightleftharpoons CH_3 \cdot COOH(aq)$$

Thus, hydrolysis results in the formation of sodium hydroxide and ethanoic acid. Since the former compound is fully dissociated and the latter only partially so, the solution will contain more HO^- than H^+ and it will be alkaline.

Similarly, Na_2CO_3, KCN, Na_2S, etc. will all give alkaline solutions, because the bases formed by hydrolysis will be fully ionised, whereas the weak acids formed (H_2CO_3, HCN, and H_2S) will be only partially ionised.

b. Salts of strong acids and strong bases

Salts of strong acids and bases give neutral solutions. For example, the situation in sodium chloride solutions is:

$$Na^+(aq) + Cl^-(aq)$$
$$H_2O(l) \rightleftharpoons HO^-(aq) + H^+(aq)$$

Hydrolysis would give sodium hydroxide and hydrochloric acid, both of which dissociate completely. The only partially dissociated substance present is H_2O, and $[H^+] = [HO^-]$.

c. Salts of weak acids and weak bases

This type is illustrated by a solution of ammonium ethanoate (ammonium acetate); the following equilibria are set up:

$$NH_4^+(aq) + CH_3 \cdot COO^-(aq) \left.\begin{array}{c} \\ \\ \end{array}\right\} \rightleftharpoons NH_3(aq) + CH_3 \cdot COOH(aq)$$

$$H_2O(l) \rightleftharpoons HO^-(aq) + H^+(aq)$$

Hydrolysis gives a weak base and a weak acid, and since they dissociate to approximately the same extent, the resultant solution is nearly neutral. Similar examples are ammonium carbonate and ammonium sulphide.

d. Salts of strong acids and weak bases

An example of this type is ammonium chloride and, in aqueous solution, the situation is:

$$Cl^-(aq) + NH_4^+(aq) \left.\begin{array}{c} \\ \\ \end{array}\right\} \rightleftharpoons NH_3(aq)$$

$$H_2O(l) \rightleftharpoons H^+(aq) + HO^-(aq)$$

Hydrolysis gives hydrochloric acid which is fully dissociated and ammonia solution which is only partially so. Consequently there will be more H^+ than HO^- present and an acidic solution is obtained. Similar examples are NH_4NO_3, $(NH_4)_2SO_4$, $FeCl_3$, $AlCl_3$, and $CuSO_4$.

ACID-ALKALI INDICATORS

Most indicators of this type are weak acids, but a few are weak bases. For a compound to act as an acid–alkali indicator, it is necessary for the anion or cation to be markedly different in colour to the undissociated molecule.

If the indicator is a weak acid it will dissociate as follows:

$$HA \rightleftharpoons H^+ + A^-$$

and
$$K_a = \frac{[H^+][A^-]}{[HA]}$$

When the indicator is added to an acid solution, $[H^+]$ will be increased and, in order for the value of K_a to be maintained, $[A^-]$ must fall. Thus, the dissociation of the indicator is supressed and the colour of the undissociated molecule predominates. In alkaline solution, $[HO^-]$ is increased and, to enable K_a and K_w to be maintained, some of the HO^- and H^+ combine and then more indicator dissociates. The colour of the anion becomes more apparent as the dissociation of the indicator increases.

Phenolphthalein is an indicator of the weak acid type:

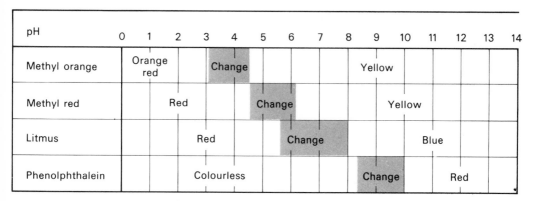

Colourless   Pink

If the indicator is a weak base of the type BOH, the dissociation is:

$$BOH \rightleftharpoons B^+ + HO^-$$

and

$$K_b = \frac{[B^+][HO^-]}{[BOH]}$$

In acid conditions, $[HO^-]$ falls and so $[B^+]$ increases and its colour predominates. In alkaline solutions, $[HO^-]$ is large and consequently $[B^+]$ is small and the colour of BOH is predominant.

Methyl orange is an indicator of this type:

$$^-O_3S-\langle\bigcirc\rangle-N{=}N-\langle\bigcirc\rangle-N(CH_3)_2 \quad \begin{matrix} H^+ \\ \rightleftharpoons \\ HO^- \end{matrix}$$

Yellow

$$^-O_3S-\langle\bigcirc\rangle-NH-N{=}\langle\bigcirc\rangle{=}\overset{+}{N}(CH_3)_2$$

Red

The weak acids and bases which act as indicators have varying strengths, and so each has a characteristic pH range over which it changes colour. Some of the common indicators and their colour ranges are shown in Figure 7.3.

pH	0	1	2	3	4	5	6	7	8	9	10	11	12	13	14
Methyl orange		Orange red			Change					Yellow					
Methyl red			Red			Change					Yellow				
Litmus				Red			Change					Blue			
Phenolphthalein				Colourless						Change		Red			

Figure 7.3 Ranges and colours of common indicators

Choice of indicators in acid–alkali titrations

For an indicator to be useful, it should change over its whole range of colour on the addition of one drop of the titrating liquid.

There are four possible types of titration to be considered: titration of a strong acid by a strong or a weak base, and titration of a weak acid by a strong or a weak base. Each type of titration gives a characteristic titration curve as illustrated in Figure 7.4 and so the indicator must be chosen to suit the conditions encountered.

Strong acid–strong base

Consider a titration between 25 cm³ of 0.1 M hydrochloric acid and 0.1 M sodium hydroxide solution. When say 24.95 cm³ of the alkali have been added, 0.05 cm³ of the acid remains and the volume will be 49.95 cm³. The H^+ concentration will therefore be 0.0001 mol litre⁻¹, and the pH will be 4. The pH at any stage in the titration may be calculated in a similar manner and some are shown in Table 7.1. It is apparent from this and from Figure 7.4(a) that near the equivalence point the pH changes very sharply on the addition of a very small

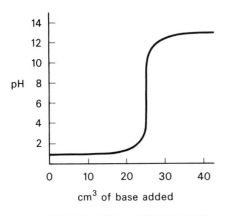

(a)   STRONG ACID - STRONG BASE

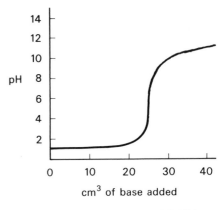

(b)   STRONG ACID - WEAK BASE

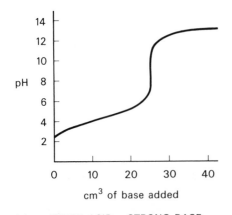

(c)   WEAK ACID - STRONG BASE

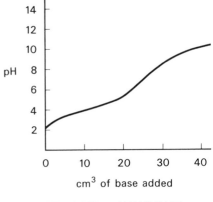

(d)   WEAK ACID - WEAK BASE

Figure 7.4   The variation of pH during titrations of 25 cm³ of 0.1 M acid with 0.1 M base

quantity of the titrating solution. Hence the equivalence point will be shown by any indicator which changes colour when the pH lies in the vertical part of the graph. Methyl orange, methyl red, litmus, and phenolphthalein all change colour near enough to this range to give the equivalence point to within the limits possible with the apparatus used.

Table 7.1. Changes in pH in a hydrochloric acid-sodium hydroxide titration ·

Volume of 0.1 M NaOH added to 25 cm³ of 0.1 M HCl	H_3O^+ /mol litre^{-1}	pH
24.00	0.002	2.7
24.90	0.0002	3.7
24.95	0.0001	4.0
24.99	0.00002	4.7
25.00	0.0000001	7.0

b. Stong acid–weak base

The type of curve obtained from a hydrochloric acid–ammonia solution titration is shown in Figure 7.4(b). The graph is similar to that obtained with strong acids and bases until the equivalence point is reached. On addition of more ammonia solution, however, the pH increases slowly since the excess weak base does not greatly increase the HO⁻ concentration. An indicator for a titration of this type must change colour within the approximate range of pH 3 to 7 and so methyl orange and methyl red would be satisfactory.

c. Weak acid–strong base

Figure 7.4(c) shows the type of curve obtained from an ethanoic acid (acetic acid)–sodium hydroxide titration. When the base is initially added to the acid, the pH rises rather faster than may have been expected because the sodium ethanoate produced undergoes hydrolysis, thus increasing the HO⁻ concentration. After pH = 7, the graph is similar to that for a strong acid–strong base titration. An indicator must be chosen such that it changes colour in the approximate range of pH 7 to 11; phenolphthalein is satisfactory.

d. Weak acid–weak base

In a titration of say ethanoic acid (acetic acid) with ammonia solution the graph takes the form shown in Figure 7.4(d). It is seen that there is no sharp change in pH and so indicators cannot be used for titrations of this type. The reaction can be followed, however, by a potentiometric titration — a titration involving the use of a pH meter (page 144) — or by a conductometric titration (page 128).

Titration of soluble carbonates with dilute mineral acids

In this type of titration a weak acid is liberated throughout, e.g.,

$$Na_2CO_3 + 2HCl \rightarrow 2NaCl + CO_2 + H_2O$$

As long as any carbonate is present, the acid is consumed and carbon dioxide is liberated. When all the carbonate has just been decomposed, the pH will be 6.3, i.e. the pH of a saturated solution of carbon dioxide. As soon as any excess acid is added, the pH will fall sharply. Methyl orange may therefore be used as indicator since it will retain its yellow colour as long as any carbonate remains and then turn pink as soon as any excess acid is added. If phenolphthalein is used as indicator, the colour change occurs at the half equivalence point i.e. when carbonate has been converted to hydrogencarbonate and the pH is about 8:

$$Na_2CO_3 + HCl \rightarrow NaHCO_3 + NaCl + H_2O$$

Determination of pH by indicator methods

The approximate pH of a given solution may be found by means of Universal Indicator. This is a mixture of indicators each of which changes colour over a particular pH range, the colour varying from red in strongly acidic solutions through orange and yellow to yellow-green at pH = 7, and through darker green and blue to violet in strongly alkaline solutions. Once the approximate pH is known, the actual pH can be found by using a specific indicator and comparing, by means of a comparator, the colour produced by the same indicator in a series of buffer solutions with a pH difference of 0.1.

The required buffer solutions may be prepared by using different weak acids or bases and their salts. Since a weak acid dissociates as follows:

$$HA \rightleftharpoons H^+ + A^-$$

then

$$K_a = \frac{[H^+][A^-]}{[HA]}$$

or

$$[H^+] = \frac{K_a[HA]}{[A^-]}$$

If the molar concentration of the acid and its salt are the same, $[H^+] = K_a$ and so preparing a buffer solution of any pH is simply a matter of selecting a weak acid or base with the required K_a value. On occasions, when suitable acids or bases are not available, it becomes necessary to alter the relative amounts of the acid or base and its salt.

Oxidation and reduction

Oxidation used to be defined as the combination of oxygen with a substance or the removal of hydrogen from a substance. Conversely, reduction was initially defined as the combination of a substance with hydrogen or the removal of oxygen from a substance. These definitions are now obsolete since they are too limiting. For example, the two reactions

$$2Mg + O_2 \rightarrow 2MgO$$

and

$$Mg + S \rightarrow MgS$$

are obviously very similar but, according to the original definitions, the former is classes as oxidation of magnesium, but the latter is not. The common factor in the reactions is that in each case the magnesium atom loses two electrons to give a Mg^{2+} ion and so it is apparent that a much broader definition of oxidation is possible.

Oxidation is now regarded as loss of electrons from an element, compound, or ion whilst *reduction* is gain of electrons by an element, compound, or ion. Obviously, if one substance gains electrons another must lose them and so every oxidation is accompanied by a reduction and *vice versa*. Reactions involving oxidation and reduction are known as *redox* reactions (short for reduction–oxidation).

Oxidising agents are electron acceptors and *reducing agents* are electron donors.

The mode of operation of some important oxidising agents is given below. These types of equation are known as *half-reactions*.

(1) Potassium dichromate(VI) in acid solution:

$$Cr_2O_7{}^{2-} + 14H^+ + 6e^- \rightarrow 2Cr^{3+} + 7H_2O$$

The half-reactions of oxy-anions, and compounds acting as oxidising agents in acid solution, may be derived using the following general procedure.

(a) Write down the formulae of the reactant and product and the relevant oxidation states (see page 170), e.g.

$$\overset{(+6)}{Cr_2O_7{}^{2-}} \qquad \rightarrow \overset{(+3)}{2Cr^{3+}}$$

(b) Add the necessary electrons required to bring about the change in oxidation number:

$$Cr_2O_7{}^{2-} + 6e^- \rightarrow 2Cr^{3+}$$

(c) Add sufficient H^+ ions to combine with the oxygen atoms being removed from the oxidising agent (to balance the charges):

$$Cr_2O_7{}^{2-} + 6e^- + 14H^+ \rightarrow 2Cr^{3+}$$

(d) Add water to balance the equation:

$$Cr_2O_7{}^{2-} + 6e^- + 14H^+ \rightarrow 2Cr^{3+} + 7H_2O$$

(2) Potassium manganate(VII) (potassium permanganate) in acid solution:

$$MnO_4{}^- + 8H^+ + 5e^- \rightarrow Mn^{2+} + 4H_2O$$

Potassium manganate(VII) in alkaline solution:

$$MnO_4{}^- + 2H_2O + 3e^- \rightarrow MnO_2 + 4HO^-$$

The half-reactions for oxy-anions acting as oxidising agents in aklaline solutions may be derived as follows.

(a) Write down the basic change with the relevant oxidation states, e.g.

$$\overset{(+7)}{MnO_4{}^-} \rightarrow \overset{(+4)}{MnO_2}$$

(b) Add the electrons necessary to bring about the change:

$$MnO_4{}^- + 3e^- \rightarrow MnO_2$$

(c) Balance the charges by adding HO^- ions:

$$MnO_4^- + 3e^- \rightarrow MnO_2 + 4HO^-$$

(d)   Add water to balance the hydrogens and oxygens:

$$MnO_4^- + 3e^- + 2H_2O \rightarrow MnO_2 + 4HO^-$$

(3)   Oxygen, sulphur, and chlorine on reaction with metals:

$$O_2 + 4e^- \rightarrow 2O^{2-}$$
$$S + 2e^- \rightarrow S^{2-}$$
$$Cl_2 + 2e^- \rightarrow 2Cl^-$$

(4)   Protons from dilute acids, e.g. metal + acid: $2H^+ + 2e^- \rightarrow H_2$

(5)   Hydrogen peroxide: $H_2O_2 + 2H^+ + 2e^- \rightarrow 2H_2O$

(6)   Manganese(IV) oxide (manganese dioxide) in acid solution:

$$MnO_2 + 4H^+ + 2e^- \rightarrow Mn^{2+} + 2H_2O$$

(7)   Concentrated sulphuric(VI) acid, e.g.

$$Cu + 2H_2SO_4 \rightarrow CuSO_4 + 2H_2O + SO_2$$

Half reaction: $SO_4^{2-} + 2e^- + 4H^+ \rightarrow SO_2 + 2H_2O$

(8)   Nitric(V) acid. This can react in several ways depending upon the concentration, temperature, etc. Examples:

Hot concentrated acid Cu + $4HNO_3$ $\rightarrow Cu(NO_3)_2 + 2H_2O + 2NO_2$
Half-reaction $NO_3^- + e^- + 2H^+$ $\rightarrow NO_2 + H_2O$

Cold dilute acid $3Cu + 8HNO_3$ $\rightarrow 3Cu(NO_3)_2 + 4H_2O + 2NO$
Half-reaction $NO_3^- + 3e^- + 4H^+$ $\rightarrow NO + 2H_2O$

Very dilute acid $4Zn + 10HNO_3$ $\rightarrow 4Zn(NO_3)_2 + 5H_2O + N_2O$
Half-reaction $2NO_3^- + 8e^- + 10H^+ \rightarrow N_2O + 5H_2O$

The common reducing agents operate as follows.

(1)   Metal and acid. Electrons are released as the metal goes into solution as ions,

e.g. $Zn \rightarrow Zn^{2+} + 2e^-$

Metal and alkali: $Al + 4HO^- \rightarrow [Al(OH)_4]^- + 3e^-$

(2)   Hydrogen, carbon, and carbon monoxide in the reduction of metal oxides:

$$H_2 + O^{2-} \rightarrow H_2O + 2e^-$$
$$C + O^{2-} \rightarrow CO + 2e^-$$
or $$C + 2O^{2-} \rightarrow CO_2 + 4e^-$$
$$CO + O^{2-} \rightarrow CO_2 + 2e^-$$

(3)   Acidified potassium iodide solution: $2I^- \rightarrow I_2 + 2e^-$

(4)   Solutions of iron(II) salts: $Fe^{2+} \rightarrow Fe^{3+} + e^-$

(5)   Hydrogen sulphide solution: $H_2S \rightleftharpoons 2H^+ + S^{2-}$
$$S^{2-} \rightarrow S + 2e^-$$

(6) Tin(II) chloride and hydrochloric acid:

$$Sn^{2+} \rightarrow Sn^{4+} + 2e^-$$

(7) Sodium thiosulphate(VI) solution:

$$2S_2O_3^{2-} \rightarrow S_4O_6^{2-} + 2e^-$$

(8) Solutions of ethanedioates (oxalates):

$$C_2O_4^{2-} \rightarrow 2CO_2 + 2e^-$$

(9) Sulphuric(IV) acid (sulphurous acid):

$$SO_2 + H_2O \rightleftharpoons H_2SO_3 \rightleftharpoons 2H^+ + SO_3^{2-}$$
$$SO_3^{2-} + H_2O \rightarrow SO_4^{2-} + 2H^+ + 2e^-$$

Combination of an oxidising agent half-reaction and a reducing agent half-reaction, with any necessary balance of electrons, gives the overall stoichiometry of the reaction. It should be stressed, however, that some oxidising and reducing agents are more powerful than others and so not all combinations lead to reaction. Some examples of redox reactions are given below; several of them are generally referred to as oxidations because the main substance under consideration is oxidised, but some other substance must obviously be reduced at the same time.

(a) Oxidation of hot ethanedioate (oxalate) solutions with potassium manganate(VII) (potassium permanganate).

$$C_2O_4^{2-} \rightarrow 2CO_2 + 2e^- \, (C_2O_4^{2-} \text{ oxidised})$$
$$MnO_4^- + 8H^+ + 5e^- \rightarrow Mn^{2+} + 4H_2O \, (MnO_4^- \text{ reduced})$$

The combined equation is obtained by multiplying the first equation by 5 and the second by 2 (to balance the electrons) and then adding, i.e.

$$5C_2O_4^{2-} + 2MnO_4^- + 16H^+ \rightarrow 10CO_2 + 2Mn^{2+} + 8H_2O$$

This equation covers all the reacting ions, but the non-reacting ions may also be added, if desired, to give the complete equation, e.g.

$$5Na_2C_2O_4 + 2KMnO_4 + 8H_2SO_4 \rightarrow 10CO_2 + 2MnSO_4$$
$$+ 8H_2O + K_2SO_4 + 5Na_2SO_4$$

(b) Oxidation of iron(II) salts with hydrogen peroxide.

$$Fe^{2+} \rightarrow Fe^{3+} + e^- \quad (Fe^{2+} \text{ oxidised})$$
$$H_2O_2 + 2H^+ + 2e^- \rightarrow 2H_2O \quad (H_2O_2 \text{ reduced})$$

i.e.
$$2Fe^{2+} + H_2O_2 + 2H^+ \rightarrow 2Fe^{3+} + 2H_2O$$

(c) Oxidation of hydrogen sulphide by chlorine.

$$H_2S \rightleftharpoons 2H^+ + S^{2-} \qquad S^{2-} \rightarrow S + 2e^- \quad (H_2S \text{ oxidised})$$
$$Cl_2 + 2e^- \rightarrow 2Cl^- \quad (Cl_2 \text{ reduced})$$

i.e.
$$S^{2-} + Cl_2 \rightarrow S + 2Cl^-$$

(d)   Oxidation of copper by hot concentrated sulphuric(VI) acid.

$$2H_2SO_4 + 2e^- \rightarrow SO_4^{2-} + 2H_2O + SO_2 \quad (H_2SO_4 \text{ reduced})$$
$$Cu \rightarrow Cu^{2+} + 2e^- \quad (Cu \text{ oxidised})$$

i.e. $$Cu + 2H_2SO_4 \rightarrow Cu^{2+} + SO_4^{2-} + 2H_2O + SO_2$$

Some substances can act as both oxidising and reducing agents depending upon the conditions. Thus, both hydrogen peroxide and nitric(III) acid (nitrous acid) behave as reducing agents in the presence of oxidising agents appreciably stronger than themselves, i.e. potassium manganate(VII) (potassium permanganate) and potassium dichromate(VI)

$$H_2O_2 + 2H^+ + 2e^- \rightarrow 2H_2O \qquad \text{(oxidising agent)}$$
$$H_2O_2 \rightarrow 2H^+ + O_2 + 2e^- \qquad \text{(reducing agent)}$$
$$HNO_2 + H^+ + e^- \rightarrow H_2O + NO \qquad \text{(oxidising agent)}$$
$$H_2O + HNO_2 \rightarrow HNO_3 + 2H^+ + 2e^- \qquad \text{(reducing agent)}$$

In all electrolyses, reduction occurs at the cathode whilst oxidation occurs at the anode. Thus, the electrolysis of molten sodium chloride is an example of an electrochemical redox reaction:

$$\text{Cathode} \quad Na^+ + e^- \rightarrow Na \quad \text{reduction}$$
$$\text{Anode} \quad Cl^- - e^- \rightarrow Cl \quad \text{oxidation}$$
$$2Cl \rightarrow Cl_2$$

ELECTROCHEMICAL CELLS

When a metal is in contact with a solution of one of its salts, an equilibrium is set up between the tendency of the metal to lose electrons and go into solution as ions and the opposing tendency of the ions in solution to gain electrons and be deposited on the metal. The metal, therefore, acquires a negative or positive charge respectively depending upon which process predominates. This means that a potential difference is set up between the metal and the solution. The metal in its salt solution is known as a *half-cell*.

Three common types of half-cell are given below.

(a)   Metal–metal ion as above.

(b)   Gas–gas ion. In this type a gas is allowed to equilibriate with its ions in solution at the surface of an inert metal conductor such as platinised platinum, e.g. the hydrogen electrode.

(c)   Redox cells. Here an inert metal conductor (platinum) is in contact with an element in two different oxidation states, e.g. Fe^{2+} and Fe^{3+} .

The combination of two half-cells will give an *electrochemical* or *voltaic cell*, e.g. the Daniell cell illustrated in Figure 7.5. Instead of a porous partition the two half-cells can be linked by a *salt bridge*, i.e. an inverted U-tube containing saturated potassium chloride solution and a porous plug each end.

If a cell is short-circuited by connecting the two poles by a conducting wire,

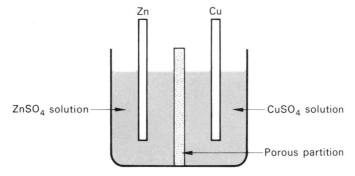

Figure 7.5   The Daniell cell

electrons flow from one pole to the other. Thus, in the Daniell cell, the zinc goes into solution as Zn^{2+} ions:

$$Zn(s) \rightarrow Zn^{2+} (aq) + 2e^-$$

and the electrons produced flow in the external circuit to the copper electrode where Cu^{2+} ions in solution accept them and are deposited:

$$Cu^{2+} (aq) + 2e^- \rightarrow Cu(s)$$

It is found that the potential difference between the two electrodes increases as the resistance in the external circuit is increased. The maximum potential difference of the cell is known as the *electromotive force* or e.m.f. and it occurs when no current is flowing. Accurate measurement of the e.m.f. of a cell is not possible with a normal voltmeter since current needs to flow to operate it. This problem may be largely overcome by using a high resistance voltmeter (a valve voltmeter) so that the current taken is very small. Alternatively, a potentiometer may be used, the principle here being that the potentiometer produces a measured adjustable voltage to just balance the current flowing in a circuit of which the cell is part. The counter-voltage applied then equals the e.m.f. of the cell under test.

The potential difference of a half-cell is dependent on the materials used, the temperature, and the concentration. Increasing the temperature increases the potential difference since it facilitates the loss of electrons from the metal and discourages the reverse reaction. Application of Le Chatelier's principle to the

$$M \rightleftharpoons M^{n+} + ne^-$$

equilibrium shows that increasing the concentration of the metal ion will repress electron loss from the metal and so lower the potential difference between the electrode and solution. The temperature and concentration are therefore standardised by taking a standard temperature of 25 °C and using molar solutions; the potential is then known as the *standard electrode potential*, $E^{\ominus}$. The absolute potential of an electrode or half-cell cannot be measured, but the potential difference between two electrodes can be measured. Hence the hydrogen electrode is taken as a reference electrode, its electrode potential being taken as zero. The potential difference (in volts) of all other electrodes are therefore obtained relative to the hydrogen electrode (Figure 7.6). In practice a secondary reference electrode, i.e. the calomel electrode (mercury in contact with a solution of potassium chloride saturated with mercury(I) chloride) is used since it is easier to set up and maintain.

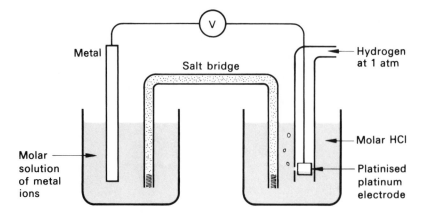

Figure 7.6   Measurement of electrode potential

The standard electrode potentials of some common elements are given in Table 7.2.

Table 7.2.   Redox potentials for some common elements

Oxidised form		Reduced form	E/volts
K^+	$+ e^-$	K	-2.92
Ca^{2+}	$+ 2e^-$	Ca	-2.87
Na^+	$+ e^-$	Na	-2.71
Zn^{2+}	$+ 2e^-$	Zn	-0.76
Fe^{2+}	$+ 2e^-$	Fe	-0.44
Pb^{2+}	$+ 2e^-$	Pb	-0.13
H^+	$+ e^-$	H	0.00
I_2	$+ 2e^-$	$2I^-$	0.54
Ag^+	$+ e^-$	Ag	0.80
Br_2	$+ 2e^-$	$2Br^-$	1.07
Cl_2	$+ 2e^-$	$2Cl^-$	1.36
F_2	$+ 2e^-$	$2F^-$	2.87

A cell is made by combining two half-cells and is represented by a *cell diagram*, for example, the cell diagram for the Daniell cell is represented by

$$Zn(s) \mid Zn^{2+}(aq) \parallel Cu^{2+}(aq) \mid Cu(s)$$

The single lines represent phase boundaries and the double line a salt bridge. By convention, the more positive electrode is put on the right hand side, and then the e.m.f. of the cell is given by

$$E_{cell} = E_{right\text{-}hand\,side\,electrode} - E_{left\text{-}hand\,side\,electrode}$$

Hence, for the Daniell cell

$$E_{cell} = 0.34 - (-0.76)$$
$$= 1.10 \text{ V}$$

The cell diagram implies the direction of the reaction, e.g. in the Daniell cell

$$Zn \rightarrow Zn^{2+} \quad \text{and} \quad Cu^{2+} \rightarrow Cu$$

or

$$Zn + Cu^{2+} \rightarrow Zn^{2+} + Cu$$

Arrangement of the elements, in order of their standard electrode potentials, gives the *electrochemical series* and so Table 7.2 constitutes part of this. The elements with the greatest negative electrode potential are at the top of the list and they have the greatest tendency to exist as positive ions in solution, i.e. they are the strongest reducing agents. The electrochemical series has a number of important uses.

(a) It shows the order of reactivity of the elements. The metals decrease in reactivity as the series is descended, whilst the reverse is the case for the non-metals.

(b) It shows which elements can displace each other from solutions of their salts, e.g. iron will displace copper from copper(II) sulphate(VI) solution since it is higher up the series. Similarly, chlorine will displace bromide ions from their solutions:

$$Cl_2(g) + 2Br^-(aq) \rightarrow 2Cl^-(aq) + Br_2(aq)$$

(c) It shows why zinc and aluminium may be used to protect the surface of iron — they are higher up the series and so will dissolve first.

(d) It can be used to predict the products discharged at the electrodes during electrolyses, but it should be remembered that other factors, such as the nature of the electrodes, etc. also affect this.

The electrochemical series has been extended to give the *redox series* which includes the standard electrode potentials of systems in which elements are present in different valency states, e.g:

$$Sn^{2+} \rightleftharpoons Sn^{4+} + 2e^-.$$

The redox series shows which reductions are energetically favourable although it must be understood that this does not imply that the rate is such as to make the reaction practicable. As a rough guide, cell reactions will proceed at a reasonable rate and go to completion if E_{cell} exceeds about 0.4 V. However, if E_{cell} is below this value the reaction tends to be slow and not to go to completion, i.e. an equilibrium mixture is formed. The use of $E^{\ominus}$ values to predict cell reactions is illustrated below.

Example Predict the cell reaction and calculate E_{cell} given that

$$Sn^{4+} + 2e^- \rightarrow Sn^{2+} \qquad E^{\ominus} = 0.15 \text{ V}$$

and for

$$Fe^{3+} + e^- \rightarrow Fe^{2+} \qquad E^{\ominus} = 0.77 \text{ V}$$

From the size of the $E^{\ominus}$ values for the half-cells it is apparent that Fe^{3+} has a greater attraction for electrons than Sn^{4+}, and so Fe^{3+} should be able to take electrons from Sn^{2+} to give Fe^{2+} and Sn^{4+}. This suggests that the half-reaction for tin should be reversed to obtain the actual equation for the process, i.e.

$$Sn^{2+} \rightarrow Sn^{4+} + 2e^-$$
$$\underline{2e^- + 2Fe^{3+} \rightarrow 2Fe^{2+}}$$

Adding:
$$Sn^{2+} + 2Fe^{3+} \rightarrow Sn^{4+} + 2Fe^{2+}$$

The cell is therefore written as

$$Sn^{2+}(aq) \mid Sn^{4+}(aq) \parallel Fe^{3+}(aq) \mid Fe^{2+}(aq)$$

$$E_{cell} = E_{R.H.S.} - E_{L.H.S.} = 0.77 - 0.15 = 0.62 \text{ V}$$

Note that the prediction applies when the temperature is 25 °C and molar solutions are used, but changing the conditions could alter the direction of the reaction.

MEASUREMENT OF pH BY ELECTRICAL METHODS

The pH of a solution can be found by immersing a hydrogen electrode in it and then completing the cell by combining it with a reference electrode. The e.m.f. of the cell is measured and this is utilised to calculate the H^+ concentration and hence the pH. (The calculation involved uses the Nernst equation which is beyond the scope of this text.) However, the method is not very convenient because the hydrogen electrode is difficult to set up and maintain and so another electrode sensitive to H^+ is preferable. A number of such electrodes exist but, to date, the glass electrode is the most successful.

The glass electrode, shown in Figure 7.7(a), consists of a glass tube terminating in a thin walled bulb containing 0.1 M hydrochloric acid solution. A platinum wire with its end plated with silver and then silver chloride passes down the tube and dips into the hydrochloric acid solution. When this electrode is placed in a solution of different pH, a potential difference is set up across the glass. The glass electrode is combined with a reference electrode which is usually a silver/silver chloride/saturated potassium chloride electrode or a calomel electrode (mercury/mercury(I) chloride/saturated potassium chloride — Figure

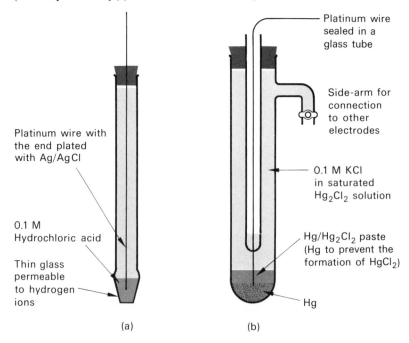

Platinum wire sealed in a glass tube

Side-arm for connection to other electrodes

Platinum wire with the end plated with Ag/AgCl

0.1 M KCl in saturated Hg_2Cl_2 solution

0.1 M Hydrochloric acid

Hg/Hg_2Cl_2 paste (Hg to prevent the formation of $HgCl_2$)

Thin glass permeable to hydrogen ions

Hg

(a)   (b)

Figure 7.7   (a) The glass electrode   (b) The calomel electrode

7.7(b)). The cell is put in the test solution and the e.m.f. measured by means of a valve or transistor voltmeter, this type being necessary because the glass bulb has a very high resistance, i.e. of the order of 10^6 to 10^8 ohms. The valve voltmeter may be calibrated in pH units by determining the e.m.f. using buffer solutions of known pH and the instrument is then termed a pH meter. In commercial pH meters the glass and reference electrode are combined in a single unit. Acid–base titrations may be followed by means of a pH meter and they are then known as *potentiometric titrations*.

Extraction of metals from their purified ores

The type of ore in which an element occurs is related to its electropositivity as indicated by Table 7.3. The electropositivity also determines the method used for extracting the metal from its purified ore.

The weakly electropositive elements are not strongly attracted to electronegative elements with the result that their oxides and sulphides are fairly readily decomposed. Thus, heating copper(I) sulphide in air partially converts it to the oxide which reacts with unchanged sulphide:

$$2Cu_2S + 3O_2 \rightarrow 2Cu_2O + 2SO_2$$
$$2Cu_2O + Cu_2S \rightarrow 6Cu + SO_2$$

Table 7.3. Relationship between $E^{\ominus}$ for an element and its type of ore

Element	$E^{\ominus}$/volts	Type of ore
Na	-2.71	Chloride
Mg	-2.37	Chloride and carbonate
Al	-1.66	Oxide
Ti	-1.60	Oxide
Zn	-0.76	Oxide and sulphide
Fe	-0.44	Oxide and sulphide
H	0.00	Oxide
Cu	0.34	Sulphide
Au	1.55	Free metal

More electropositive elements such as iron have their oxides reduced by heating with carbon or carbon monoxide, e.g.

$$Fe_2O_3 + 3CO \rightarrow 2Fe + 3CO_2$$
$$Fe_3O_4 + 4C \rightarrow 3Fe + 4CO$$

In the case of zinc, the reaction with carbon monoxide is reversible and so carbon has to be the reducing agent.

Titanium is quite strongly electropositive and has a strong affinity for oxygen. Carbon will reduce the oxide at high temperatures but, under these conditions, the titanium reacts with the carbon to form carbides. A more powerful reducing agent is therefore required and so the chloride is heated with sodium in an inert atmosphere of argon.

$$TiO_2 + 2C + 2Cl_2 \rightarrow TiCl_4 + 2CO$$
$$TiCl_4 + 4Na \rightarrow Ti + 4NaCl$$

Aluminium, magnesium, and sodium are too electropositive for their ores to be reduced by the usual reducing agents, and electrolytic reduction becomes necessary. The chlorides are generally used because their melting points tend to be lower than those of other salts. However, since aluminium chloride is covalent and will not conduct electricity, the oxide is used; molten sodium hexafluoro-aluminate(III) (cryolite, Na_3AlF_6) is employed as the solvent.

Questions

7.1 (a) Define (i) saturated solution, (ii) solubility product of a binary electrolyte.

(b) Derive an expression for the solubility product of a salt A_xB_y.

(c) Explain why the clear filtrate from a cold saturated solution of lead chloride becomes turbid on the addition of dilute (2 M) hydrochloric acid.

(d) The solubility of silver chloride is 1.87×10^{-3} g litre^{-1} of water at 25 °C.

  (i) What is the solubility product of silver chloride at this temperature?

  (ii) What weight of silver chloride would dissolve in 10 litres of 0.01 M hydrochloric acid at 25 °C?   [J.M.B.]

7.2 Define solubility product and quote two electrolytes to which the term can be applied and two to which it cannot. (Compounds named below in this question must not be included.)

  The solubilities of silver chloride and silver chromate(VI) (Ag_2CrO_4) are 2.009×10^{-3} g litre^{-1} and 3.207×10^{-2} g litre^{-1} respectively at 25 °C. Calculate

(a) the solubility product of each;

(b) the concentration in mol litre^{-1} of silver ions needed to precipitate silver chloride from a neutral 0.1 M solution of chloride ions;

(c) the concentration in mol litre^{-1} of silver ions needed to precipitate silver chromate(VI) from a neutral 0.005 M solution of chromate(VI) ions. What bearing have these results on the volumetric estimation of chloride ion?
                                    [J.M.B.]

7.3 Explain what is meant by solubility product and indicate the conditions under which the principle can be applied.

  The solubility product of calcium fluoride is 4.0×10^{-11} mol^3 litre^{-3} at 25 °C. Calculate the solubility (in g litre^{-1}) of calcium fluoride in (a) water and (b) a 0.01 M solution of sodium fluoride.

7.4 Briefly explain to which of the following compounds the concept of solubility product may be applied: $CuSO_4$, $MgCl_2$, $Fe(OH)_3$, CCl_4, and $Zn(NO_3)_2$.

  The solubility of strontium hydroxide, $Sr(OH)_2$, is 4.097 g litre^{-1} at 25 °C. Calculate:

(a) the solubility product of strontium hydroxide,

(b) the mass of strontium hydroxide which will dissolve in 250 cm^3 of 0.5 M sodium hydroxide solution.

7.5 (a) In each of the following reactions one of the reactants is behaving as a base. In each case state the species you believe to be the base.

  (i) $HSO_4^- + HNO_2 \rightarrow H_2NO_2^+ + SO_4^{2-}$

  (ii) $H_2PO_4^- + HCO_3^- \rightarrow HPO_4^{2-} + H_2O + CO_2$

  (iii) $CH_3CO_2H + HNO_3 \rightarrow CH_3CO_2H_2^+ + NO_3^-$

  (iv) $HBr + HCl \rightarrow Br^- + H_2Cl^+$

(b) (i) Using one of the systems given above as an example, give and explain briefly the Brønsted–Lowry theory of acids and bases.

  (ii) According to the theory, water is able to function both as an acid and as a base. Give one reaction in which water functions as an acid and another in which it functions as a base.

(c) Comment on the fact that a mixture of pure nitric acid, HNO_3, and pure chloric(VII) acid, $HClO_4$, contains the ions $H_2NO_3^+$ and ClO_4^-.

[University of London]

7.6 State Ostwald's dilution law.

The degree of dissociation of 0.02 M benzenecarboxylic acid $C_6H_5 \cdot COOH$, is 0.056 at 25 °C. Calculate (a) the dissociation constant of the acid, (b) the hydrogen ion concentration in the 0.02 M solution, and (c) the degree of dissociation in a 0.01 M solution.

7.7 Define (a) dissociation constant of an acid, (b) pH, and (c) buffer solution. Calculate (i) the approximate degree of dissociation and (ii) the pH of a solution of methanoic acid, HCOOH, containing 4.0 g litre^{-1} at 25 °C.

What would be the pH of a solution which is 0.1 M with respect to both methanoic acid and sodium methanoate?

The dissociation constant of methanoic acid is 1.6×10^{-4} mol litre^{-1} at 25 °C.

7.8 Define pH.

Explain why 0.1 M hydrochloric acid has pH = 1, but 0.1 M ethanoic acid (acetic acid) has a pH of approximately 3. Show, with the aid of diagrams, how these facts influence the choice of indicator when these solutions are titrated with sodium hydroxide solution.

Calculate the pH of

(a) 0.1 M ammonia solution,

(b) the solution obtained when 24.9 cm^3 of 0.1 M sodium hydroxide is added to 25.0 cm^3 of 0.1 M hydrochloric acid.

K_b for ammonia solution = 1.8×10^{-5} mol litre^{-1}.

7.9 (a) What do you understand by the Lewis theory of acids and bases?

(b) What is the difference between a strong and a weak acid?

How could the strength of an acid be determined experimentally?

(c) What is a buffer solution? Explain its mode of action. Give one example of its use. [J.M.B.]

7.10 You are given a sample of monobasic acid HA and are asked to obtain a quantitative measure of its strength in aqueous solution. Describe as fully as you can ONE experimental method which would enable you to do this, indicating how the result may be calculated from the measurements made.

[University of London]

7.11 The 'strength' of acids is customarily expressed in pK-units where pK denotes the negative logarithm to the base 10 of the dissociation constant of an acid.

For an investigation of the strength of hydrofluoric acid in aqueous solution, the following procedure was adopted: 25.0 cm^3 of a dilute hydrofluoric acid solution was added to 50.00 cm^3 of a 1.000 M sodium hydroxide solution contained in a laboratory flask. The excess sodium hydroxide was then back-titrated with 1.000 M sulphuric acid solution, requiring 6.25 cm^3 of the latter. In a separate experiment the pH of the hydrofluoric acid solution was found to be 1.66.

(a) From the above information, calculate the dissociation constant and the pK value of hydrofluoric acid.

(b) Briefly comment upon the procedure adopted for the titration.

(c) Suggest a method that might be used for the determination of the pH of the above solution.

(d) How does the strength of hydrofluoric acid compare with that of the other hydrogen halide acids? Suggest reasons for any differences you may infer.

[University of London]

7.12 'Electron transfer forms the basis of all redox reactions.' Make clear what you understand by this statement. Consider the statement in connection with the following reagents:

(a) hydrogen,

(b) hydrogen peroxide,

(c) sulphur dioxide,

(d) acidified sodium nitrate(III) (sodium nitrite).

   If possible give two reactions in each case which illustrate a different type of reaction. [University of London]

7.13 Define oxidation in terms of electron transfer.

Rewrite each of the following molecular equations first as a single ionic equation then as a pair of ion–electron equations, and show in each case which species is oxidised and which is reduced.

(a) $Fe_2(SO_4)_3 + 2KI \rightarrow 2FeSO_4 + I_2 + K_2SO_4$

(b) $2KMnO_4 + 16HCl \rightarrow 2KCl + 2MnCl_2 + 5Cl_2 + 8H_2O$

(c) $3Cu + 8HNO_3 \rightarrow 3Cu(NO_3)_2 + 2NO + 4H_2O$

(d) $4Zn + 10HNO_3 \rightarrow 4Zn(NO_3)_2 + NH_4NO_3 + 3H_2O$

(e) $3CdS + 8HNO_3 \rightarrow 3Cd(NO_3)_2 + 3S + 2NO + 4H_2O$

[J.M.B.]

7.14 Explain the process of oxidation and reduction in terms of electron transfer. Interpret the following oxidation–reduction reactions:

(a) $2Na + 2H_2O \rightarrow 2NaOH + H_2$

(b) $2H_2SO_4 \rightarrow H_2S_2O_8 + H_2$

(c) $2Na_2S_2O_3 + I_2 \rightarrow Na_2S_4O_6 + 2NaI$

It was found that, when 1.713 g of sodium ethanedioate (sodium oxalate) was dissolved in water and made up to 250 cm^3, 25 cm^3 of this solution, after acidification, reacted with 23.9 cm^3 of a potassium manganate(VII) (potassium permanganate) solution. Calculate the molarity of the latter solution.

7.15 Define (i) oxidation and (ii) reduction. Illustrate your definitions by reference to the electrolysis of copper(II) sulphate(VI) solution using copper electrodes.

Explain why iron, coated with a layer of tin, will rust rapidly if it is scratched, whereas iron coated with zinc does not rust in this way. Describe briefly, with the aid of a diagram, how you would determine the standard electrode potential of zinc at room temperature. How would you use this method to construct an electrochemical series? What use can be made of such a series?

Electrode potentials: $Zn^{2+}|Zn$ −0.762 V,

$Fe^{2+}|Fe$ −0.441 V,

$Sn^{2+}|Sn$ −0.136 V.

[J.M.B.]

7.16 Use the redox potentials given to decide what products, if any, would be obtained by the action of potassium dichromate(VI) in hydrochloric acid on (i) iron(II) chloride, (ii) tin, (iii) zinc, (iv) cerium(III) chloride.

Standard redox potentials in volts:

$Zn^{2+}|Zn$ −0.76, $Fe^{2+}|Fe$ −0.44, $Cr^{3+}|Cr^{2+}$ −0.41, $Ce^{4+}|Ce^{3+}$ +1.45,

$Fe^{3+}|Fe^{2+}$ +0.76, $\frac{1}{2}Cr_2O_7{}^{2-}|Cr^{3+}$ +1.33, $Sn^{2+}|Sn$ −0.14,

$Sn^{4+}|Sn^{2+}$ +0.15

[J.M.B.]

8 Chemical Kinetics

All chemical reactions proceed at a definite rate, the rate being controlled by the experimental conditions, e.g. the concentration of the reactants, the temperature, and the presence of a catalyst or light of a suitable wavelength. Some reactions are so slow that, for all practical purposes, no change occurs, e.g. the combination of nitrogen and hydrogen at room temperature. On the other hand, other reactions are so fast as to appear to be instantaneous, e.g. the reaction between halides and silver nitrate(V). Between these extremes, however, there are many reactions which proceed at measurable rates and the study of these, and the factors which affect them, is known as chemical kinetics. Important factors affecting the rate are discussed below.

1 CONCENTRATION

It was seen in Chapter 6 that the rate of a chemical reaction is proportional to the concentration of the reactants (the law of mass action). Since the reactants are continually being used up, it follows that the rate of the reaction will fall steadily as the reaction proceeds, i.e. the rate is continually changing. A plot of reaction rate against time will take the form shown in Figure 8.1. The rate of a reaction is usually measured in terms of the amount of reactants used up, or products formed, in a given time.

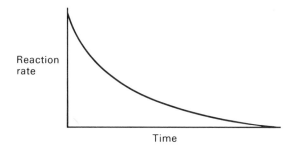

Figure 8.1   Variation of reaction rate with time

In the reaction

$$N_2O_5 \rightarrow N_2O_4 + \tfrac{1}{2}O_2$$

it is found that the rate is proportional to the concentration of the dinitrogen pentoxide, i.e.

$$\text{rate} \propto [N_2O_5]$$

or
$$\text{rate} = k[N_2O_5] \qquad (8.1)$$

where k is a proportionality constant known as the *velocity constant* or *rate constant*. Plotting rate against concentration will give a straight line graph passing through the origin and the gradient will be equal to k. The rate is in terms of decrease in the concentration of the dinitrogen pentoxide in unit time and so its units will be mol litre^{-1}s^{-1}. Rearranging equation (8.1) gives

$$k = \frac{\text{rate}}{[N_2O_5]}$$

and substituting the units in this gives

$$k = \frac{\text{mol litre}^{-1}\text{s}^{-1}}{\text{mol litre}^{-1}}$$

Hence the units of k are s^{-1}. The value of k is dependent upon the temperature at which the reaction is carried out.

In the reaction

$$CH_3I + NaOH \rightarrow CH_3OH + NaI$$

it is found experimentally that the rate is proportional to the concentration of the iodomethane (methyl iodide) and of the sodium hydroxide, i.e.

$$\text{rate} \propto [CH_3I] = k_1[CH_3I]$$
$$\text{rate} \propto [NaOH] = k_2[NaOH]$$

Combining these equations gives

$$\text{rate} \propto [CH_3I][NaOH]$$

or $$\text{rate} = k[CH_3I][NaOH] \qquad (8.2)$$

where k is numerically equal to $k_1 \times k_2$

A plot of rate against the concentration of iodomethane multiplied by the concentration of sodium hydroxide will give a straight line graph with gradient equal to k. The units of k may be found by rearranging equation (8.2) and substituting the various units, i.e.

$$k = \frac{\text{rate}}{[CH_3I][NaOH]}$$

$$= \frac{\text{mol litre}^{-1}\text{s}^{-1}}{\text{mol litre}^{-1} \times \text{mol litre}^{-1}}$$

$$= \text{litre mol}^{-1}\text{s}^{-1}$$

The reaction

$$2HI \rightarrow H_2 + I_2$$

is somewhat similar to the previous example; it is found that

$$\text{rate} \propto [HI][HI]$$

i.e. $$\text{rate} = k[HI]^2$$

If in the general reaction

$$aA + bB \rightarrow \text{products}$$

it is found that the rate is proportional to $[A]^x$ and $[B]^y$, then the rate expression is

$$\text{rate} = k[A]^x[B]^y \qquad (8.3)$$

The overall *order* of the reaction is said to be $x + y$ whilst the order with respect to A is x and with respect to B is y. The order of a reaction is, therefore, the sum of the powers of the concentration of the reactants which appear in the rate equation. Some reactions have simple orders, e.g. the reaction

$$H_2 + I_2 \rightleftharpoons 2HI$$

is a second order reaction. However, fractional orders are possible, i.e. for the reaction

$$CH_3 \cdot CHO \xrightarrow{450\ °C} CH_4 + CO$$

the rate is found to be proportional to $[CH_3 \cdot CHO]^{1.5}$, and so the order is 1.5. Pseudo orders are also observed, thus the hydrolysis of ethyl ethanoate (ethyl acetate)

$$CH_3 \cdot COOC_2H_5 + H_2O \rightleftharpoons CH_3 \cdot COOH + C_2H_5OH$$

appears to be first order because a large excess of water is used and its concentration remains nearly constant.

It should be noted that orders of reaction are determined from experimental measurements and that neither they nor the rate expression can be determined from the equation for the reaction.

Many reactions proceed via a series of steps, and the overall rate by which they proceed will be controlled by the slowest stage which is known as the *rate determining step*. The number of molecules, atoms, radicals, or ions taking part in the rate determining step is known as the *molecularity* and numerically it may or may not be the same as the order of the reaction.

The treatment of results from kinetic measurements is illustrated in the examples below.

Example   Dinitrogen pentoxide undergoes thermal decomposition according to the equation

$$2N_2O_5 \rightarrow 4NO_2 + O_2$$

From the following data, obtained at 45 °C, show that the reaction is first order and determine the rate constant.

Time/min	Conc. $N_2O_5/ 10^{-3}$ mol litre^{-1}
0	22.90
10	16.27
20	12.29
30	9.35
40	6.89
50	4.88
60	3.68
70	2.74
80	2.16
90	1.85

The first step is to plot a graph of concentration against time (Figure 8.2). Tangents are then drawn at a number of points on the curve and the gradients of the tangents give the rates of the reaction as those points, e.g.

Conc. N_2O_5 in 10^{-3} mol litre^{-1}	Rate in 10^{-3} mol litre^{-1} min^{-1}
16.27	0.483
6.89	0.209
3.68	0.121
2.74	0.078

A graph is then plotted of reaction rate against concentration (Figure 8.3), and it is seen that, as expected for a first order reaction, the points fall close to a straight line. The gradient of this graph gives the value of the rate constant, k, i.e.

$$k = 3.0 \times 10^{-2} \, min^{-1}$$

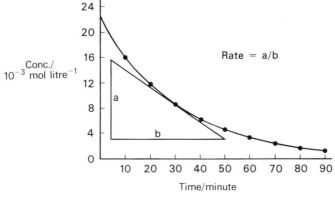

Figure 8.2   Plot of concentration of N_2O_5 against time

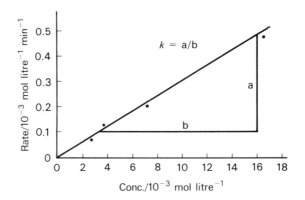

Figure 8.3   First order rate plot for thermal decomposition of N_2O_5

It will be seen from the concentration–time graphs of first order reactions that the time taken for the initial concentration to fall by one half is the same as the time taken for it to fall from one half to one quarter, and from one quarter to one eighth, etc. The *half-life*, $t_{\frac{1}{2}}$, for a first order reaction is therefore constant and further, it is independent of the initial concentration. Radioactive decay is a typical first order reaction.

Example   When 0.02 mole of bromoethane (ethyl bromide) and of sodium hydroxide in solution were mixed and then 20 cm^3 samples were taken, at regular intervals, and titrated against 0.05 M hydrochloric acid, the following results were obtained.

Time/min	Titre/cm³	Time/min	Titre/cm³
0	32.0	100	8.8
25	18.0	150	6.2
50	13.1	200	4.8
75	10.8	250	4.0

Show that the reaction is second order. Note that the titres are directly proportional to the concentration of sodium hydroxide.

$$C_2H_5Br + NaOH \rightarrow C_2H_5OH + NaBr$$

If the reaction is second order, then

$$rate \propto [C_2H_5Br] \, [NaOH]$$

and so a plot of rate against $[C_2H_5Br] \, [NaOH]$ should give a straight line passing through the origin. The rate at various times may be found by plotting a concentration–time graph (Figure 8.4), drawing tangents to the curve, and then finding their gradients.

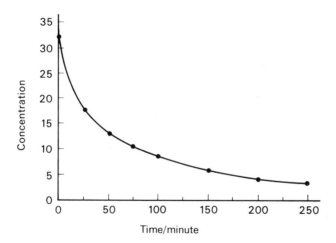

Figure 8.4   Plot of [NaOH] against time

The following results are obtained:

Time/min	Rate	[NaOH] [C₂H₅Br] i.e. titre²
25	0.285	324
50	0.128	172
75	0.089	117
100	0.071	77
200	0.022	23

The plot of rate against $[NaOH] \, [C_2H_5Br]$ is shown in Figure 8.5 and since it is a straight line graph the reaction is second order.

It will be seen from the concentration–time graph for a second order reaction (Figure 8.4) that the half-life is not constant, and that it is dependent on the concentrations at the start of the half-life.

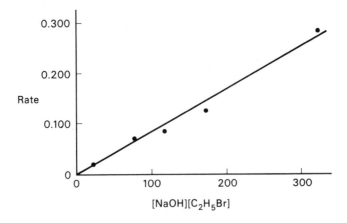

Figure 8.5   Second order rate plot for the reaction between NaOH and C_2H_5Br

Methods of following rates of reaction

There are a number of methods of determining the extent to which a reaction has progressed, some of the most widely applicable ones being:

(a)   taking a sample, quenching if necessary, and then titrating,

(b)   measuring the volume of gas evolved,

(c)   measuring the intensity of the colour of the solution by means of a colorimeter,

(d)   measuring the electrical resistance of the solution (electrolytic conductivity measurements).

a.   This method is used to follow the acid catalysed hydrolysis of methyl methanoate (methyl formate)

$$HCOOCH_3 + H_2O \xrightarrow{H^+} HCOOH + CH_3OH$$

Thus, the ester is added to dilute hydrochloric acid solution and then a sample is immediately removed and added to a large excess of cold water so that the reaction is effectively stopped. The acid is titrated with sodium hydroxide solution using phenolphthalein as indicator. The process is repeated at regular intervals and finally after about 48 hours when reaction is complete. In each case, the concentration of acid (hydrochloric acid plus methanoic acid) is proportional to the sodium hydroxide titre.

b.   Diazonium salt solutions (page 343) decompose on warming e.g.

$$C_6H_5 \cdot N{=}N^+Cl^- + H_2O \rightarrow C_6H_5OH + N_2 + HCl$$

and the reaction may be followed by measuring the volume of nitrogen evolved. The apparatus used is illustrated in Figure 8.6.

Once the solution has had time to reach the bath temperature, the tap is closed and the time noted. The volume is then read at various intervals after the levels have been equalized.

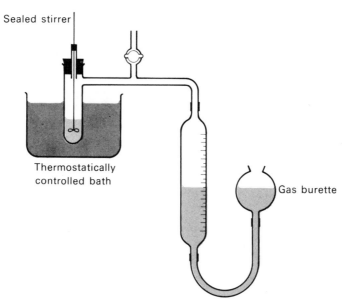

Figure 8.6   Apparatus for measuring the volume of gas evolved in a reaction

c.   Colorimetric methods may be used to follow the reaction between aqueous solutions of propanone (acetone) and iodine:

$$CH_3 \cdot CO \cdot CH_3 + I_2 \rightarrow CH_3 \cdot CO \cdot CH_2I + HI$$

The colorimeter (Figure 8.7) consists basically of a light source, various filters, and a light sensitive cell in conjunction with a meter. The filters are selected so that a band of light of the wavelengths most strongly absorbed by the solution is used. This light passes through the solution on to a light sensitive cell which gives an e.m.f. proportional to the intensity of light falling on it. When the colorimeter has been calibrated, the concentration of the unreacted iodine can be found from the intensity of the light emerging from the solution.

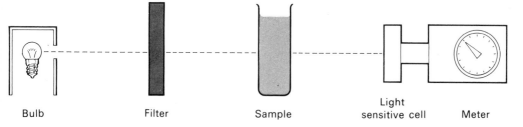

Bulb   Filter   Sample   Light
                                                          sensitive cell   Meter

Figure 8.7   The essentials of a colorimeter

d.   Electrolytic conductivity measurements (page 127) can be used to study the alkaline hydrolysis of ethyl ethanoate (ethyl acetate):

$$CH_3 \cdot COOC_2H_5 + NaOH \rightarrow CH_3 \cdot COONa + C_2H_5OH$$

since the fast moving HO^- ion is replaced by the less effective $CH_3 \cdot COO^-$ ion.

A similar example is the hydrolysis of bromoethane (ethyl bromide) with potassium hydroxide solution:

$$C_2H_5Br + KOH \rightarrow C_2H_5OH + KBr$$

2 TEMPERATURE

It has been recognised for a long time that the rate of a reaction is greatly affected by temperature. In fact, it is generally found that a temperature rise of about 10 °C approximately doubles the rate of reaction. Two theories — the collision and transition state theories — have been developed to explain this. The former theory is concerned with collisions between molecules whilst the latter concentrates on what happens once molecules have collided.

a. The collision theory

This is based on the kinetic theory and assumes that molecules must collide before they can react. However, calculations show, and experiments confirm, that often only one collision in about every 10^{14} leads to reaction. It was suggested, therefore, that only molecules having more than a certain minimum amount of kinetic energy, known as the *activation energy*, react when they collide. The activation energy varies considerably between reactions.

It was noted on page 34 that the kinetic energy of molecules is continually changing as a result of collisions. Further, at any given time, a small fraction of the molecules has relatively high kinetic energies. The proportion of high energy molecules is increased by raising the temperature. Thus, Figure 8.8 shows that, at temperature T_2, there is a greater proportion of molecules with kinetic energy in excess of the arbitrary value E than there is at the lower temperature T_1. At higher temperatures, therefore, there are more molecules of the reactants with energy in excess of the activation energy and so the rate of reaction will be greater.

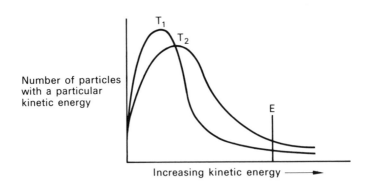

Figure 8.8   Distribution on kinetic energy of molecules at different temperatures

Using the collision theory, it is possible to calculate a theoretical value for the rate constant for various reactions. A comparison of the theoretical values with the experimental results shows that there is agreement in the case of some simple reactions. However, with many reactions the predicted and observed rates differ by many powers of ten. One reason for the discrepancies is that in some reactions a steric factor is involved, i.e. the molecules must collide in the correct relative positions. This is known as *collision geometry*. Effective and ineffective collisions are illustrated in Figure 8.9.

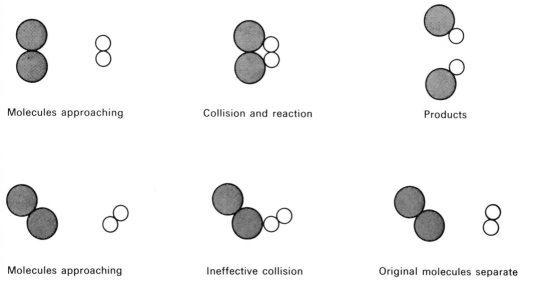

| Molecules approaching | Collision and reaction | Products |

| Molecules approaching | Ineffective collision | Original molecules separate |

Figure 8.9   Effective and ineffective collision of molecules

When steric factors are involved, the calculated rate constant is larger than the experimental value. The reverse is the case, however, in reactions which occur via a free radical mechanism (see page 160). The collision theory fails with reactions involving heterogeneous catalysis since the reaction rates of these do not depend upon the number of collisions between the molecules of the reactants.

b. The transition state theory

This theory suggests that as molecules collide and reaction takes place, they are momentarily in a less stable state than either the reactants or the products, i.e. the atoms are rearranging themselves. As the atoms are separated, the potential energy of the system increases and this results in an energy barrier between the reactants and products. An *activated complex* or *transition state* is formed. This is illustrated by the reaction between hydrogen and iodine giving hydrogen iodide:

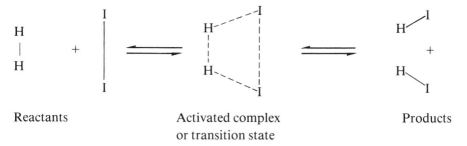

| Reactants | Activated complex or transition state | Products |

(N.B. The reaction is bimolecular because two molecules are required to form the activated complex.)

The products are formed only if the colliding molecules have sufficient energy to overcome the energy barrier, the energy required being the activation energy. The higher the activation energy the lower the number of effective collisions.

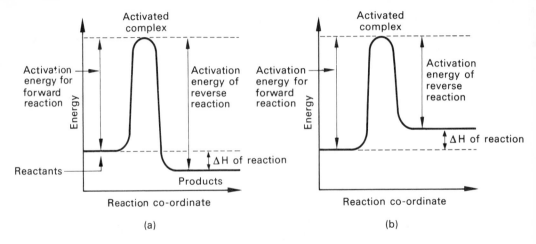

Figure 8.10   Energy changes in (a) exothermic and (b) endothermic reactions

Figure 8.10 gives a pictorial representation of the courses of exothermic and endothermic reactions.

The transition state theory, unlike the collision theory, may be used to predict reaction rates from a knowledge of the molecular structure of the reactants. However, the mathematical development of the theory is complicated and beyond the scope of this text.

The Arrhenius equation

The relationship between the rate constant and the temperature for a given reaction is expressed by the *Arrhenius equation*

$$k = A e^{-E/RT} \tag{8.4}$$

where A is a constant for the reaction and is known as the Arrhenius factor, E, is the activation energy, R, is the gas constant, and T is the kelvin temperature. The exponential factor, $e^{-E/RT}$, provides for the marked effect that increasing the temperature has on the proportion of high-kinetic-energy molecules.

Taking log to the base ten of equation (8.4) gives

$$\lg k = \lg A - \frac{E}{2.303RT}$$

Since this is an equation of the type $a = c + bx$, a plot of values of $\lg k$ at different temperatures against $1/T$ will give a straight line graph with slope $-E/2.303\,R$ and intercept $\lg A$. Hence the activation energy and the Arrhenius factor may be determined.

The Arrhenius equation is very similar to the mathematical expression for the Collision Theory of Reaction Rates, the latter being

$$k = Z^0 e^{-E/RT}$$

where Z^0 is known as the collision number. The collision number is the number of collisions per unit volume per unit time divided by the product of the number of molecules of each reactant per unit volume. When steric factors, etc. are involved (see above) the equation is modified to

$$k = PZ^0e^{-E/RT}$$

where P is called the *probability* or *steric* factor and is the ratio of the observed rate constant to the value calculated from the collision theory.

3 CATALYSTS

A catalyst is a substance which affects the rate of a chemical reaction, but which can be completely recovered unchanged at the end of the reaction. During the reaction, however, the catalyst may be changed physically or it may undergo temporary chemical change. Catalysts operate by providing an alternative route for the reaction. If the alternative route has a lower activation energy, the catalyst is said to be a *positive catalyst*, whereas one which leads to a higher activation energy is known as a *negative catalyst* or *inhibitor*. Note that catalysts do not affect the position of equilibrium, they merely alter the rate at which equilibrium is achieved.

The amount of catalyst required varies greatly. Thus, the reaction

$$14H^+(aq) + Cr_2O_7^{2-}(aq) + 6I^-(aq) \rightarrow 2Cr^{3+}(aq) + 3I_2(aq) + 7H_2O(l)$$

is catalysed by minute traces of Cu^{2+} whereas the Friedel-Crafts reaction

$$C_6H_6(l) + CH_3 \cdot COCl(l) \xrightarrow{\text{AlCl}_3} C_6H_5 \cdot CO \cdot CH_3(l) + HCl(g)$$

requires just over one mole of aluminium chloride for every mole of acid chloride used.

Some catalysts have their activity impaired or destroyed by the presence of other substances, for example, traces of arsenic poison the platinum catalyst in the contact process,

$$2SO_2(g) + O_2(g) \xrightarrow{\text{Pt,450 °C}} 2SO_3(g)$$

Conversely, traces of some substances, which have no catalytic activity, can make some catalysts more active; such substances are said to be *promoters*. Molybdenum is the catalyst promoter in the Haber process,

$$N_2(g) + 3H_2(g) \xrightarrow{\text{550 °C, 250 atm}} 2NH_3(g)$$

In some reactions one of the products catalyses the reaction and this is known as *autocatalysis*, e.g. the Mn^{2+} ion, formed in the oxidation of ethane-1,2-dioic acid (oxalic acid) by acidified potassium manganate(VII) (potassium permanganate), catalyses the reaction

$$2MnO_4^-(aq) + 16H^+(aq) + 5C_2O_4^{2-}(aq) \rightarrow 2Mn^{2+}(aq) + 8H_2O(l) + 10CO_2(g)$$

There are two main types of catalysis, i.e. homogeneous and heterogeneous catalysis. However, enzymes (page 258) are very important catalysts which do not readily fit into either of the above categories because they are present as colloidal solutions.

a. Homogeneous catalysis

In homogeneous catalysis, the catalyst and the reactants are in the same phase. A typical example is the acid catalysed reaction between organic acids and alcohols (see page 332), e.g.

$$CH_3 \cdot COOH(l) + C_2H_5OH(l) \underset{\text{(e.g. H}_2\text{SO}_4)}{\overset{H^+}{\rightleftharpoons}} CH_3 \cdot COOC_2H_5(l) + H_2O(l)$$

An example in the gas phase is the catalysis of the reaction between oxygen and sulphur dioxide by nitrogen oxide. For simplicity, this complex reaction may be represented as:

$$2NO(g) + O_2(g) \rightarrow 2NO_2(g)$$
$$NO_2(g) + SO_2(g) \rightarrow SO_3(g) + NO(g)$$

b. Heterogeneous catalysis

Two phases are present in this type of catalysis. Generally the process involves the reaction of two gases at the surface of a solid catalyst, e.g. the hydrogenation of alkenes using nickel as catalyst (see page 301). In this reaction it is thought that the hydrogen is adsorbed as hydrogen atoms on the surface of the metal and that this also attracts the π electrons of the alkene. The π electrons are then attacked by the hydrogen atoms, the resultant alkane is expelled, and the process is repeated. The catalyst is finely divided to give a large surface area.

The catalytic decomposition of hydrogen peroxide with manganese(IV) oxide is an example of heterogeneous catalysis of a reaction in solution.

4 LIGHT

Methane reacts with chlorine under the influence of ultraviolet light to give chlorinated compounds. The function of the light is to split up the chlorine molecules into reactive chlorine atoms.

$$Cl-Cl \xrightarrow{\text{u.v.}} 2Cl \cdot$$
$$CH_4 + Cl \cdot \rightarrow CH_3 \cdot + HCl$$
$$CH_3 \cdot + Cl_2 \rightarrow CH_3Cl + Cl \cdot \text{ etc.}$$

Bromine behaves similarly. Reactions such as this are quite common, particularly in organic chemistry, and they are referred to as *photochemical reactions*. They are said to proceed via a free radical mechanism, a *free radical* being an atom or group with an unpaired electron.

The reaction between hydrogen and chlorine in sunlight is similar to the above reaction:

$$Cl_2 \xrightarrow{\text{u.v.}} 2Cl \cdot$$
$$H_2 + Cl \cdot \rightarrow HCl + H \cdot$$
$$H \cdot + Cl_2 \rightarrow HCl + Cl \cdot$$

Kinetics and mechanism

Kinetics provide a major starting point for formulating mechanisms. However, it must be stressed that a mechanism is a theoretical explanation of the experimental (kinetic) facts.

Investigation of the alkaline hydrolysis of bromoethane (ethyl bromide):

$$C_2H_5Br + HO^- \rightarrow C_2H_5OH + Br^-$$

shows that

$$\text{rate} = k[C_2H_5Br]\,[HO^-]$$

Hence, both the bromoethane and hydroxide ion are involved in the rate determining step. The mechanism is thought to be

$$HO^- + C_2H_5Br \rightarrow [HO \cdots C_2H_5 \cdots Br]^- \rightarrow C_2H_5OH + Br^-$$

Dotted lines indicate partial bonds.

The reaction is a bimolecular second order reaction.

On the other hand, the hydrolysis of 2-bromo-2-methylpropane (*t*-butyl bromide) has a rate equation of

$$\text{rate} = k[(CH_3)_3CBr]$$

and so it is independent of the alkali concentration (i.e. the reaction is zero order with respect to HO^-). The inference here is that only the 2-bromo-2-methylpropane is involved in the rate determining step. It is thought that the hydrolysis is a two-stage process, i.e.

$$(CH_3)_3CBr \xrightarrow{\text{slow}} (CH_3)_3C^+ + Br^- \xrightarrow[\text{fast}]{HO^-} (CH_3)_3COH$$

This hydrolysis is a unimolecular first order reaction.

Questions

8.1 In the presence of hydrochloric acid, *N*- chloro-*N*-phenylethanamide (*N*-chloro-acetanilide) (A) is changed to its isomer, *N*-(4-chlorophenyl)ethamide (4-chloroacetanilide) (B):

| A | B |

The progress of the reaction can be followed because A liberates iodine from potassium iodide solution whereas B does not react. The iodine can be estimated by titration with standard sodium thiosulphate(VI) solution.

The table shows the volume, x, of a sodium thiosulphate(VI) solution needed at various times in the course of a particular experiment, in each case for a fixed volume of reaction mixture (x measures the amount of A left at each titration).

t/min	0	15	30	45	60	75
x/cm^3	24.5	18.1	13.3	9.7	7.1	5.2

(a) Using suitable scales, plot a graph of x against t.

(b) From your graph, read off the time taken
  (i) for half of the original A,

(ii) for three quarters of the original A, to have been transformed into B. What does a comparison of these two figures tell us about the order of the reaction?

(c) Describe in outline how you would investigate the effect of changing acid concentration on the rate of reaction.

(d) Discuss briefly the effect that an increase in temperature would have on the rate of the reaction.

(e) Does the fact that a reaction is of first order necessarily mean that it is unimolecular?
Explain.   [University of London]

8.2   Explain the terms (a) order of reaction, (b) molecularity, (c) half-life, and (d) rate of reaction.
Hydrolysis of 2-bromo-2-methylpropane at 25 °C gave the following results:

Time/h	Conc./mol litre^{-1}	Time/h	Conc./mol litre^{-1}
0.1	0.105	14.5	0.051
2.5	0.092	20.0	0.037
3.5	0.088	27.0	0.026
7.0	0.074	35.5	0.016
11.0	0.061	45.0	0.010

(i) Determine by graphical means, the order of reaction with respect to the 2-bromo-2-methylpropane.
(ii) The rate constant.
(iii) The rate of reaction when the initial concentration has fallen by half.

8.3   Discuss, in terms of the collision theory, the effect of temperature on the rate of a reaction. The rate constant of a reaction varies with temperature as follows:

Temperature/ °C	25	45	65
Rate constant/s^{-1}	3.45×10^{-1}	4.16×10^{-1}	4.88×10^{-3}

Use these data to determine the activation energy of the reaction.

8.4 (a) Sketch a graph, with labelled axes, to illustrate the distribution of molecular velocities in a gas. On the same graph indicate how the distribution changes with a small increase in temperature.

(b) Use a knowledge of the kinetic theory of gases and the information given in part (a) to describe what happens when a volatile liquid, in equilibrium with its vapour, is heated to a slightly higher temperature.

(c) Draw a diagram which illustrates the changes which occur during the course of the chemical reaction

$$A + B \rightarrow C + D$$

On the diagram, indicate the activation energy and the enthalpy change (ΔH) for the reaction. Assume that the reaction is exothermic, i.e. ΔH is negative.

(d) Show how the concepts involved in part (a) of this question are of value in understanding why the rate of a chemical reaction increases with temperature.
[J.M.B.]

8.5   What are the essential features of a catalyst? Outline an experiment in which the rate of a catalysed reaction is followed.
Indicate what experiments you would carry out to show that the reaction of your choice is a catalytic one.
Discuss, with a suitable example in each case, TWO main theories of catalytic action which have been proposed. In your account consider the effect of a catalyst, if any, on the rate of reaction, the route of the reaction, and the position of equilibrium.
The replacement of hydrogen in a hydrocarbon by bromine is sometimes described as being 'catalysed by bright sunlight'. Explain the action of sunlight and comment on this description.
[J.M.B.]

9 Periodicity

The outer electron configuration of atoms is a periodic function. For example, lithium, sodium, potassium, etc. all have an electronic configuration with one electron in the outer shell, and on moving to the right across the periodic table, the number of electrons in the outer shell increases from one to eight. It is to be expected, therefore, that any physical properties connected with electron arrangement will also exhibit this periodicity. Such properties include ionisation energy, electron affinity, covalent radii, ionic radii, oxidation state, atomic volume, melting points, boiling points, and enthalpies (heats) of fusion and vaporisation.

1 IONISATION ENERGY (IONISATION POTENTIAL)

Ionisation energy is the enthalpy change when a mole of electrons is removed from a mole of gaseous atoms to give positively charged ions. Energy is always absorbed in this process because work has to be done in overcoming the attractive force of the positively charged nucleus for the negatively charged electrons. The energy is measured in kilojoules per mole (or electronvolts where $1\,eV = 1.602 \times 10^{-19}\,J$).

The ionisation energy, I, of an element may be determined from its atomic emission spectrum (Chapter 1). Thus, the convergency limit represents the point at which an electron becomes free from the attraction of the nucleus. The energy associated with this frequency of radiation is given by the formula $E = h\nu$ (page 10).

Alternatively, an electron impact method may be used; the apparatus required is shown in Figure 9.1. The valve is filled with a monatomic gas at low pressure and then the cathode is heated by means of an electric current. This produces a stream of electrons in the valve and they are accelerated by means of a potential applied between the grid and the cathode. The electrons pass through the grid and move towards the anode, but they are repelled from this by applying a small voltage to make it negative. (N.B. The valve anode is not being used as an anode in this determination — it is negatively charged.) Thus, at this stage no current flows between the anode and cathode.

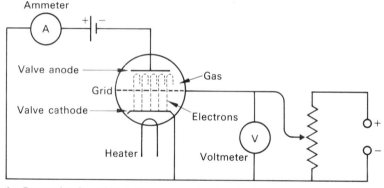

Figure 9.1   Determination of ionisation energy by the electron impact method

If an electron of low energy collides with an atom of the gas in the valve, the collision is elastic, i.e. the electron simply bounces off and no energy is transferred to the atom. However, if the energy of the electron is increased by increasing the potential difference applied to the grid, a state is reached where the collisions become inelastic and some of the electron's energy is absorbed by the atom which consequently assumes an excited state. The excitation potential is characteristic of a particular atom.

Several excitation potentials may be detected, but finally an electron is dislodged from the atom or, in other words, ionisation occurs. The positive ions formed are attracted to the valve anode and so, under these conditions, current flows between the cathode and anode and the ammeter indicates this accordingly. A plot of anode current against applied voltage takes the form shown in Figure 9.2.

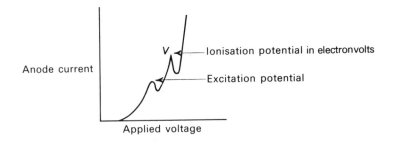

Figure 9.2   Variation of anode current with applied voltage in the electron impact method of determining ionisation potential

The energy attained by an electron when it is accelerated through a potential difference of   1 volt = $1.6 \times 10^{-19} \times 1$ joule   (1.6×10^{-19} coulomb is the electron charge). Thus, the energy attained by   1 mole of electrons = $1.6 \times 10^{-19} \times 6.02 \times 10^{23}$ J (6.02×10^{23} mol^{-1} is the Avogadro constant) = 96.3 kJ.

It follows from this that the energy possessed by one mole of electrons, when it ionises one mole of the gas is $96.3 \times V$ kJ (where V is the voltage applied to the grid — see Figure 9.2). This corresponds to the ionisation energy of the gas.

A plot of ionisation energy against atomic number is shown in Figure 9.3 and the periodicity is clearly evident. The individual values are given in Figure 9.4.

The magnitude of the ionisation energy of an element is determined by the attraction the nucleus has for the electron removed. This attraction is dependent upon the nuclear charge and also on how well the other electrons shield it. Thus, the *effective nuclear charge*, Z^*, is less than the true nuclear charge. For example, beryllium has a much larger Z^* value than lithium because its extra electron is in the same shell as lithium's, but the nuclear charge is one more ($I_{Li} = 520$ kJ mol^{-1}, $I_{Be} = 900$ kJ mol^{-1}). The ionisation energy therefore increases steadily across each period. The minor breaks shown in the plot of ionisation energy against atomic number for the 2nd and 3rd periods (Figure 9.5) are due to the relatively high shielding power of filled or exactly half filled sub-shells. Boron has a lower ionisation energy than beryllium because its extra electron is shielded by the full $2s$ orbital (Be, $1s^2 2s^2$) and oxygen is lower than nitrogen because its extra electron is shielded by the half-filled $2p$ orbitals (N, $1s^2 2s^2 2p^3$).

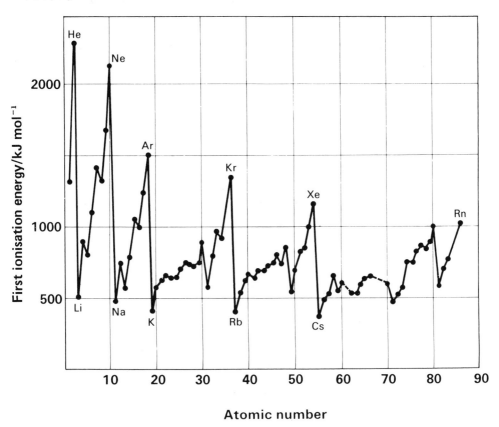

Figure 9.3   Variation of ionisation energies of the elements with atomic number

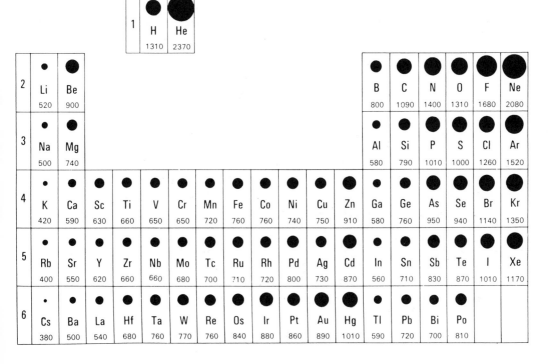

Figure 9.4   Ionisation energies of the elements/kJ mol⁻¹

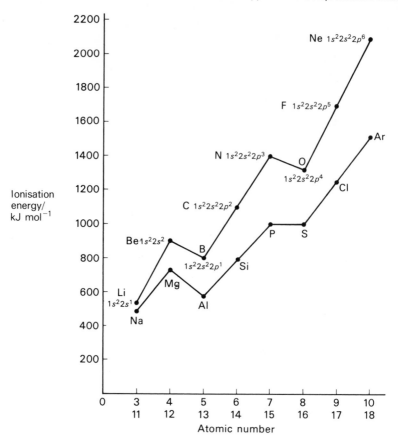

Figure 9.5   Variation of ionisation energy with atomic number for the 2nd and 3rd periods of the periodic table

Ionisation energy generally decreases steadily down a group because the outer electron becomes progressively further from the nucleus and so less tightly held. However, some exceptions do occur immediately after the transition elements. For instance, in Group III, the ionisation energies for indium ($Z = 49$) and thallium ($Z = 81$) are 560 and 590 kJ mol^{-1} respectively. This is explained by the fact that, of the 32 extra electrons in thallium, 14 have gone into an inner shell (the $4f$). Thus, the increase in atomic size is not paralleled by the increase in nuclear charge.

The ionisation energies referred to above are first ionisation energies, but it is possible to remove further electrons and so get 2nd, 3rd, 4th, etc. ionisation energies. The values of these successive ionisation energies are interesting since they confirm quantitatively the observed fact that completed shells show increased stability. The successive ionisation energies for potassium are illustrated in Figure 3.2.

Obviously, the size of the ionisation energy exerts a considerable influence on the type of bonding an atom undergoes. In the case of aluminium, the first three ionisation energies are 580, 1800, and 2700 kJ mol^{-1} respectively, and so the formation of an Al^{3+} ion (with the noble gas structure) requires 5080 kJ mol^{-1}. Hence ionic aluminium compounds are not formed unless this large energy requirement is compensated for by an even larger lattice energy. As a result, most aluminium compounds, other than the fluoride, are covalent or have appreciable covalent character.

It is perhaps pertinent at this stage to stress the distinction between ionisation energy and standard electrode potential (page 141). Both apply to the formation of ions, the difference being the state of the reactants and products as shown by the equations below:

$$M(s) \rightarrow M^+(aq) + e^- \quad \text{electrode potential}$$

$$M(g) \rightarrow M^+(g) + e^- \quad \text{ionisation energy}$$

Note that, as a consequence of this, the order of the electrochemical series is not necessarily the same as the order of ionisation energies.

2 ELECTRON AFFINITY

Electron affinity is the enthalpy change when a mole of gaseous atoms acquires a mole of electrons to give a mole of negatively charged ions, e.g.

$$F(g) + e^- \rightarrow F^-(g)$$

The general trend is for electron affinity to increase from left to right across the periodic table. This is because, as the nuclear charge increases, the extra electrons enter the same electron shell and so are attracted more strongly. Thus, atoms have a progressively greater attraction for electrons as the group number increases. Some electron affinities are given below, the values being in kJ mol^{-1} (the values vary considerably according to the source from which they are obtained):

C	-120	N	≈ 0	O	-141	F	-333
		P	-67	S	-200	Cl	-364
						Br	-342
						I	-295

Most electron affinities are exothermic because the orbitals of an electron theoretically extend to infinity and so the nuclear charge is never completely balanced. However, when an electron is added to an atom with a full or exactly half full outer sub-shell (e.g. as with nitrogen), the electron affinity is low or endothermic since, as noted previously (pages 18 and 164), these structures show enhanced stability. These values reinforce the deductions made about stability from ionisation energies.

The reason for fluorine having a lower electron affinity than chlorine is discussed on page 215.

Second electron affinities are possible, the values for O$^-$ and S$^-$ being $+791$ and $+649$ kJ mol^{-1} respectively. These are always endothermic on purely electrostatic grounds.

Electron affinities are generally obtained indirectly from thermochemical cycles because direct measurement is difficult.

3 ATOMIC RADII

According to the quantum theory, the orbitals of an atom extend to infinity and so there is no definite size to the atom. In order to assign a size, it is necessary to make some arbitrary decision as to where the boundary of an atom is; this can be done in various ways.

In the case of a diatomic molecule of an element, the atomic or covalent radius is taken as half the internuclear distance (determined by X-ray or electron diffraction). The covalent radii of the elements in the second period of the periodic table are:

Li	Be	B	C	N	O	F
0.123	0.106	0.088	0.077	0.070	0.066	0.064 nm

The values decrease from left to right since, as the nuclear charge increases, the electrons enter the same shell and so are attracted more strongly. The radii increase on descending a group (e.g.: F, 0.064; Cl, 0.099; Br, 0.111; and I, 0.128 nm) because, although the nuclear charge increases considerably, the extra electrons are entering shells further from the nucleus.

In any period of the periodic table, the alkali metals have by far the largest atoms, and this is reflected in their low densities, etc. The reason for this is that their outer electron has started a new shell and there is only one electron per atom involved in bonding. The atomic radii start to rise towards the end of each transition series because as the *d* electrons become paired they tend to repel each other rather than take part in the bonding. A plot of atomic radius against atomic number for some elements is shown in Figure 9.6.

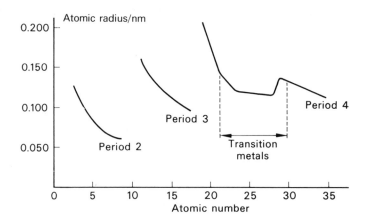

Figure 9.6   Plot of atomic radius against atomic number

Half the distance between adjacent atoms in a metal is known as the metallic radius. In the case of lithium vapour, Li_2, the covalent radius is 0.123 nm, but the metallic radius is 0.152 nm. This difference is to be expected since, in the molecule, the atoms are pulled together by their attraction for the electrons shared in the bond between them. On the other hand the valency electrons of metals are delocalised throughout the lattice (page 65) and consequently a more open structure is adopted. As expected, metallic radii decrease across the periodic table and increase down it.

Another way of expressing atomic size is to determine how closely the nuclei of the atoms, in two different molecules, come to one another during a molecular collision and then halve this distance. The result is known as the van der Waals radius and this value is always larger than the covalent radius (for example, the van der Waals and covalent radii of fluorine are 0.135 and 0.064 nm respectively) but they follow the same trends in the periodic table. The collision distances can be derived from a comparison of the pressure–volume relationships of the real gas compared with the results expected if the gas behaved ideally (real and ideal gases are discussed on page 35).

Van der Waals and covalent radii are illustrated in Figure 9.7.

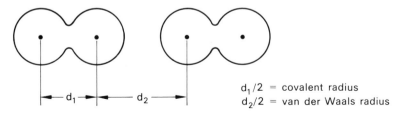

$d_1/2$ = covalent radius
$d_2/2$ = van der Waals radius

Figure 9.7   Covalent and van der Waals radii

4 IONIC RADII

Ionic radii, like covalent radii, are found by measuring internuclear distances by X-ray diffraction, etc. However, this gives the sum of the ionic radii of two different ions and certain assumptions must be made about their relative sizes in order to apportion the distance between them. The size of an ion varies to some extent with its environment and so average values are used.

The trends of ionic radii down a group are similar to those for covalent radii, e.g.

Ion	radius/nm	Ion	radius/nm
Li^+	0.068	F^-	0.133
Na^+	0.098	Cl^-	0.181
K^+	0.133	Br^-	0.196
Rb^+	0.148	I^-	0.219
Cs^+	0.167		

Cations have smaller radii than the corresponding atoms since their nuclear charge outweighs the electronic charge and consequently the electrons are attracted more strongly. This is illustrated by the following figures:

Na 0.157 nm, Na^+ 0.098 nm;   Mg 0.140 nm, Mg^{2+} 0.065 nm

Anions, on the other hand, have larger radii than their atoms because the protons are exceeded by the electrons and the latter repel one another, e.g.

S 0.104 nm, S^{2-} 0.190 nm;   F 0.064 nm, F^- 0.133 nm

The variation of ionic radii with nuclear charge may be illustrated by the *isoelectronic series* (a series with the same number of electrons) below; the nuclear charge is given in brackets:

N^{3-}	0.171 nm	(7)	Na^+	0.098 nm	(11)
O^{2-}	0.146 nm	(8)	Mg^{2+}	0.065 nm	(12)
F^-	0.133 nm	(9)	Al^{3+}	0.045 nm	(13)

5 OXIDATION STATE OR OXIDATION NUMBER

The use of valency to describe the combining power of an atom in a compound can lead to uncertainty or ambiguity. For example, it may have been thought that nickel exhibits a valency of four in $Ni(CO)_4$, but, in fact, its valency here is zero because none of its electrons are involved in the bonding. To avoid difficulties such as this, the concept of oxidation number or oxidation state has been developed.

The oxidation number is the number of electrons which must be added to a positive ion or removed from a negative ion to give a neutral atom. Thus, the oxidation number of the Al^{3+} ion is $+3$ and of the S^{2-} ion, -2. In the case of covalent compounds, it is assumed that the electrons actually go to the most electronegative atom. Hence the oxidation number of nitrogen in NH_3 is -3.

A few points need to be noted.

(a)   The oxidation number of an uncombined element is taken as zero.

(b)   The algebraic sum of the oxidation numbers of the elements in a compound is zero, e.g.

$$\overset{+2\ -2}{MgO}, \qquad \overset{+1\ -2\ +1}{NaOCl}$$

(c)   The signs of oxidation numbers are relative to the elements with which they are combined, e.g. the oxidation numbers of phosphorus in Na_3P and P_2O_5 are -3 and $+5$ respectively.

(d)   In ions, the algebraic sum of the oxidation numbers of the atoms forming the ion is the same as the charge on the ion. Consider the NO_3^- ion. The total oxidation number of the oxygens is -6 and since the sum of the oxidation numbers in the ion is -1, the oxidation number of the nitrogen must be $+5$.

(e)   The oxidation number of oxygen is always -2 except in peroxides, $(-O-O-)^{2-}$, and in oxygen difluoride, F_2O, where it is -1 and $+2$ respectively.

The oxidation number chart for chlorine compounds (Figure 9.8) will aid familiarity with oxidation numbers

Oxidation number	Examples
$+7$	$HClO_4$, $KClO_4$, Cl_2O_7
$+6$	Cl_2O_6
$+5$	$HClO_3$, $KClO_3$
$+4$	ClO_2
$+3$	ClF_3, $HClO_2$
$+2$	—
$+1$	$HOCl$, $NaOCl$
0	Cl_2
-1	HCl, $MgCl_2$, CCl_4

Figure 9.8   Oxidation number chart for chlorine

Oxidation number varies in a periodic manner across the periodic table. The most oxidation numbers tend to be observed with the transition elements, particularly in the centre of the series where the number of unpaired d electrons is at a maximum. The oxidation numbers of Groups I and II are invariably $+1$ and $+2$ respectively. Figure 9.9 shows the oxidation numbers of the elements when combined with chlorine.

Oxidation numbers are useful as a means of deciding whether oxidation or reduction has occurred in a reaction. For example, in the reaction

$$2KMnO_4 + 8H_2SO_4 + 5(COONa)_2 \rightarrow$$
$$K_2SO_4 + 2MnSO_4 + 5Na_2SO_4 + 8H_2O + 10CO_2$$

the oxidation number of the manganese has fallen from $+7$ to $+2$ and so the MnO_4^- ion has obviously been reduced.

The modern method of nomenclature (the Stock notation) of inorganic compounds uses oxidation numbers to avoid the use of endings such as -ous and -ic to denote different valency states of the elements. Thus, the names ferrous and ferric chloride have been replaced by iron(II) and iron(III) chloride respectively.

												H (1)					
Li (1)	**Be** (2)											**B** (3)	**C** (4)	**N** (3)			
Na (1)	**Mg** (2)											**Al** (3)	**Si** (4)	**P** (3,5)	**S** (1,2,4)		
K (1)	**Ca** (2)	**Sc** (3)	**Ti** (2,3,4)	**V** (2,3,4)	**Cr** (2,3)	**Mn** (2)	**Fe** (2,3)	**Co** (2)	**Ni** (2)	**Cu** (1,2)	**Zn** (2)	**Ga** (2,3)	**Ge** (2,4)	**As** (3)	**Se** (1,4)	**Br** (1)	
Rb (1)	**Sr** (2)	**Y** (3)	**Zr** (2,3,4)	**Nb** (3,4,5)	**Mo** (2,3,4,5)		**Ru** (3,4)	**Rh** (3)	**Pd** (2)	**Ag** (1)	**Cd** (2)	**In** (1,2,3)	**Sn** (2,4)	**Sb** (3,5)	**Te** (2,4)	**I** (1,3)	
Cs (1)	**Ba** (2)	**La** (3)		**Ta** (2,3,4,5)	**W** (2,4,5,6)	**Re** (3,5)	**Os** (2,3,4)	**Ir** (1,2,3,4)	**Pt** (1,2,3,4)	**Au** (1,3)	**Hg** (1,2)	**Tl** (1,3)	**Pb** (2,4)	**Bi** (2,3)			

Figure 9.9   Oxidation states of the elements in their chlorides

Other physical properties exhibiting periodicity

Periodicity is exhibited by several other physical properties, for example, atomic volume, melting point, boiling point, enthalpy (latent heat) of fusion and vaporisation; these are illustrated in Figures 9.10 to 9.17. However, the trends here are not so clear-cut since these properties do not concern isolated gaseous atoms, and chemical structure may be a predominant factor. These properties should be examined in conjunction with Figure 9.18 which shows the states of the various elements. Thus, the very high melting point of carbon followed by the very low melting point of nitrogen is readily understood when it is seen that the former element exists as a macromolecule whilst the latter exists as diatomic molecules.

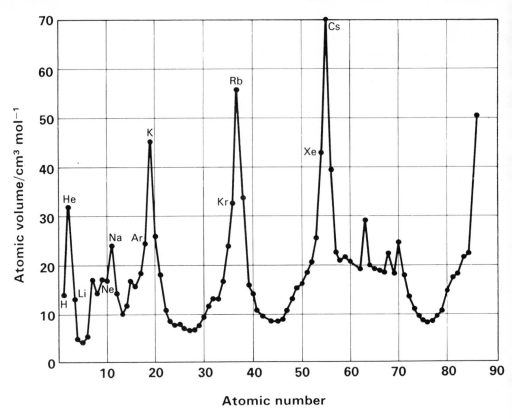

Figure 9.10   Variation of atomic volume with atomic number

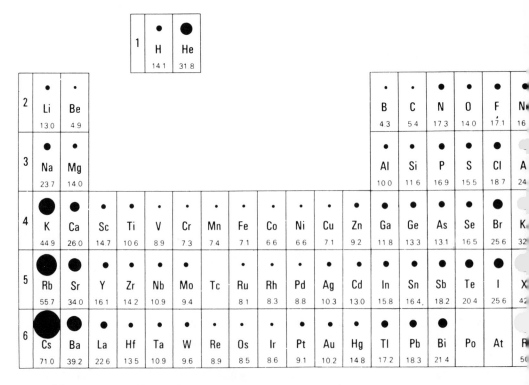

Figure 9.11   Atomic volumes of the elements/cm³ mol⁻¹

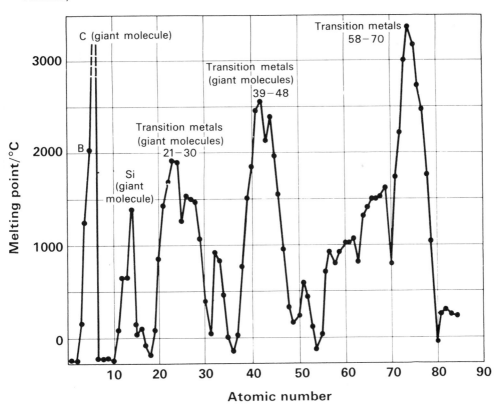

Figure 9.12   Variation of melting points of the elements with atomic number

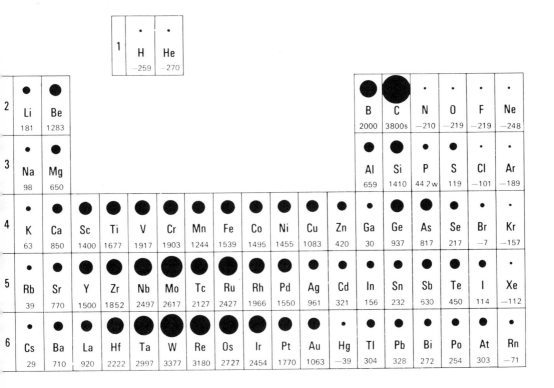

Figure 9.13   Melting points of the elements / °C

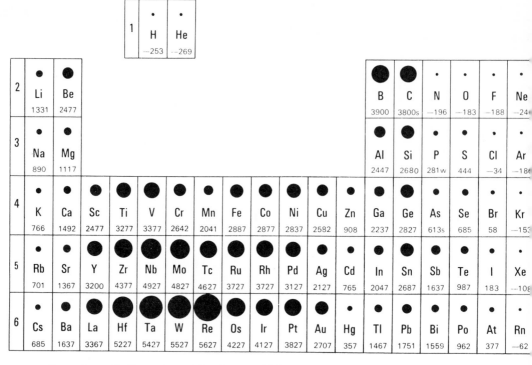

Figure 9.14   Boiling points of the elements / °C

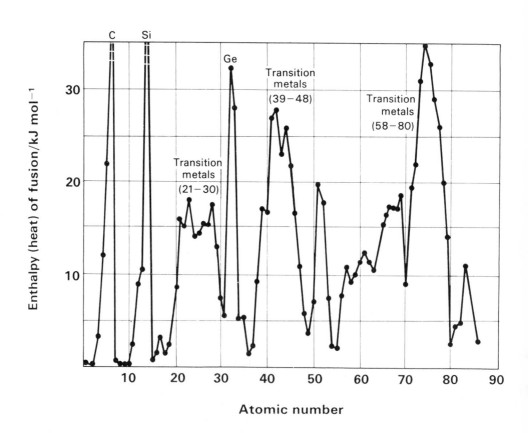

Figure 9.15   Variation of enthalpy (latent heat) of fusion with atomic number

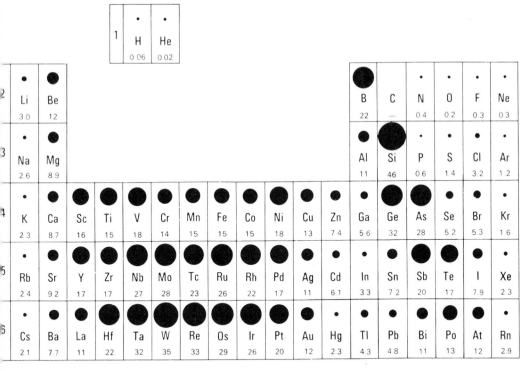

Figure 9.16   Enthalpies (latent heats) of fusion of the elements/kJ mol⁻¹

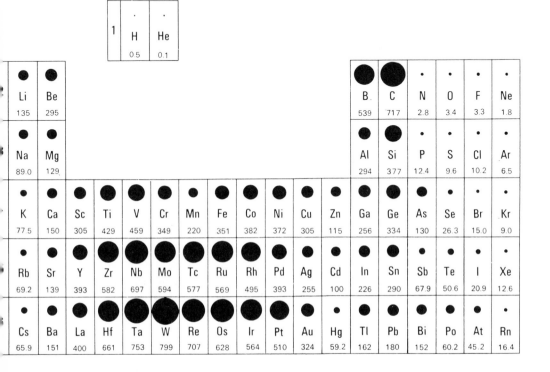

Figure 9.17   Enthalpies (latent heats) of vaporisation of the elements/kJ mol⁻¹

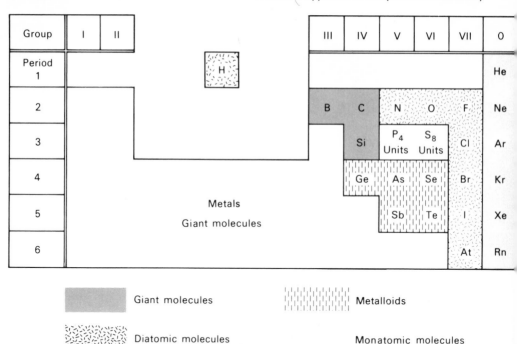

Figure 9.18   States of the elements

Periodicity of chemical properties

The arrangement of the elements in the periodic table is such that the elements in each group have similar electronic structures and so similar properties. The variation in properties down Groups I, II, IV, and VII is discussed in Chapter 10.

On travelling from left to right across the periodic table, the elements gradually change from metals to non-metals, the dividing line in the 2nd and 3rd periods being indicated below by the dotted line:

Element	Li	Be	B	C	N	O	F	Ne
Electronic structure	2.1	2.2	2.3	2.4	2.5	2.6	2.7	2.8

Element	Na	Mg	Al	Si	P	S	Cl	Ar
Electronic structure	2.8.1	2.8.2	2.8.3	2.8.4	2.8.5	2.8.6	2.8.7	2.8.8

The elements to the left of the line are electropositive — they tend to lose electrons and form positive ions. The electropositivity decreases, however, on going from left to right as indicated by their ionisation energies. On the other side of the line, the electronegativity gradually increases on moving from left to right, i.e. the elements have an increasing tendency to gain electrons and form negative ions. The elements in the centre (B, C, Si, N, and P) form covalent bonds whereas the most electronegative elements can form either covalent or ionic compounds. The noble gases are unreactive because they have stable electronic structures.

As explained on page 189, the head elements in each group tend to show some properties uncharacteristic of the group as a result of their smaller size. The variation in chemical properties across the periodic table is, therefore, best illustrated by the elements in the third period (sodium to chlorine). The oxides, chlorides, and hydrides typify the changing properties.

(a) Oxides

Formula	Na_2O Na_2O_2	MgO	Al_2O_3	SiO_2	P_4O_6 P_4O_{10}	SO_2 SO_3	Cl_2O
State at 20 °C	Solids	Solid	Solid	Solid	Solids	Gas Liquid	Gas
Bonding	Ionic	Ionic	Ionic	Covalent	Covalent	Covalent	Covalent
Type of oxide	Basic	Basic	Amphoteric	Acidic	Acidic	Acidic	Acidic

Disodium oxide (sodium monoxide) reacts vigorously with water giving sodium hydroxide:

$$Na_2O(s) + H_2O(l) \rightarrow 2NaOH(aq)$$

whilst the peroxide also gives hydrogen peroxide:

$$Na_2O_2(s) + 2H_2O(l) \rightarrow 2NaOH(aq) + H_2O_2(aq)$$

Magnesium oxide is less basic; it is insoluble in water, but reacts with acids to give salts:

$$MgO(s) + H_2SO_4(aq) \rightarrow MgSO_4(aq) + H_2O(l)$$

Aluminium is a weak metal and so its oxide is amphoteric, i.e. it can behave as an acid and as a base. It is insoluble in water.

$$Al_2O_3(s) + 3H_2SO_4(aq) \rightarrow Al_2(SO_4)_3(aq) + 3H_2O(l)$$
$$Al_2O_3(s) + 2NaOH(aq) + 3H_2O(l) \rightarrow 2NaAl(OH)_4(aq)$$

Silicon(IV) oxide (silica) is weakly acidic; it reacts with boiling alkalis giving silicates(IV):

$$SiO_2(s) + 2NaOH(aq) \rightarrow Na_2SiO_3(aq) + H_2O(l)$$

Phosphorus(III) oxide (phosphorus trioxide) reacts slowly with cold water giving phosphonic acid (phosphorous acid):

$$P_4O_6(s) + 6H_2O(l) \rightarrow 4H_3PO_3(aq)$$

Phosphorus(V) oxide (phosphorus pentoxide) reacts vigorously with cold water giving polyphosphoric(V) acid (metaphosphoric acid):

$$P_4O_{10}(s) + 2H_2O(l) \rightarrow 4HPO_3(aq)$$

Sulphur dioxide and sulphur(VI) oxide (sulphur trioxide) react with water to give sulphuric(IV) acid (sulphurous acid) and sulphuric(VI) acid respectively:

$$SO_2(g) + H_2O(l) \rightarrow H_2SO_3(aq)$$
$$SO_3(l) + H_2O(l) \rightarrow H_2SO_4(aq)$$

Dichlorine oxide yields chloric(I) acid (hypochlorous acid) when passed into water:

$$Cl_2O(g) + H_2O(l) \rightarrow 2HOCl(aq)$$

(b) Chlorides

Formula	NaCl	MgCl$_2$	AlCl$_3$	SiCl$_4$	PCl$_3$	PCl$_5$	S$_2$Cl$_2$
State at 20 °C	Solid	Solid	Solid	Liquid	Liquid	Solid	Liquid
Bonding	Ionic	Ionic	Covalent	Covalent	Covalent	Ionic	Covalent
Effect on moist air	None	None	Fumes	Fumes	Fumes	Fumes	Fumes

Sodium chloride dissolves in water, but does not react with it. On the other hand, magnesium chloride undergoes some hydrolysis:

$$MgCl_2(s) + 2H_2O(l) \rightleftharpoons Mg(OH)_2(aq) + 2HCl(aq)$$

Aluminium chloride undergoes reversible hydrolysis:

$$AlCl_3(s) + 6H_2O(l) \rightleftharpoons [Al(H_2O)_6]^{3+}(Cl^-)_3(aq)$$

whilst the hydrolysis of silicon tetrachloride is complete:

$$SiCl_4(l) + 4H_2O(l) \rightarrow SiO_2 \cdot 2H_2O(s) + 4HCl(aq)$$

Phosphorus trichloride is instantly hydrolysed by cold water to phosphonic acid (phosphorous acid):

$$PCl_3(l) + 3H_2O(l) \rightarrow H_3PO_3(aq) + 3HCl(aq)$$

The pentachloride reacts vigorously with water, initially giving phosphorus trichloride oxide (phosphorus oxychloride) but this is then hydrolysed to phosphoric(V) acid (orthophosphoric acid):

$$PCl_5(s) + H_2O(l) \rightarrow POCl_3(l) + 2HCl(g)$$
$$POCl_3(l) + 3H_2O(l) \rightarrow H_3PO_4(aq) + 3HCl(aq)$$

(c) Hydrides

Formula	NaH	MgH$_2$	(AlH$_3$)$_n$	SiH$_4$	PH$_3$	H$_2$S	HCl
State at 20 °C	Solid	Solid	Solid	Gas	Gas	Gas	Gas
Bonding	Ionic	Ionic	Covalent	Covalent	Covalent	Covalent	Covalent

The hydrides of sodium, magnesium, and aluminium all react with water with the evolution of hydrogen, e.g.

$$MgH_2(s) + 2H_2O(l) \rightarrow Mg(OH)_2(s) + 2H_2(g)$$

Silane dissolves in water, undergoing slow hydrolysis, the reaction being catalysed by HO$^-$ ions,

$$SiH_4(g) + 4H_2O(l) \rightarrow Si(OH)_4(s) + 4H_2(g)$$

Phosphine is slightly soluble in water giving a practically neutral solution, but hydrogen sulphide is a little more soluble and the solution is weakly acidic:

$$H_2S(g) + H_2O(l) \rightleftharpoons H_3O^+(aq) + HS^-(aq)$$

$$\big\Updownarrow H_2O$$

$$H_3O^+(aq) + S^{2-}(aq)$$

Hydrogen chloride is very soluble in water and its solution is strongly acidic:

$$HCl(g) + H_2O(l) \rightarrow H_3O^+(aq) + Cl^-(aq)$$

Diagonal relationships

On travelling from left to right across the periodic table, the electropositivity gradually decreases whilst on descending a group it increases. Hence, elements diagonally below one another have similar electropositivities and so similar properties. These diagonal relationships are particularly evident with the elements shown below:

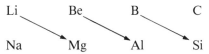

Thus, the ions of lithium and magnesium are extensively hydrated due to their small size, their carbonates and nitrates(V) undergo ready decomposition on heating, their carbonates and phosphates(V) are insoluble in water, and both metals combine with nitrogen at high temperatures. In all these reactions, lithium differs from the rest of the elements in Group I.

Beryllium and aluminium have many similar properties, e.g. they both react with acids and alkalis:

$$Be(s) + 2HCl(aq) \rightarrow BeCl_2(aq) + H_2(g)$$
$$2Al(s) + 6HCl(aq) \rightarrow 2AlCl_3(aq) + 3H_2(g)$$
$$Be(s) + 2NaOH(aq) + 2H_2O(l) \rightarrow Na_2Be(OH)_4(aq) + H_2(g)$$
$$2Al(s) + 2NaOH(aq) + 6H_2O(l) \rightarrow 2NaAl(OH)_4(aq) + 3H_2(g)$$

They both form salts which are extensively hydrolysed in solution. Their oxides have very high melting points, and their hydroxides are amphoteric. Both beryllium and aluminium chlorides are low-melting-deliquescent solids which fume in moist air, dissolve in water giving acidic solutions, dissolve in organic solvents, and do not conduct electricity in the fused state.

Similar relationships are found with boron and silicon. They are both non-metals, they have similar properties, and their compounds are covalent. Their oxides are acidic and macromolecular. They form a number of hydrides, whilst their chlorides are covalent and readily hydrolysed.

Questions

9.1 (a) For the sodium atom, the first ionisation energy (I_1) is 493 kJ and the second ionisation energy (I_2) is 4560 kJ.
  (i) Give the equations which describe the changes taking place during the measurements of I_1 and I_2.
  (ii) Explain why the value of I_2 is very much greater than that of I_1.

(b) The following diagram indicates the first ionisation energies of the elements from hydrogen to neon.

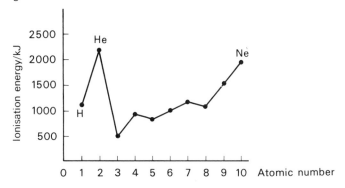

Comment upon the important features of the diagram showing how it may be interpreted, assuming a knowledge of the extranuclear structure of the atoms involved.

(c) Using a knowledge of electronic structure and bonding explain why beryllium chloride is soluble in organic solvents. [J.M.B.]

9.2 Define ionisation energy and briefly outline two methods by which it may be determined.

Discuss the variation of ionisation energy down and across the periodic table.

9.3 (a) The first four ionisation energies of an element are 740, 1500, 7700, and $10\,500\,kJ\,mol^{-1}$ respectively. Suggest which group of the periodic table this element is in and explain the relative sizes of the values.

(b) Define electron affinity and explain how it varies down Group VII (the halogens).

(c) Account for the diagonal relationship shown by some first and second row elements.

9.4 Discuss the periodicity of physical properties of the elements.

9.5 Determine the oxidation numbers of the named element in the following reactions and so decide if oxidation or reduction has occurred.

(a) Chlorine in $6NaOH + 3Cl_2 \rightarrow 5NaCl + NaClO_3 + 3H_2O$
(b) Phosphorus in $PBr_3 + 3H_2O \rightarrow H_3PO_3 + 3HBr$
(c) Chromium in $K_2CrO_4 + H_2SO_4 \rightarrow K_2SO_4 + H_2Cr_2O_7$
(d) Chlorine in $HOCl + 2KBr + HCl \rightarrow 2KCl + Br_2 + H_2O$
(e) Copper in $2CuSO_4 + 4KCN \rightarrow 2CuCN + C_2N_2 + 2K_2SO_4$
(f) Manganese in $10FeSO_4 + 2KMnO_4 + 8H_2SO_4 \rightarrow 5Fe_2(SO_4)_3 + K_2SO_4 + 2MnSO_4 + 8H_2O$

9.6 (a) Define the terms *first ionisation energy, second ionisation energy*, and *electron affinity*. In each case illustrate your answer with an electronic equation.

(b) Sketch a graph showing the relative first ionisation energies of the elements lithium to neon.

(c) The table below gives the first, second, third, and fourth ionisation energies (in $kJ\,mol^{-1}$) of five elements A, B, C, D, and E. These letters are not the usual symbols for the elements.

	First	Second	Third	Fourth
A	800	2400	3700	25 000
B	900	1800	14 800	21 000
C	500	4600	6900	9500
D	1090	2400	4600	6200
E	1310	3400	5300	7500

(i) Which element forms an ionic chloride of the type XCl?
(ii) Which element is an alkaline earth metal?
(iii) Which element is likely to have a maximum valency of 3?
Briefly explain the basis of your choices.

(d) Explain why the first electron affinity of the sulphur atom has a negative value $(-199.5\,kJ\,mol^-)$ whilst the second electron affinity has a positive value $(+ 648.5\,kJ\,mol^{-1})$. [University of London]

9.7 Many of the oxides of the elements may be classified as acidic or basic depending on the nature of the solution when the oxide is dissolved in water. Discuss the trend from basic to acidic oxides across the third period (Na to S) of the Periodic table paying particular attention to (a) the trend in the value of the first ionisation energy, (b) the change from metallic to non-metallic character of the elements and (c) the trend found in bond types across this period. [J.M.B.]

9.8 Describe and explain, by using the oxides, chlorides, and hydrides as examples, the variation in chemical properties across the series Na, Mg, Al, Si, P, S, and Cl.

10 A Study of Some of the Elements and Groups in the Periodic Table

HYDROGEN

Hydrogen is similar in some ways to the alkali metals, e.g. it has one electron in its outer shell and forms positive ions (although this needs far more energy than Li, Na, etc). It is also similar to the halogens, e.g. its molecules are diatomic, it has one electron short of the noble gas structure, it forms a negative ion, H^- (although it has far less tendency to do this than F, Cl, etc), and it forms covalent bonds by sharing electrons. As a matter of convenience, it is often placed at the top of Group I or at the tops of Groups I and VII but it is best regarded in a class of its own.

The enthalpy (heat) of atomisation of hydrogen ($\frac{1}{2}H_2(g) \rightarrow H(g)$) is 218 kJ mol^{-1}. This is a high value for a single bonded diatomic molecule, the corresponding figures for fluorine and chlorine, for example, being 79 and 121 kJ mol^{-1} respectively. Even the double bonded oxygen molecule only has an enthalpy of atomisation of 249 kJ mol^{-1}. The strength of the H—H bond is due to the fact that the electrons in the bond are very close to the two nuclei and so firmly held in position.

The ionisation energy ($H(g) \rightarrow H^+(g) + e^-$) is also very high as illustrated by the following figures:

H, 1310; Li, 520; O, 1310; and Xe, 1170 kJ mol^{-1}.

Again the reason is the closeness of the electron to the nucleus.

The proton has a radius of approximately 0.0001 nm whereas most cations are at least 10^3 times as large, and so it has intense polarising power. This is reflected in a hydration energy ($H^+(g) + H_2O(l) \rightarrow H^+$ (aq)) of −1075 kJ mol^{-1}, and, in fact, H^+ cannot exist in the unhydrated state in aqueous solution. The energy involved in the formation of the hydrated hydrogen ion from the hydrogen molecule is

$$\Delta H \text{ atomisation} \quad + \Delta H \text{ ionisation} \quad + \Delta H \text{ hydration}$$

i.e. $\quad \Delta H(\frac{1}{2}H_2 \rightarrow H) \quad + \Delta H(H \rightarrow H^+ + e^-) + \Delta H(H^+ + H_2O \rightarrow H^+ \text{ (aq)})$

$$218 \quad + \quad 1310 \quad + \quad (-1075)$$

$$= 453 \text{ kJ mol}^{-1}$$

Hydrides

Salt-like or ionic hydrides are formed when the most electropositive elements (Groups I and II) are heated with hydrogen, e.g.

$$2Na(s) + H_2(g) \xrightarrow{\quad 400\,°C \quad} 2Na^+H^-(s)$$
$$Ca(s) + H_2(g) \xrightarrow{\quad 400\text{--}500\,°C \quad} Ca^{2+}(H^-)_2(s)$$

The lattice energy (page 76) offsets the ionisation energy of the metal, etc. The Born–Haber cycle for sodium hydride is shown in Figure 10.1 and its enthalpy (heat) of formation is seen to be $-57\ kJ\ mol^{-1}$.

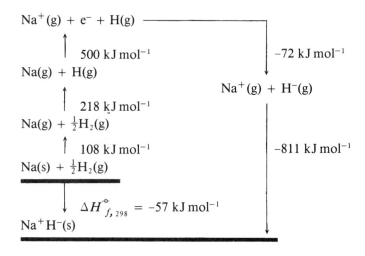

Figure 10.1   Born–Haber cycle for sodium hydride

The ionic hydrides are hydrolysed by water, e.g.

$$Na^+H^-(s) + H_2O(l) \rightarrow Na^+OH^-(aq) + H_2(g)$$

On electrolysis they give hydrogen at the anode:

Cathode: $Na^+ + e^- \rightarrow Na$
Anode: $H^- - e^- \rightarrow H;\quad 2H \rightarrow H_2$

Covalent hydrides are formed with the Group IV, V, VI, and VII elements. C—H, N—H, O—H, and F—H bonds are strong, but the bond dissociation enthalpies decrease down each group as the size of the atom and the metallic character increases. The covalent hydrides are generally gaseous at room temperature, but some are dimeric, e.g. B_2H_6, or polymeric, e.g. $(AlH_3)_n$.

The hydrides in any particular group would be expected to have gradually increasing boiling points as the relative molecular mass increases, but there are some notable exceptions as shown by Table 10.1 and Figure 10.2.

Anomalies occur in the case of NH_3, H_2O, and HF owing to hydrogen bonding (page 37) — the very small electronegative nitrogen, oxygen, and fluorine atoms

Table 10.1. Boiling points of Groups IV, V, VI and VII hydrides

Compound	Boiling point/°C	Compound	Boiling point/°C
CH_4	−161	NH_3	−33
SiH_4	−112	PH_3	−90
GeH_4	−90	AsH_3	−55
SnH_4	−52	SbH_3	−17
PbH_4	−13	BiH_3	+22
H_2O	+100	HF	+20
H_2S	−60	HCl	−85
SeH_2	−41	HBr	−67
TeH_2	−2	HI	−35

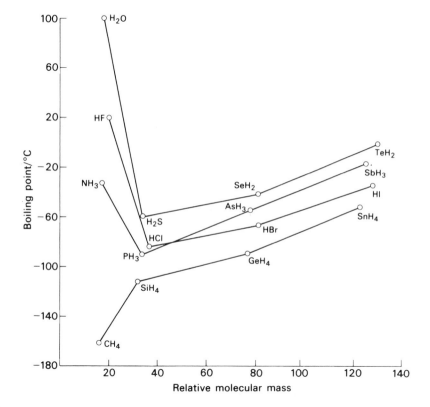

Figure 10.2   Effect of hydrogen bonding on the boiling points of some hydrides

polarise the molecules and cause intermolecular attractions, e.g. liquid hydrogen fluoride contains zig-zag chains:

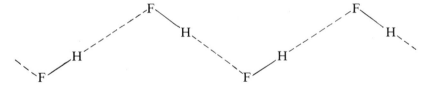

Energy is needed to break these hydrogen bonds before the liquid will boil, and so higher boiling points result. Hydrogen bonds are intermediate in strength between covalent bonds and van der Waals forces as shown by the following figures: covalent bonds 150–500 kJ mol^{-1}, hydrogen bonds 30–100 kJ mol^{-1}, and van der Waals forces approximately 20 kJ mol^{-1}.

The effect of hydrogen bonding is reflected in high enthalpies (latent heats) of fusion and vaporisation for these compounds. Thus, $\Delta H_{fus.}$ for CH_4 and SiH_4 are 0.92 and 0.67 kJ mol^{-1} respectively, but for NH_3 (hydrogen bonds present) and PH_3 the figures are 5.65 and 1.13 kJ mol^{-1} respectively.

A number of hydrides are known in which hydrogen is present in a complex anion, e.g. lithium tetrahydridoaluminate(III) (lithium aluminium hydride), $LiAlH_4$ and sodium tetrahydridoborate(III) (sodium borohydride), $NaBH_4$. These can be prepared as follows:

$$4LiH + AlCl_3 \xrightarrow{\text{ether}} LiAlH_4 + 3LiCl$$
$$4NaH + BF_3 \xrightarrow{\text{ether}} NaBH_4 + 3NaF$$

Unlike the salt-like hydrides, they are strong reducing agents at room temperature. They find considerable application, particularly in organic chemistry, e.g. reduction of carboxylic acids and aldehydes to primary alcohols, and ketones to secondary alcohols.

GROUPS I (Li-Cs) AND II (Be-Ba)

These elements have one and two electrons respectively in their outer shell and their chemistry is dominated by the tendency to lose the outer electron(s) and form positive ions. The ionisation energies are as follows.

Element	I_1/kJ mol^{-1}	Element	$(I_1 + I_2)$/kJ mol^{-1}
Li	520	Be	2700
Na	500	Mg	2240
K	420	Ca	1690
Rb	400	Sr	1650
Cs	380	Ba	1500

Ionisation energy decreases and reactivity increases down each group since the outer electrons are further from the nucleus and more readily lost.

The elements in these groups are known as *s*-block elements because their valency electrons are in *s* orbitals.

The lattice energy of their compounds decreases down the groups due to a decrease in charge density, but it is considerably greater in Group II because there is double the charge, e.g. the lattice energies for sodium chloride and magnesium chloride are –769 and –2502 kJ mol^{-1} respectively.

The hydration energy, e.g. $M^+(g) + nH_2O(l) \rightarrow M^+(aq)$ decreases down each group since, the lower the charge density, the smaller the attraction for the slightly negatively charged oxygens in water molecules. Again the values for the alkaline earths are greater (Table 10.2).

Table 10.2　Hydration energies

Ion	$\Delta H^{\ominus}$/kJ mol$^-$	Ion	$\Delta H^{\ominus}$/kJ mol^{-1}
Li^+	–499	Be^{2+}	–2385
Na^+	–390	Mg^{2+}	–1891
K^+	–305	Ca^{2+}	–1561
Rb^+	–281	Sr^{2+}	–1414
Cs^+	–248	Ba^{2+}	–1273

This order of hydration energies results in salts with fewer molecules of water of crystallisation as the groups are descended, e.g.

$$Na_2CO_3 \cdot 10H_2O \quad MgSO_4 \cdot 7H_2O$$
$$K_2CO_3 \cdot 2H_2O \quad CaSO_4 \cdot 2H_2O$$
$$SrSO_4$$

Solubilities are also affected. (N.B. Solubility is controlled by two opposing factors: lattice energy and hydration energy.) Low solubility will result if the lattice energy exceeds the hydration energy. Hydration energy overrides lattice energy in the case of the Group II sulphates(VI) and its decrease down the group is illustrated by $MgSO_4$ being soluble, $CaSO_4$ slightly soluble, $SrSO_4$ insoluble, and $BaSO_4$ insoluble.

The metals react with water with increasing vigour down the groups $(M + H^+OH^- \rightarrow M^+ + OH^- + H, \quad 2H \rightarrow H_2)$. Beryllium does not react with water, magnesium reacts with boiling water and calcium reacts with cold water. The basic strength of the hydroxides increases down the groups since there is less attraction between M^+ or M^{2+} and HO^- as the charge density falls.

The reactivity trend of the elements is indicated by their standard electrode potentials (page 141) as well as by the ionisation energies, but the figure for lithium is anomalous due to the high hydration energy of Li^+.

Metal $E_{298}^{\ominus}$/volts		Metal $E_{298}^{\ominus}$/volts	
Li	−3.03	Be	−1.85
Na	−2.71	Mg	−2.37
K	−2.92	Ca	−2.87
Rb	−2.93	Sr	−2.89
Cs	−3.02	Ba	−2.90

Oxides

Controlled oxidation of the Group I metals gives the monoxides, peroxides, $(O^- - O^-)$, and superoxides ($[O \dot{\dot{\cdot}} O]^-$) shown below but superoxides are not formed with the Group II elements.

$$Li \xrightarrow{O_2} Li_2O \qquad\qquad\qquad\qquad Be \xrightarrow{O_2} BeO$$

$$Na \xrightarrow{O_2} Na_2O \xrightarrow{O_2} Na_2O_2 \qquad\qquad Mg \xrightarrow{O_2} MgO$$

$$K \xrightarrow{O_2} K_2O \xrightarrow{O_2} K_2O_2 \xrightarrow{O_2} KO_2 \qquad Ca \xrightarrow{O_2} CaO$$

$$Rb \xrightarrow{O_2} Rb_2O \xrightarrow{O_2} Rb_2O_2 \xrightarrow{O_2} RbO_2 \quad Sr \xrightarrow{O_2} SrO \xrightarrow{O_2} SrO_2$$

$$Cs \xrightarrow{O_2} Cs_2O \xrightarrow{O_2} Cs_2O_2 \xrightarrow{O_2} CsO_2 \quad Ba \xrightarrow{O_2} BaO \xrightarrow{O_2} BaO_2$$

The superoxides contain a three-electron bond and, since they contain an odd number of electrons, they are paramagnetic (page 59).

All the Group I oxides are strongly basic and react vigorously with water giving the hydroxide:

$$O^{2-}(s) + H_2O(l) \rightarrow 2HO^-(aq)$$
$$O^- - O^-(s) + 2H_2O(l) \rightarrow 2HO^-(aq) + H_2O_2(aq)$$
$$2(O \dot{\dot{\cdot}} O)^-(s) + 2H_2O(l) \rightarrow 2HO^-(aq) + H_2O_2(aq) + O_2(g)$$

The peroxides and superoxides are powerful oxidising agents and may be used, for example, to oxidise chromium(III) salts to chromates(VI), CrO_4^{2-} (see page 251) or Mn^{2+} to manganese(IV) oxide (see page 253).

Beryllium oxide is covalent and amphoteric and is not attacked by water. The other Group II oxides are ionic and become more basic as the group is descended, i.e. as the metallic character increases.

Chlorides

All the chlorides may be formed by direct combination of the elements under the influence of heat. They are white solids which are soluble in water.

The standard enthalpies (heats) of formation of the chlorides may be obtained from Born–Haber cycles. The relevant data, expressed in kJ mol^{-1}, are given below and the cycle for magnesium chloride is illustrated in Figure 10.3

	ΔH_{at}	I_1		ΔH_{at}	$I_1 + I_2$
Li	161	520	Be	326	2700
Na	108	500	Mg	149	2240
K	90	420	Ca	177	1690
Rb	82	400	Sr	164	1650
Cs	78	380	Ba	175	1500

Chlorine: ΔH_{at}, 121; electron affinity, -364
Lattice energies:

LiCl	-849		$BeCl_2$	-3052
NaCl	-781		$MgCl_2$	-2545
KCl	-710		$CaCl_2$	-2176
RbCl	-685		$SrCl_2$	-2156
CsCl	-648		$BaCl_2$	-2049

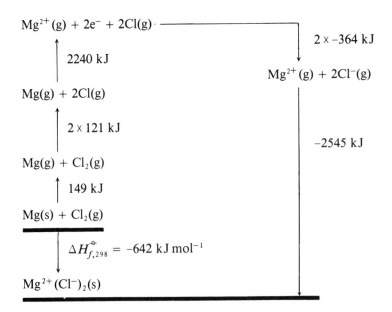

Figure 10.3   Born–Haber cycle for magnesium chloride

The standard enthalpies (heats) of formation of the Group II chlorides are as follows:

$BeCl_2$, -512; $MgCl_2$, -642; $CaCl_2$, -795; $SrCl_2$, -828; and $BaCl_2$, -860 kJ mol^{-1}. The order reflects the greater reactivity of the metals as the group is descended. Differences in crystal structure make the trend in the Group I chlorides less clear cut.

Carbonates and hydrogencarbonates (bicarbonates)

Carbonates, M_2CO_3 are known for all the Group I metals. They are white solids which, with the exception of lithium carbonate, are readily soluble in water. Owing to its low solubility, lithium carbonate can be precipitated by adding sodium carbonate solution to a lithium salt solution or by action of carbon dioxide on lithium hydroxide solution:

$$2LiCl(aq) + Na_2CO_3(aq) \rightarrow Li_2CO_3(s) + 2NaCl(aq)$$
$$2LiOH(aq) + CO_2(g) \rightarrow Li_2CO_3(s) + H_2O(l)$$

The other carbonates are too soluble to be precipitated by passing carbon dioxide through their hydroxide solution, but saturating the solution with carbon dioxide leads to the precipitation of the less soluble hydrogencarbonates, e.g.

$$2KOH(aq) + CO_2(g) \rightarrow K_2CO_3(aq) + H_2O(l)$$
$$K_2CO_3(aq) + H_2O(l) + CO_2(g) \rightarrow 2KHCO_3(s)$$

Heating the hydrogencarbonates gives the anhydrous carbonates, e.g.

$$2KHCO_3(s) \rightarrow K_2CO_3(s) + H_2O(g) + CO_2(g)$$

The reactions of the carbonates are generally similar since the carbonate ion is involved in each case. Lithium carbonate decomposes on strong heating:

$$Li_2CO_3(s) \rightarrow Li_2O(s) + CO_2(g)$$

but the others are thermally stable. (The relative instability of lithium carbonate can be explained in terms of the gain in electrostatic attraction when the very small cation, Li^+, is combined with the smaller oxide ion instead of the much larger carbonate ion.) Solutions of carbonates are alkaline, as are to a lesser extent the hydrogencarbonates, owing to hydrolysis:

$$CO_3{}^{2-}(aq) + H_2O(l) \rightleftharpoons HO^-(aq) + HCO_3{}^-(aq)$$
$$HCO_3{}^-(aq) + H_2O(l) \rightleftharpoons HO^-(aq) + H_2CO_3(aq)$$

The soluble carbonates (Na to Cs) can be titrated with acids using methyl orange as indicator:

$$CO_3{}^{2-}(aq) + 2H^+(aq) \rightarrow H_2O(l) + CO_2(g)$$

The Group II metal carbonates are all insoluble and thermally decomposed to the oxide and carbon dioxide, but with increasing difficulty down the group. They are prepared by adding sodium carbonate to solutions of their salts. In the case of magnesium, this results in precipitation of the basic carbonate,

$$Mg(OH)_2 \cdot 3MgCO_3 \cdot 3H_2O$$

because the CO_3^{2-} ion is partially hydrolysed to give HO^- and so carbonate and hydroxide are precipitated together. Magnesium carbonate is prepared by addition of sodium hydrogencarbonate to a magnesium salt solution followed by boiling.

The hydrogencarbonates of the Group II elements are unstable and only occur in solution. They are made by the action of carbon dioxide on the carbonate in the presence of water, e.g.

$$CaCO_3(s) + H_2O(l) + CO_2(g) \rightarrow Ca(HCO_3)_2(aq)$$

The above reaction accounts for temporary hardness in water from limestone areas.

Hydroxides

The Group I hydroxides are more important than those of Group II and are prepared by electrolysis of the chloride solution, e.g.

$$2NaCl(aq) + 2H_2O(l) \xrightarrow{\text{electrolysis}} 2NaOH(aq) + Cl_2(g) + H_2(g)$$

They are white solids and, with the exception of the less soluble lithium hydroxide, are deliquescent.

Sodium and potassium hydroxides are widely used as sources of HO^- ions both in preparative and quantitative work. If a metal forms an insoluble hydroxide, this is usually prepared by adding HO^- to an aqueous solution of a salt of the metal, e.g.

$$Cu^{2+}(aq) + 2HO^-(aq) \rightarrow Cu(OH)_2(s)$$
$$Fe^{3+}(aq) + 3HO^-(aq) \rightarrow Fe(OH)_3(s)$$

In a few cases the oxide is precipitated, the hydroxide being unstable, e.g.

$$Hg^{2+}(aq) + 2HO^-(aq) \rightarrow HgO(s) + H_2O(l)$$
$$2Ag^+(aq) + 2HO^-(aq) \rightarrow Ag_2O(s) + H_2O(l)$$

Sodium and potassium hydroxides are used in the preparation of alcohols from halogenoalkanes (alkyl halides) and for hydrolysing esters, etc. They neutralise acids by their HO^- ions accepting protons:

$$H_3O^+(aq) + HO^-(aq) \rightarrow 2H_2O(l)$$

and are therefore important in volumetric analysis. The ability to accept protons also explains the liberation of ammonia from ammonium salts by alkali metal and calcium hydroxides:

$$NH_4^+(aq) + HO^-(aq) \rightarrow NH_3(g) + H_2O(l)$$

Flame tests

A number of the Group I and II elements give characteristic flame colorations and this is a useful aid in their detection. Thus, a platinum wire moistened with concentrated hydrochloric acid is dipped into the substance and then held in a non-luminous flame. The following results are obtained.

Element	*Flame coloration*
Sodium	Persistent golden-yellow
Potassium	Lilac (crimson when viewed through blue glass)
Calcium	Brick-red
Strontium	Crimson
Barium	Yellow-green

The concentrated hydrochloric acid is used to form the metal chlorides since these are generally more volatile than the other salts.

Summary of the general characteristics of the groups

The general properties of the groups are summarised in Table 10.3.

Table 10.3. Summary of the Properties of Group I and II Elements and their Compounds

	Group I (Alkali metals)	Group II (Alkaline earths)
Elements	Electropositive	Electropositive
Hydroxides	Strong bases, stable to heat except LiOH	Sparingly soluble, not very strong bases, give oxide on heating
Carbonates	Thermally stable except Li_2CO_3	Evolve carbon dioxide on heating
Hydrogencarbonates (Bicarbonates)	Exist in solid state except $LiHCO_3$	Obtained only in solution
Nitrates (V)	Evolve oxygen on strong heating, e.g. $2NaNO_3 \rightarrow 2NaNO_2 + O_2$ $LiNO_3$ gives the oxide	Evolve nitrogen dioxide and oxygen on heating, e.g. $2Ca(NO_3)_2 \rightarrow 2CaO + 4NO_2 + O_2$
Nitrides	Not formed directly except by lithium $6Li + N_2 \rightarrow 2Li_3N$	Form nitrides on heating in nitrogen
Salts of strong acids	Not hydrolysed	Not usually hydrolysed but some Be and Mg exceptions
Salts of weak acids	Hydrolysed in solution: e.g. $NaCN + H_2O \rightleftharpoons NaOH + HCN$	Not hydrolysed at room temperature

Differences of the head elements

The head element of a group often shows abnormal properties due to the very small size of its atoms. Lithium, for example, differs from the other alkali metals in that it forms a nitride directly and its hydroxide, carbonate, and nitrate(V) decompose on heating to give the oxide. All the above reactions are similar to those of magnesium (see diagonal relationships — page 179).

Many lithium compounds have low solubility, e.g. LiOH, LiF, and Li_2CO_3 and this can often be ascribed to the high lattice energy, but there are exceptions — the chloride, bromide, iodide, nitrate(V), and sulphate(VI) are all soluble (hydration energy affects solubility as well as lattice energy).

Several beryllium and lithium compounds are covalent or partially covalent, e.g. beryllium chloride; even beryllium oxide and fluoride have some covalent character as may be expected from Fajan's rules. (Covalent character is favoured by small size and high charge of the cation — i.e. high polarising power — and large size and high charge on the anion — i.e. high polarisability.) The high ionisation energies also encourage covalency.

Beryllium oxide and hydroxide are amphoteric and dissolve in alkalis to give beryllates, e.g.

$$Be(OH)_2(s) + 2NaOH(aq) \rightarrow Na_2Be(OH)_4(aq)$$

GROUP III - BORON AND ALUMINIUM

The elements in Groups III to VII are known as p-block elements since their outer p electrons are involved in bonding.

Boron and aluminium are the first two elements in the third group and they have the electronic structures:

$$B \quad 2 \cdot 3 \quad \text{i.e.} \quad 1s^2 2s^2 2p^1$$

and $\qquad$ Al $\quad 2 \cdot 8 \cdot 3 \quad$ i.e. $\quad 1s^2 2s^2 2p^6 3s^2 3p^1$

The ionisation energies are:

$$B \quad I_1 \quad 800; \quad I_2 \quad 2400; \quad I_3 \quad 3700 \text{ kJ mol}^{-1}.$$
$$Al \quad I_1 \quad 580; \quad I_2 \quad 1800; \quad I_3 \quad 2700 \text{ kJ mol}^{-1}.$$

The main group valency is three since one of the outer s electrons is promoted to the p orbital. However, the sums of the first three ionisation energies of each element are large and this indicates that it is difficult to form the M^{3+} ions. Boron, in fact, does not form the simple ion B^{3+}; it is a non-metal and its compounds are covalent. It has just a trace of metallic character in that it forms a sulphate(VI) and a phosphate(V). The compound $B(OH)_3$ is usually written as H_3BO_3 because it is a weak monobasic acid, boric(III) acid.

Aluminium is the most abundant metal and the third most abundant element. It can form the simple Al^{3+} ion but this is probably present only in the fluoride. The ion does occur in a number of complexes. For example, in aqueous solution it gives the hexaaquaaluminium(III) ion, $[Al(H_2O)_6]^{3+}$. The small Al^{3+} ion has a high charge density and so the hydration is strongly exothermic. The high hydration energy is the reason why aluminium, despite the large amount of energy required to form Al^{3+}, has a standard electrode potential of -1.66 V.

Aluminium generally forms covalent bonds but these may have some ionic character. Covalency is to be expected because the Al^{3+} ion is very small and so has intense polarising power. Aluminium hydroxide is amphoteric, e.g:

$$Al(OH)_3(s) + 3HNO_3(aq) \rightarrow Al(NO_3)_3(aq) + 3H_2O(l)$$
$$Al(OH)_3(s) + NaOH(aq) \rightarrow NaAl(OH)_4(aq)$$

A valency of one is sometimes encountered with aluminium (and with the lower members of the group). Thus, the covalent compounds AlCl and AlBr are formed by the aluminium just utilising its outer p electron.

Neither boron nor aluminium attain the noble gas electronic structure when they share their valency electrons and so the resulting compounds tend to form complexes or to polymerise. Examples of their complexes are $NaBH_4$, HBF_4, $LiAlH_4$, and Na_3AlF_6. Aluminium hydride is polymeric, $(AlH_3)_n$.

Trihalides

The trihalides of boron may be prepared by direct combination of the elements or by heating the oxide with carbon and the halogen, e.g.

$$B_2O_3(s) + 3C(s) + 3Cl_2(g) \rightarrow 2BCl_3(g) + 3CO(g)$$

The fluoride and chloride are gases at room temperature whilst the bromide is a liquid and the iodide a solid.

All the boron halides are completely hydrolysed by water giving boric(III) acid, e.g.

$$BCl_3(g) + 3H_2O(l) \rightarrow H_3BO_3(aq) + 3HCl(aq)$$

They are attacked by water because the boron is electron deficient, i.e. it has only six electrons in its outer shell, e.g.

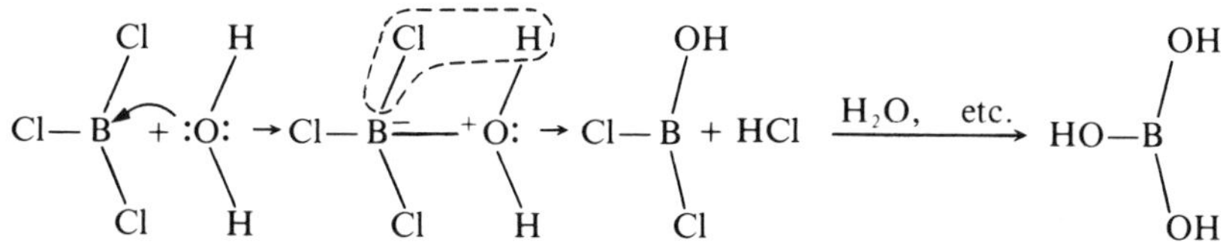

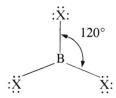

Hydrolysis of boron trifluoride also gives some hydrogen tetrafluoroborate(III) (fluoroboric acid), HBF_4 since the hydrogen fluoride reacts with some unhydrolysed reactant:

$$BF_3(g) + HF(g) \rightarrow HBF_4(g)$$

The boron trihalides have the trigonal shape

However, the bonds are shorter than expected, and it is thought that they exist as resonance hybrids (pages 65-6), lone pairs on the halogens being donated to the vacant p orbital of the boron:

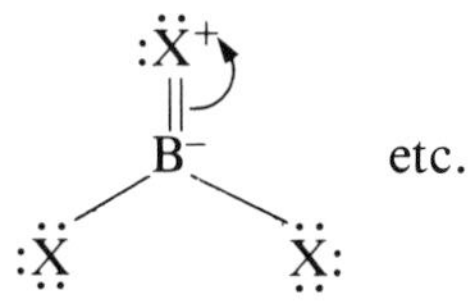

The aluminium halides can be prepared by direct combination of the elements. The chloride, bromide, and iodide may also be prepared by heating aluminium with the hydrogen halide, e.g.

$$2Al(s) + 6HCl(g) \rightarrow 2AlCl_3(s) + 3H_2(g)$$

Aluminium fluoride is ionic and sparingly soluble in water. However, the other halides are covalent and readily hydrolysed, e.g.

$$AlCl_3(s) + 6H_2O(l) \rightarrow [Al(H_2O)_6]^{3+}(Cl^-)_3(aq)$$

The chloride, bromide, and iodide exist as dimers in non-polar solvents and also in the vapour state provided that the temperature is not too high. Dissociation of Al_2Cl_6 into $AlCl_3$, for example, starts at about 400 °C and is complete at 800 °C. In the AlX_3 molecule, the aluminium has only six electrons in its outer shell. However, in the dimer it has the noble gas electronic structure because one of the halogens in each molecule acts as a donor, i.e.

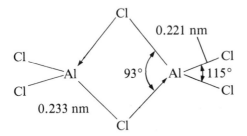

The bonds round each Al approximate to the tetrahedral structure.

As a result of boron and aluminium having incomplete outer electron shells, these compounds act as Lewis acids, i.e. lone pair acceptors. They therefore react with Lewis bases so forming addition compounds such as $\bar{B}F_3 \leftarrow \overset{+}{N}H_3$ and $\bar{A}lCl_3 \leftarrow \overset{+}{N}H_3$.

Boron has a covalency maximum of four because the second shell can only hold eight electrons. It exhibits this covalency in the $[BF_4]^-$ ion as in hydrogen tetrafluoroborate(III) (fluoroboric acid), HBF_4. Aluminium, however, has d orbitals available in its outer shell and so its covalency maximum rises to six. This is exemplified by the $[AlF_6]^{3-}$ ion as in sodium hexafluoroaluminate(III) (cryolite), Na_3AlF_6. Three d orbitals in the aluminium remain unoccupied because there is not room for more than six F^- ions round the small Al^{3+} ion. With the larger halogens, the covalency of aluminium is restricted to four and so the $[AlCl_4]^-$ ion is formed; $[AlCl_6]^{3-}$ does not occur.

Hydrolysis of aluminium chloride

Addition of water to aluminium chloride gives $[Al(H_2O)_6]^{3+}(aq)$ and $Cl^-(aq)$. The complex cation is acidic since Al^{3+} is small and highly polarising and it weakens the O—H bonds.

$$[Al(H_2O)_6]^{3+} \rightleftharpoons [Al(H_2O)_5OH]^{2+} + H^+$$

This can be suppressed by adding H^+. Proton loss will eventually give hydrated aluminium hydroxide, $Al(H_2O)_3(OH)_3$, which is uncharged and insoluble. It is polymeric due to bridging co-ordinate bonds involving H_2O or HO^-. Further H^+ can be removed by adding HO^-, i.e. amphoteric behaviour occurs.

$$[Al(H_2O)_3(OH)_3]_n \rightleftharpoons [Al(H_2O)_2(OH)_4]_n^- + nH^+$$

The exact nature of the final product, the aluminate(III) ion, is not certain, but it is not the AlO_2^- ion sometimes stated. It is now generally represented as $Al(OH)_4^-$ with or without the two molecules of water or as $[Al(OH)_6]^{3-}$.

Aluminium is discussed further in Chapter 11.

GROUP IV

The electronic structures of the elements are as follows:

Carbon	C	2·4
Silicon	Si	2·8·4
Germanium	Ge	2·8·18·4
Tin	Sn	2·8·18·18·4
Lead	Pb	2·8·18·32·18·4

Each element has four electrons in its outer shell and so is four electrons short of the noble gas structure. Carbon and silicon are non-metals, germanium is metalloid, whilst tin and lead are weak metals (e.g. they form amphoteric oxides and their compounds have appreciable covalent character).

The relevant ionisation energies, expressed in $kJ\,mol^{-1}$, are:

	C	Si	Ge	Sn	Pb
I_1	1090	790	760	710	720
I_2	2400	1600	1500	1400	1500
I_3	4600	3200	3300	2900	3100
I_4	6200	4400	4400	3900	4100

Valency

The main valency is four and here the elements form covalent bonds. However, Sn^{4+} and Pb^{4+} occur in complexes such as $Na_2[Sn(OH)_6]$ and $Na_2[Pb(OH)_6]$.

Divalent ions are formed by germanium, tin, and lead. The Ge(II) compounds have appreciable covalent character and, to a lesser extent, so do Sn(II) compounds. The importance of the divalent state increases down the group. Thus, Ge(II) compounds are very readily oxidised, i.e. they are strong reducing agents:

$$Ge(II) - 2e^- \rightarrow Ge(IV)$$

Tin(II) compounds also are readily oxidised:

$$Sn(II) - 2e^- \rightarrow Sn(IV)$$

However, the +2 oxidation state of lead is more stable than the +4 state. Lead(IV) compounds are, in fact, powerful oxidising agents:

$$Pb(IV) + 2e^- \rightarrow Pb(II)$$

The relative stabilities of the +2 and +4 oxidation states of tin and lead are illustrated by the following standard electrode potentials:

$$Sn^{4+}(aq) + 2e^- \rightarrow Sn^{2+}(aq); \quad E^\ominus = 0.15\ V$$
$$Pb^{4+}(aq) + 2e^- \rightarrow Pb^{2+}(aq); \quad E^\ominus = 1.69\ V$$

The valency of carbon is four rather than two because one of the $2s$ electrons is promoted to the vacant $2p$ orbital (see page 287):

$$1s \qquad 2s \qquad 2p \qquad \rightarrow \qquad 1s \qquad 2s \qquad 2p$$

Hence four unpaired electrons are available for sharing with other atoms. The energy released in forming the two extra bonds compensates for that required to promote the electrons to the higher energy level. With the heavy metals, however, the bonds are weaker since the bonding electrons are farther from the nucleus. Consequently, the energy gain by forming the extra bonds is less and so the tetravalent state is less stable. The reluctance of the outer pair of s electrons in heavy atoms to enter into bond formation is known as the *inert pair effect*.

Catenation

This is the ability of an element to form bonds between its own atoms and so produce long chains. It is particularly evident with carbon and gives rise to the very large number of organic compounds. However, the tendency rapidly decreases down the group as illustrated by the observed hydrides of the elements:

C	Si	Ge	Sn	Pb
Unlimited length	$Si \rightarrow Si_6$	GeH_4	SnH_4	PbH_4
of chain e.g. poly(ethene)	i.e. SiH_4 to Si_6H_{14}	Ge_2H_6		
$+CH_2-CH_2+_n$	c.f. alkanes	Ge_3H_8		

The reason for the decreases in catenation lies in the relative bond strengths:

C—C	Si—Si	Ge—Ge	Sn—Sn	
346	175	168	155	$kJ\,mol^{-1}$

The C$\rightarrow$C bond is strong because carbon atoms have a small radius and so the electrons in the bond are close to the nuclei and are firmly held in position. As the size of the atoms increases so the strength of the bond decreases.

Silicon–oxygen bonds are much stronger than silicon–silicon bonds because of the smaller size of the oxygen atom. Long —Si—O— chains can occur and silicates are abundant in nature. The —Si—O— chains are also present in silicones and these compounds are stable, readily made, and are very important as insulators, lubricants, and water repellants. A silicone may be prepared as follows:

$$n\, Cl-\underset{CH_3}{\overset{CH_3}{Si}}-Cl \xrightarrow{H_2O} n\, HO-\underset{CH_3}{\overset{CH_3}{Si}}-OH \xrightarrow[(-H_2O)]{Heat} \left[-\underset{CH_3}{\overset{CH_3 \cdot}{Si}}-O- \right]_n$$

A further difference between carbon and the other elements is that it can form multiple bonds with itself and also with other elements. This is illustrated by the following examples:

$$CH_2{=}CH_2, \ CH{\equiv}CH, \ CH_3-\underset{O}{\overset{\|}{C}}-CH_3, \ \text{and} \ H-C{\equiv}N.$$

Allotropy

Carbon exists in two allotropic forms, namely diamond and graphite. In diamond, each carbon is bonded covalently to four others in the tetrahedral configuration (Figure 2.28(a)), the C—C bond length being 0.154 nm. Diamond is a macromolecule, it has the highest melting point of any element (3500 °C +), and is a non-conductor of electricity since all its valency electrons are paired. In graphite, each carbon is joined covalently to three others so that a layer lattice of hexagonal rings is produced (Figure 2.28(b)). The layers are held together by weak van der Waals forces. Graphite conducts electricity along the layers since only three of the four valency electrons of each carbon are paired. The unpaired electron from each carbon atom is free to move along the layer.

Silicon and germanium do not exhibit allotropy. Silicon exists as a grey shiny solid or a brown solid (different crystal size) and germanium has a grey metallic appearance. They both have the diamond structure and are semiconductors.

Three allotropes of tin are known; grey, white, and rhombic:

$$\text{grey tin} \; \overset{13.2\ ^\circ C}{\rightleftharpoons} \; \text{white tin} \; \overset{161\ ^\circ C}{\rightleftharpoons} \; \text{rhombic tin}$$

Stable below 13.2 °C	Stable between 13.2 and 161 °C	Stable between 161 °C and the m.p.

This is an example of *enantiotropy* because each allotrope is stable over a definite range of temperature and there is a specific transition temperature. (*Monotropy* is when more than one allotrope can exist under a given set of conditions, but only one is stable, e.g. red phosphorus is stable at room temperature whilst white phosphorus is metastable.) White tin is the allotrope normally encountered, it has a typical metallic lattice and its change into grey tin (the non-metallic form) with the diamond lattice becomes rapid only when the temperature is lowered to about –50 °C.

Lead does not exhibit allotropy and it has a cubic metallic lattice.

General properties

Carbon and silicon are fairly inert at room temperature, but are more reactive on heating, e.g.

$$C + O_2 \rightarrow CO \quad \text{or} \quad CO_2$$

depending on the available oxygen and the temperature (more CO is formed at higher temperatures). This ease of oxidation at high temperatures makes hot carbon a powerful reducing agent, e.g. it is used to reduce zinc oxide:

$$ZnO(s) + C(s) \rightarrow Zn(s) + CO(g)$$

Heating the elements with sulphur gives the sulphide: CS_2 (at 1500 °C), SiS_2, GeS_2, SnS_2, and PbS, but the last three are better prepared by passing hydrogen sulphide through acidic solutions of $GeCl_4$, $SnCl_4$, and $PbCl_2$ respectively.

Carbon is attacked by hot oxidising acids, e.g.

$$C(s) + 2H_2SO_4(conc) \rightarrow CO_2(g) + 2H_2O(l) + 2SO_2(g)$$
$$C(s) + 4HNO_3(conc) \rightarrow CO_2(g) + 2H_2O(l) + 4NO_2(g)$$

Silicon is resistant to all acids except hydrogen fluoride, but germanium, tin, and lead are oxidised by concentrated nitric(V) acid:

$$3Ge(s) + 4HNO_3(conc) \longrightarrow 3GeO_2(s) + 2H_2O(l) + 4NO(g)$$
$$3Sn(s) + 4HNO_3(conc) \longrightarrow 3SnO_2(s) + 2H_2O(l) + 4NO(g)$$
$$Pb(s) + 4HNO_3 \xrightarrow[\text{conc.}]{\text{Dil. or}} Pb(NO_3)_2(aq) + 2H_2O(l) + 2NO_2(g)$$

There is little reaction between tin and cold concentrated sulphuric(VI) acid, but, on heating, tin(IV) sulphate(VI) is formed:

$$Sn(s) + 4H_2SO_4(conc) \rightarrow Sn(SO_4)_2(s) + 4H_2O(l) + 2SO_2(g)$$

Lead does not react with dilute sulphuric(VI) acid because a thin protective layer of the insoluble sulphate(VI) is formed, but its reaction with the hot concentrated acid gives lead(II) sulphate(VI):

$$Pb(s) + 2H_2SO_4(conc) \rightarrow PbSO_4(s) + 2H_2O(l) + SO_2(g)$$

Carbon is resistant to alkali, but silicon dissolves in concentrated solutions giving a silicate(IV)

$$Si(s) + 2KOH(aq) + H_2O(l) \rightarrow K_2SiO_3(aq) + 2H_2(g)$$

Germanium dissolves in alkali; with potassium hydroxide solution it gives potassium hexahydroxygermanate(IV):

$$Ge(s) + 2KOH(aq) + 4H_2O(l) \rightarrow K_2Ge(OH)_6(aq) + 2H_2(g)$$

Tin and lead dissolve very slowly in hot concentrated solutions of alkalis. With potassium hydroxide, for example, they give the hexahydroxystannate(II) and plumbate(II) respectively:

$$Sn(s) + 4KOH(aq) + 2H_2O(l) \rightarrow K_4Sn(OH)_6(aq) + H_2(g)$$
$$Pb(s) + 4KOH(aq) + 2H_2O(l) \rightarrow K_4Pb(OH)_6(aq) + H_2(g)$$

If air is present, tin gives the hexahydroxystannate(IV) rather than the hexahydroxystannate(II):

$$[Sn(OH)_6]^{4-} - 2e^- \rightarrow [Sn(OH)_6]^{2-}$$

Hydrides

A preparation of methane is given on page 297. SiH_4, GeH_4, and SnH_4 can be prepared by reducing the tetrachloride with lithium tetrahydridoaluminate(III) (lithium aluminium hydride) in ethoxyethane (ether) solution, e.g.

$$SiCl_4 + LiAlH_4 \rightarrow LiCl + AlCl_3 + SiH_4(g)$$

Traces of PbH_4 are formed by electrolytic reduction of lead(II) salts at a lead cathode.

There are large differences chemically, e.g. hydrolysis:

CH_4 — methane — unaffected by alkali,

SiH_4 — silane — (spontaneously flammable in air) violently hydrolysed by HO^-,

GeH$_4$ — germane — stable in up to 30% alkali,
SnH$_4$ — stannane — stable in up to 15% alkali.

Three factors contribute to the difference in behaviour between methane and silane.

(a) The relative electronegativities:

Atom	Electro-negativity	Bond polarisation	Result
H	2.1		
C	2.5	$\overset{\delta-}{C}$—$\overset{\delta+}{H}$	No attraction for HO$^-$
Si	1.8	$\overset{\delta+}{Si}$—$\overset{\delta-}{H}$	Attraction for HO$^-$

(b) The relative C—H, C—O, Si—H, and Si—O bond energies:

C—H   413 kJ mol^{-1} C—O   358 kJ mol^{-1}
Si—H   318 kJ mol^{-1} Si—O   368 kJ mol^{-1}

C—H bonds are therefore stronger than C—O, but Si—H is weaker than Si—O.

(c) The availability of *d* orbitals. Carbon has no *d* orbitals available and so has a covalency maximum of four, but silicon can utilise its 3*d* orbitals to give a covalency maximum of six. The final product of hydrolysis is orthosilicic acid.

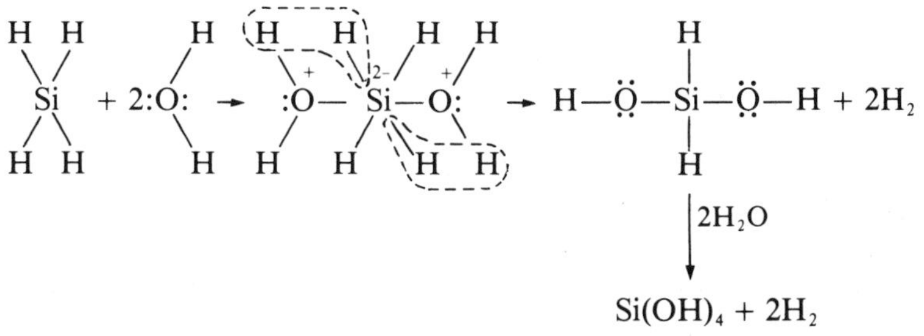

Oxides

(a) Dioxides are formed by all the elements and their physical properties and basicity are as follows:

CO$_2$	colourless gas	(monomolecular)	} acidic, covalent
SiO$_2$	colourless solid	(macromolecular)	
GeO$_2$	white		
SnO$_2$	white		amphoteric, some ionic character
PbO$_2$	purple-brown		

The first four oxides may be prepared by heating the element with oxygen, but the similar process with lead gives lead(II) oxide. Lead(IV) oxide is made by the action of dilute nitric(V) acid on dilead(II) lead(IV) oxide (red lead):

$$Pb_3O_4(s) + 4HNO_3(aq) \rightarrow PbO_2(s) + 2Pb(NO_3)_2(aq) + 2H_2O(l)$$

The expected structure of carbon dioxide would be O=C=O. However, the carbon-oxygen bond length is 0.116 nm and this compares with the 0.122 nm in the carbonyl group of ketones and a calculated figure of 0.115 nm for a C≡O bond. It is therefore thought that carbon dioxide exists as a resonance hybrid:

$$\ddot{O}{=}C{=}\ddot{O} \leftrightarrow \overset{+}{O}{\equiv}C{-}\overset{-}{\ddot{O}}{:} \leftrightarrow {:}\overset{-}{\ddot{O}}{-}C{\equiv}\overset{+}{O}$$

$$(\ddot{O}{:}\,{:}C{:}\,{:}\ddot{O} \leftrightarrow \overset{+}{O}{:}{:}C{:}\overset{-}{\ddot{O}}{:} \leftrightarrow {:}\overset{-}{\ddot{O}}{:}C{:}{:}\overset{+}{O})$$

Silicon(IV) oxide (silica or silicon dioxide) has a structure similar to diamond except that the silicon atoms are linked via oxygen atoms,

$$
\begin{array}{c}
\text{Si} \\
| \\
\text{O} \\
| \\
\text{Si} \\
\end{array}
$$

It occurs naturally in several different forms, e.g. quartz and kieselguhr. When prepared by hydrolysis of silicon tetrafluoride or chloride, hydrated silicon(IV) oxide (silica) in the form of a gel is produced. Heating the gel gives anhydrous SiO_2 but this absorbs water from the atmosphere to reform the gel; silica gel is an important drying agent. Silicon(IV) oxide is chemically inert and is attacked only by hydrogen fluoride and by alkali:

$$SiO_2(s) + 4HF(g) \rightarrow 2H_2O(l) + SiF_4(g)$$
$$SiO_2(s) + 2NaOH(aq) \rightarrow Na_2SiO_3(aq) + H_2O(l)$$

Germanium(IV) oxide is amphoteric, e.g. it reacts with acids to give the corresponding Ge(IV) salts and with sodium hydroxide giving Na_2GeO_3, disodium germanate(IV) or Na_4GeO_4, tetrasodium germanate(IV).

Tin(IV) oxide is amphoteric. With acids it gives tin(IV) salts and with alkali a hexahydroxystannate(IV) is formed, e.g.

$$SnO_2(s) + 2NaOH(aq) + 2H_2O(l) \rightarrow Na_2Sn(OH)_6(aq)$$

Lead(IV) oxide loses oxygen on heating:

$$2PbO_2(s) \rightarrow 2PbO(s) + O_2(g)$$

It is soluble in concentrated hydrochloric acid:

$$PbO_2(s) + 6HCl(conc) \rightarrow H_2PbCl_4(aq) + 2H_2O(l) + Cl_2(g)$$

(acting as an oxidising agent), and in concentrated alkali it gives sodium hexahydroxyplumbate(IV) (sodium plumbate):

$$PbO_2(s) + 2NaOH(aq) + 2H_2O(l) \rightarrow Na_2Pb(OH)_6(aq)$$

(b) Monoxides of all the elements are known. Carbon monoxide is prepared by dehydrating methanoic acid (formic acid) with concentrated sulphuric(VI) acid at room temperature:

$$HCOOH(l) \xrightarrow{H_2SO_4} H_2O(l) + CO(g)$$

It is a colourless gas which combines with chlorine under the influence of ultraviolet light giving carbonyl chloride (phosgene):

$$CO(g) + Cl_2(g) \xrightarrow{u.v.} COCl_2(g)$$

whilst on heating with sulphur it gives carbonyl sulphide, COS. Carbon monoxide is an important industrial reducing agent, e.g.

$$Fe_2O_3(s) + 3CO(g) \xrightarrow{heat} 2Fe(s) + 3CO_2(g)$$

It is slightly soluble in water giving a neutral solution. However, acidic character is evident with sodium hydroxide solution, the product being sodium methanoate (sodium formate):

$$NaOH(aq) + CO(g) \xrightarrow{150\,°C/press.} HCOONa(aq)$$

It exists as a resonance hybrid:

$$:C \rightarrow \ddot{O}: \leftrightarrow :C = \ddot{O} \leftrightarrow :C \equiv O:$$
$$(:C : \ddot{O}: \leftrightarrow :C: :\ddot{O} \leftrightarrow :C : :O:)$$

and the lone pair on the carbon can form co-ordinate bonds with transition metals, e.g. with nickel it forms tetracarbonylnickel(0) (nickel carbonyl), a colourless liquid:

$$Ni(s) + 4CO(g) \xrightarrow{60\,°C} Ni(CO)_4(l)$$

Silicon(II) oxide may be prepared by heating the dioxide with silicon. It is unstable and unimportant.

Germanium(II) oxide is prepared by hydrolysis of the dichloride:

$$GeCl_2(s) + H_2O(l) \rightarrow GeO(s) + 2HCl(g)$$

The product is the hydrated form and is yellow, but the anhydrous product, produced by reduction of the dioxide, is black. Tin(II) oxide is prepared by heating tin(II) ethanedioate (tin(II) oxalate):

$$SnC_2O_4(s) \rightarrow SnO(s) + CO_2(g) + CO(g)$$

whilst lead(II) oxide is prepared by heating the nitrate(V):

$$2Pb(NO_3)_2(s) \rightarrow 2PbO(s) + 4NO_2(g) + O_2(g)$$
$$\text{red or yellow}$$

GeO, SnO, and PbO all react with acids to give the corresponding salts and with sodium hydroxide to give sodium germanate(II), stannate(II), and plumbate(II) (sodium germanite, stannite, and plumbite) respectively. Lead(II) oxide, in contrast to the other two, shows no reducing properties.

(c) Dilead(II) lead(IV) oxide (red lead).

This is prepared by heating lead(II) oxide in air at 400 °C:

$$6PbO(s) + O_2(g) \rightarrow 2Pb_3O_4(s)$$

It contains Pb(II) and Pb(IV), i.e. $Pb_2^{(II)}[Pb^{(IV)}O_4]$ and behaves as $2PbO \cdot PbO_2$. On heating above 400 °C it loses oxygen:

$$2Pb_3O_4(s) \rightarrow 6PbO(s) + O_2(g)$$

As stated above, dilute nitric(V) acid reacts with dilead(II) lead(IV) oxide to give lead(II) nitrate(V) and lead(IV) oxide. It oxidises warm concentrated hydrochloric acid:

$$Pb_3O_4(s) + 14HCl(conc) \rightarrow 3H_2PbCl_4(aq) + 4H_2O(l) + Cl_2(g)$$

Dilead(II) lead(IV) oxide is used in paints for preventing the corrosion of iron.

Tetrahalides

All the tetrahalides are known except the bromide and iodide of lead and their absence is to be expected since Pb(IV) is an oxidising agent and bromides and iodides are readily oxidised. Lead(IV) fluoride has appreciable ionic character, but the remainder are covalent and volatile. The thermal stability of halides generally decreases with increasing size of the central atom, but here CBr_4 and CI_4 are exceptions, compared with the corresponding silicon, germanium, and tin compounds, probably because four large atoms are crowded round a very small one. The instability of Pb(IV) compounds is again illustrated by lead(IV) chloride decomposing into lead(II) chloride and chlorine at about 100 °C.

Tetrachloromethane (carbon tetrachloride) is prepared by the reaction between chlorine and boiling carbon disulphide:

$$CS_2(l) + 3Cl_2(g) \rightarrow CCl_4(l) + S_2Cl_2(l)$$

or chlorine and methane under the influence of ultraviolet light:

$$CH_4(g) + 4Cl_2(g) \xrightarrow{\text{u.v.}} CCl_4(l) + 4HCl(g)$$

The tetrachlorides, bromides, and iodides of silicon, germanium, and tin may be prepared by heating the element with the halogen, but this gives the dichloride with lead. Some lead(IV) chloride is obtained when lead(IV) oxide is dissolved in concentrated hydrochloric acid at 0 °C. However, better yields are obtained if the lead(IV) oxide is dissolved in excess cold concentrated hydrochloric acid, in the presence of chlorine, and the resultant hydrogen hexachloroplumbate(IV) (hexachloroplumbic acid) is treated with concentrated sulphuric(VI) acid:

$$PbO_2(s) + 6HCl(conc) \xrightarrow{0\,°C/Cl_2} H_2PbCl_6 + 2H_2O(l)$$

$$H_2PbCl_6 \xrightarrow{H_2SO_4,\ \text{room temp.}} PbCl_4(l) + 2HCl$$

The tetrachlorides and bromides are all liquids but the iodides are solids.

CX_4 is resistant to hydrolysis since the covalency maximum of carbon is restricted to four (there are no d orbitals in the second shell) but the covalency maximum of silicon and germanium is six and of tin and lead is eight, and so their tetrahalides are susceptible to hydrolysis, e.g. silicon tetrachloride is hydrolysed to hydrated silicon(IV) oxide (orthosilicic acid).

$$SiCl_4(l) + 4H_2O(l) \rightarrow SiO_2 \cdot 2H_2O(s) + 4HCl(g)$$

The mechanism is similar to that for the hydrolysis of silane (see above). Similarly, hydrolysis of GeX_4 and SnX_4 gives the hydrated dioxides. Hydrolysis of silicon tetrafluoride differs somewhat in that some of the hydrogen fluoride liberated reacts with unhydrolysed reactant giving hydrogen hexafluorosilicate(IV) (hexafluorosilicic acid):

$$SiF_4(g) + 2HF(aq) \rightarrow H_2SiF_6(aq)$$

The ease of hydrolysis decreases down the group from silicon to tin as the metallic character increases. This is illustrated by the hydrolysis of SiX_4 being complete whilst the hydrolyses of GeX_4 and SnX_4 are reversible and can be repressed by halogen acid, e.g.

$$GeCl_4(l) + 4H_2O(l) \rightleftharpoons Ge(OH)_4(s) + 4HCl(aq)$$

Solutions of tin(IV) chloride in concentrated hydrochloric acid can be obtained due to the formation of hydrogen hexachlorostannate(IV) (hexachlorostannic acid), H_2SnCl_6.

Lead(IV) chloride is rapidly hydrolysed (Pb(IV) compounds are not very stable) giving the dioxide and some dichloride. It dissolves in concentrated hydrochloric acid giving hydrogen hexachloroplumbate(IV), H_2PbCl_6.

Dihalides

All the dihalides of germanium, tin, and lead are known. (CF_2, CCl_2, and CBr_2 can be prepared, and so can the corresponding silicon compounds, but their life is only a fraction of a second.)

GeX_2 and SnX_2 may be prepared by heating the tetrahalide with the free metal. Dissolving tin in hydrochloric acid and crystallising the product gives the dihydrate, $SnCl_2 \cdot 2H_2O$.

Lead dihalides are not very soluble in water, and are obtained by adding halogen acid or an alkali metal halide solution to a lead(II) salt solution, i.e. the nitrate(V) or the ethanoate, e.g.

$$Pb(NO_3)_2(aq) + 2HCl(aq) \rightarrow 2HNO_3(aq) + PbCl_2(s) \quad \text{(White)}$$
$$Pb(NO_3)_2(aq) + 2KI(aq) \rightarrow 2KNO_3(aq) + PbI_2(s) \quad \text{(Golden yellow)}$$

The dihalides are more ionic and so less volatile than the tetrahalides. Ge(II) halides are polymeric:

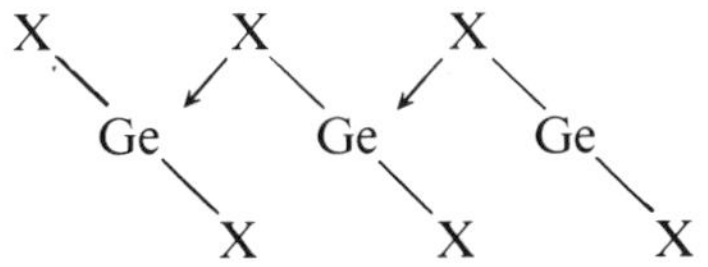

GeX_2 and SnX_2 are reducing agents, but the lower stability of GeX_2 is illustrated by its hydrolysis being accompanied by oxidation:

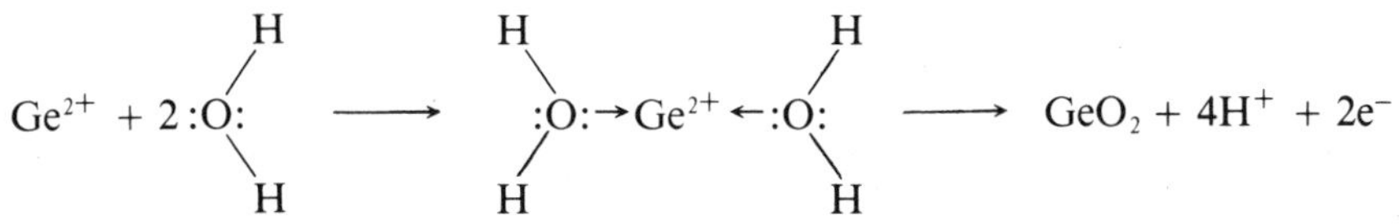

whereas with SnX_2 there is no oxidation and the hydrolysis is reversible.

Tin(II) chloride in hydrochloric acid forms the complex acid, hydrogen tetrachlorostannate(II) (chlorostannous acid), H_2SnCl_4, and diluting this solution gives $[Sn(H_2O)_6]^{2+}$ which hydrolyses to give insoluble basic salts such as Sn(OH)Cl. Lead(II) halides similarly form complexes in halogen acid, i.e. H_2PbX_4, but on dilution they decompose to PbX_2; no basic salts are formed since lead is more metallic. Lead(II) salts in solution contain the hydrated ion $[Pb(H_2O)_6]^{2+}$.

A summary of the properties of the Group IV elements is given in Table 10.4.

GROUP V — NITROGEN

Nitrogen has atomic number seven and is the first member of Group V. Its electronic structure is $1s^2 2s^2 2p^3$, i.e. there are three unpaired electrons. It can have a valency of three by accepting three electrons to give the nitride ion, N^{3-}, as in Li_3N or by sharing three electrons and forming three covalent bonds as in NH_3. It achieves its covalency maximum of four by sharing three electrons, and using the remaining lone pair to form a co-ordinate bond as in NH_4^+ and $H_3N \rightarrow AlCl_3$. The donor properties of nitrogen are attributable to its small size and so its high electron density. Phosphorus has far less tendency to donate a lone pair because the atom is much larger: N atomic radius 0.070 nm, P, 0.110 nm.

Oxidation numbers (page 170) from -3 to $+5$ are possible, e.g.

$+5$	NO_3^-	0	N_2
$+4$	N_2O_4	-1	NH_2OH
$+3$	NO_2^-	-2	N_2H_4
$+2$	NO	-3	NH_3
$+1$	N_2O		

The nitrogen molecule is diatomic and the bond length is indicative of a triple bond. It is inert at room temperature, its sole reaction being slow formation of the nitride with lithium. The inertness is a result of the very high dissociation energy of the molecule, i.e. 945 kJ mol^{-1} compared, for example, to 242 kJ mol^{-1} for the chlorine molecule. This difficulty in utilising the triple bond electrons can also be illustrated by the ionisation energy of the nitrogen molecule, $N_2(g) \rightarrow N_2^+(g) + e^-$, being 1505 kJ mol^{-1} which is virtually the same as the ionisation energy of argon. Many nitrogen compounds are endothermic as shown by the enthalpy (heat) of formation of the following gaseous oxides:

$$N_2O, \quad +82.0; \quad NO, \quad +90.4; \quad \text{and} \quad NO_2, \quad +33.2 \text{ kJ mol}^{-1}.$$

Dinitrogen oxide (nitrous oxide)

Dinitrogen oxide may be prepared by heating ammonium nitrate(V):

$$NH_4NO_3(s) \rightarrow 2H_2O(l) + N_2O(g)$$

However, since the reaction can easily get out of control, a mixture of ammonium chloride and sodium nitrate(V) is generally used.

Table 10.4. Summary of the Properties of the Group IV Elements and their Compounds

	Carbon	Silicon	Germanium	Tin	Lead
Elements	Non-metal	Non-metal	Metalloid	Weak metal	Weak metal
Allotropes	Diamond, graphite	None	None	Grey, white, rhombic	None
Most stable valency	4	4	4	4	2
Hot concentrated NaOH	No reaction	Forms Na_2SiO_3	Forms $Na_2Ge(OH)_6$	Forms $Na_2Sn(OH)_4$ or $Na_2Sn(OH)_6$ in the presence of air	Forms $Na_2Pb(OH)_4$
Chain length of hydrides	Unlimited	Up to 6	Up to 3	No chains	No chains
Oxides MO	Stable, covalent neutral, reducing agent	Unstable, covalent, reducing agent	Unstable covalent, amphoteric, reducing agent	Fairly unstable, partially ionic, amphoteric, reducing agent	Stable, ionic, amphoteric
Oxides MO_2	Stable, acidic, monomolecular	Stable, acidic, macromolecular	Stable, amphoteric, macromolecular	Stable, amphoteric, macromolecular	Decomposes on heating, amphoteric, macromolecular
Halides MX_2	Extremely unstable	Extremely unstable	Covalent, hydrolyse with oxidation	Partially ionic, reversible hydrolysis, reducing agent	Stable, ionic, not hydrolysed
Halides MX_4	Covalent, very stable, not hydrolysed	Covalent, rapidly hydrolysed	Covalent, reversible hydrolysis	Covalent, reversible hydrolysis	PbF_4 partially ionic, $PbCl_4$ covalent unstable, rapidly hydrolysed, no $PbBr_4$ or PbI_4

Its structure is similar to that of carbon dioxide with which it is *isoelectronic* (it has the same number of electrons), i.e. a linear resonance hybrid:

$$:N \overset{\cdot\cdot}{\equiv} N = \overset{\cdot\cdot}{O}: \longleftrightarrow :N \equiv N \rightarrow \overset{\cdot\cdot}{\underset{\cdot\cdot}{O}}:$$

$$(:\underset{\cdot\cdot}{N}::N::\underset{\cdot\cdot}{O}: \longleftrightarrow :N::N:\underset{\cdot\cdot}{O}:)$$

It is diamagnetic (weakly repelled by a magnetic field — see page 59) since all its electrons are paired.

It is a colourless gas, slightly soluble in water and the resultant solution is neutral. Decomposition into its elements takes place above about 550 °C and so it can act as an oxidising agent at these temperatures, e.g. it can support the combustion of sulphur and phosphorus because their flames are hot enough to decompose it:

$$S(s) + 2N_2O(g) \rightarrow SO_2(g) + 2N_2(g)$$

Nitrogen oxide (nitric oxide)

Nitrogen oxide is prepared by the action of 50% nitric(V) acid on copper:

$$3Cu(s) + 8HNO_3(aq) \rightarrow 3Cu(NO_3)_2(aq) + 4H_2O(l) + 2NO(g)\,(+ \text{ some } NO_2)$$

It is collected over water to remove any nitrogen dioxide.

It is a resonance hybrid and effectively has a three-electron bond:

$$:\overset{\cdot}{N} = \overset{\cdot\cdot}{O}: \longleftrightarrow :\overset{\cdot\cdot}{N} \overset{\cdot\cdot}{\equiv} \overset{\cdot}{O}:$$

$$(:\overset{\cdot}{N}: \ :\overset{\cdot\cdot}{O}: \longleftrightarrow :\overset{\cdot\cdot}{N}: \ :\overset{\cdot}{O}:)$$

The unpaired electron gives rise to paramagnetism (page 59).

It is a colourless gas, slightly soluble in water. The thermal stability is greater than that of dinitrogen oxide, and so it will support the combustion of magnesium and phosphorus, but not of sulphur. The unpaired electron results in greater reactivity, e.g. it reacts spontaneously with air or oxygen giving the dioxide and with halogen (except iodine) giving a nitrosyl halide:

$$2NO(g) + Cl_2(g) \rightarrow 2NOCl(g)$$

Oxidation with acidified potassium manganate(VII) (potassium permanganate) gives nitric(V) acid.

$$6KMnO_4(aq) + 10NO(g) + 9H_2SO_4(aq) \rightarrow 3K_2SO_4(aq) + 6MnSO_4(aq)$$
$$+ 10HNO_3(aq) + 4H_2O(l)$$

It can lose one electron to give the nitrosyl cation (nitrosonium ion), NO^+, and so form salts such as $(NO)^+(BF_4)^-$, nitrosyl tetrafluoroborate(III) (nitrosyl fluoroborate) and $(NO)^+(HSO_4)^-$, nitrosyl hydrogensulphate(VI). Donation of electrons is also possible, e.g.

$$FeSO_4(aq) + NO(g) \rightarrow FeSO_4 \cdot NO(aq)$$

or, more correctly

$$[Fe(H_2O)_6]^{2+}(aq) + NO(g) \rightarrow [Fe(H_2O)_5NO]^{2+}(aq) + H_2O(l)$$

This is essentially the reaction in the brown ring test for nitrates(V). In this test, concentrated sulphuric(VI) acid is run down the side of a test tube containing an acidified solution of the nitrate(V) and iron(II) sulphate(VI). A brown ring is formed at the junction:

$$NaNO_3(aq) + H_2SO_4(aq) \rightarrow NaHSO_4(aq) + HNO_3(aq)$$
$$6FeSO_4(aq) + 2HNO_3(aq) + 3H_2SO_4(conc) \rightarrow 3Fe_2(SO_4)_3(aq) + 4H_2O(l) + 2NO(g)$$
$$FeSO_4(aq) + NO(g) \rightarrow FeSO_4 \cdot NO(aq) \text{ (brown)}$$

Similar replacement occurs when nitrogen oxide (nitric oxide) is passed into acidified potassium hexacyanoferrate(III) (potassium ferricyanide), the product being the pentacyanonitrosylferrate(II) (nitroprusside):

$$[Fe(CN)_6]^{3-}(aq) + NO(g) \rightarrow [Fe(CN)_5NO]^{2-}(aq) + CN^-(aq)$$

Sodium pentacyanonitrosylferrate(II) is used in the detection of sulphides in aqueous solutions.

Nitrogen oxide can also gain an electron:

$$Na(s) + NO(g) \xrightarrow{\text{liquid } NH_3} Na^+NO^-(s) \text{ sodium nitrosyl.}$$

Nitrogen dioxide and dinitrogen tetraoxide

Nitrogen dioxide is prepared by heating lead(II) nitrate(V) (which unlike most nitrates(V) contains no water of crystallisation) and it is condensed by an ice/salt mixture.

$$2Pb(NO_3)_2(s) \rightarrow 2PbO(s) + 4NO_2(g) + O_2(g)$$

Colourless crystals are formed at $-10\ °C$ when the dimer, N_2O_4, is present. As the crystals are melted, dissociation starts and a pale brown liquid is formed which darkens gradually and at $22.4\ °C$ it boils. On heating further, the gas becomes red-brown and then almost black at $150\ °C$ when dissociation is complete. Above $150\ °C$ the colour starts to fade as decomposition occurs and at $600\ °C$ this is complete and a colourless mixture results:

$$2NO_2(g) \rightleftharpoons 2NO(g) + O_2(g)$$

The nitrogen dioxide molecule is an angular resonance hybrid and the odd electron makes it paramagnetic.

Both the N—O bonds are of equal length.

Dinitrogen tetraoxide is diamagnetic and planar, and a possible structure would be:

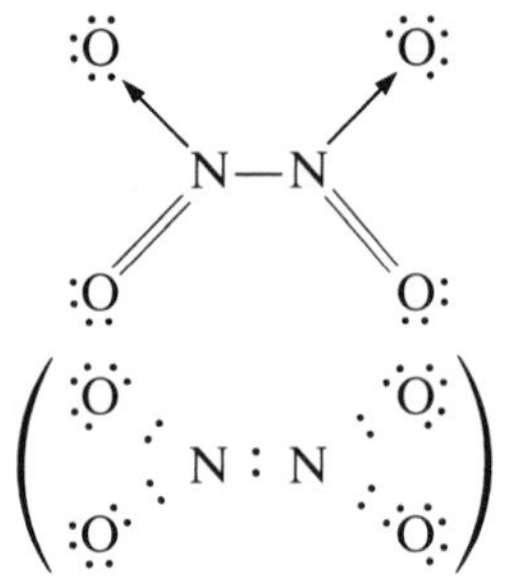

(This structure, however, does not account for the fact that liquid dinitrogen tetraoxide oxidises the alkali metals to their nitrates(V), e.g.

$$Na(s) + N_2O_4(s) \rightarrow NaNO_3(s) + NO(g)$$

The inference here is that it has some ionic character, $NO_3^-NO^+$.)

Nitrogen dioxide behaves as a mixed anhydride of nitric(III) (nitrous) and nitric(V) acids:

$$2NO_2(g) + H_2O(l) \rightarrow HNO_2(aq) + HNO_3(aq)$$

The nitric(III) acid (nitrous acid) decomposes rapidly under these conditions and so the overall reaction is:

$$3NO_2(g) + H_2O(l) \rightarrow 2HNO_3(aq) + NO(g) \text{ (oxidised by air to } NO_2)$$

A mixture of nitrate(V) and nitrate(III) (nitrite) is obtained with alkalis:

$$2NO_2(g) + 2HO^-(aq) \rightarrow NO_2^-(aq) + NO_3^-(aq) + H_2O(l)$$

Nitrogen dioxide is a powerful oxidising agent and can be reduced to nitrogen by burning phosphorus, carbon, and sulphur or by heating iron and copper in it, whereas hydrogen sulphide reduces it only to nitrogen oxide (nitric oxide):

$$H_2S(g) + NO_2(g) \rightarrow S(s) + H_2O(l) + NO(g)$$

Similar reactions are:

$$CO(g) + NO_2(g) \rightarrow CO_2(g) + NO(g)$$
$$SO_2(g) + H_2O(l) + NO_2(g) \rightarrow H_2SO_4(aq) + NO(g)$$

On the other hand, powerful oxidising agents such as hydrogen peroxide and acidified manganate(VII) (permanganate) oxidise it to nitric(V) acid:

$$2NO_2(g) + H_2O_2(aq) \rightarrow 2HNO_3(aq)$$

Ammonia

Ammonia is prepared in the laboratory by heating an ammonium salt with an alkali, e.g.

$$2NH_4Cl(s) + Ca(OH)_2(s) \rightarrow CaCl_2(s) + 2H_2O(l) + 2NH_3(g)$$

and industrially by the Haber process:

$$N_2(g) + 3H_2(g) \xrightarrow[\text{Fe catalyst}]{550\,°C, \text{ up to } 1000 \text{ atm}} 2NH_3(g)$$

It is very soluble in water; one volume of water at 0 °C dissolves 1300 volumes of ammonia and at 20 °C dissolves 700 volumes of ammonia, but this is all evolved on boiling. Concentrated ammonia solution contains 35% NH_3 and has a specific gravity of 0.880 and for this reason is often known as 'eight-eighty ammonia'.

The ammonia molecule has the trigonal pyramidal shape (page 68):

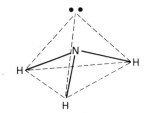

The lone pair on the nitrogen attracts the slightly positively charged hydrogens in adjacent molecules, i.e. hydrogen bonds are formed giving ammonia a relatively high boiling point and a high enthalpy (latent heat) of vaporisation. Hydrogen bonds also occur with water and this accounts for the high solubility:

$$H\overset{\delta-}{-}O\overset{\delta+}{-}H \text{ (l)} + :NH_3(g) \rightleftharpoons H-O-H\text{-----}NH_3(aq)$$

The ammonia molecule can, to a small extent, abstract a proton from the water:

$$H-O-H\text{-----}NH_3(aq) \rightleftharpoons HO^-(aq) + NH_4^+(aq)$$

Covalent NH_4OH, in fact, does not exist. The ammonia solution (often called ammonium hydroxide) is a weak base due to the slight dissociation of the hydrate, $NH_3 \cdot H_2O$. The dissociation is repressed by adding an ammonium salt. In qualitative inorganic analysis, for example, the hydroxides of iron(III), aluminium, and chromium(III) are precipitated by the addition of ammonium chloride and ammonia solution, but under these conditions the solubility products of zinc, cobalt, nickel, and manganese hydroxides are not exceeded.

Ammonia is a strong electron donor (Lewis base) and can form compounds by donation of electrons, e.g. $H_3N: \rightarrow AlCl_3$ and by dipole–ion attraction, e.g. $CaCl_2 \cdot 8NH_3$ (ammonia cannot be dried with calcium chloride) where the Ca^{2+} attracts the negative end of the ammonia dipoles.

As stated above, ammonia solution is a weak base and so it will precipitate metal hydroxides from solutions of their salts if the concentration of HO^- is sufficient to exceed the solubility product (page 120) of the metal hydroxides. However, it is quite often found that the metal hydroxide redissolves in excess ammonia solution due to the formation of ammine complexes in which the water of the hydrated metal ion has been replaced by ammonia, e.g:

$$[Cu(NH_3)_4(H_2O)_2]^{2+}, \quad [Ni(NH_3)_6]^{2+}, \quad [Co(NH_3)_6]^{3+}, \quad \text{and} \quad [Ag(NH_3)_2]^+$$

(see transition elements).

Liquid ammonia is slightly dissociated in a similar manner to water:

$$2NH_3 \rightleftharpoons NH_4^+ + NH_2^- \quad (c.f. \quad 2H_2O \rightleftharpoons H_3O^+ + HO^-)$$

Ammonium salts and amides dissolved in liquid ammonia therefore behave as acids and bases respectively, e.g.

$$NH_4Cl \; + \; KNH_2 \; \rightarrow \; KCl + \; 2NH_3$$

| Acid | Base | Salt | Solvent |

A COMPARISON OF NITROGEN AND PHOSPHORUS, THEIR HYDRIDES AND CHLORIDES

It was seen above that the covalency maximum of nitrogen is restricted to four, i.e. one co-ordinate and three covalent bonds. Phosporus, however, has d orbitals available in its outer shell and, as a result of this, it can exhibit a covalency of five as well as the group valency of three. The covalency of five is accomplished by promotion of a $3s$ electron to the $3d$ level:

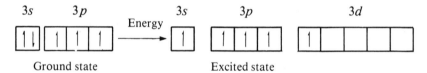

| | Ground state | | | Excited state |

Hybridisation (page 287) will give five identical sp^3d orbitals which can form covalent bonds by overlapping with orbitals from five other atoms.

Phosphorus, like nitrogen, can also attain the noble gas electronic structure by gaining three electrons and forming the phosphide ion, P^{3-}, as in sodium phosphide, Na_3P.

In contrast to nitrogen, phosphorus exhibits allotropy, the white and the red allotropes being the best known. The properties of these allotropes are summarised in Table 10.5.

Table 10.5. Comparison of the properties of white and red phosphorus

Property	White phosphorus	Red phosphorus
Density	1.82 g cm⁻³	2.20 g cm⁻³
Melting point	44 °C	Sublimes at 431 °C
Solubility in benzene, etc.	Soluble	Insoluble
Ignition temp. in moist air	About 30 °C	About 300 °C
Reaction with chlorine	Spontaneous ignition	Ignites on heating
Reaction with hot alkali	Forms phosphine — see below	No reaction

The nitrogen molecules are diatomic but phosphorus exists in benzene solution, etc. and in the vapour state as P_4 molecules.

Hydrides

Phosphine, PH_3, can be prepared by the action of sodium hydroxide solution on phosphonium iodide (c.f. the preparation of ammonia — page 206)

$$PH_4I(s) + NaOH(aq) \rightarrow NaI(aq) + H_2O(l) + PH_3(g)$$

Alternatively, white phosphorus may be boiled with concentrated sodium hydroxide solution:

$$4P(s) + 3NaOH(aq) + 3H_2O(l) \rightarrow 3NaH_2PO_2(aq) + PH_3(g)$$
Sodium phosphinate
(hypophosphite)

The product from this reaction, however, is contaminated with hydrogen and diphosphine, P_2H_4.

Phosphine is considerably less basic than ammonia because phosphorus atoms are appreciably larger than nitrogen atoms. It has a lower melting and boiling point because, unlike ammonia, it does not form hydrogen bonds. The properties of ammonia and phosphine are compared in Table 10.6.

Table 10.6 Comparison of ammonia with phosphine

Property	Ammonia	Phosphine
Appearance	Colourless gas	Colourless gas
Odour	Pungent	Fishy
Solubility in water	Very soluble giving an alkaline solution	Slightly soluble giving a neutral solution
Toxicity	Poisonous	Very poisonous
Reaction with acids	Ready formation of ammonium salts	Forms phosphonium salts with difficulty
Salts in water	Stable and soluble	Unstable and hydrolysed
Formation of complexes	Many ammines are known	Very few complexes formed

Chlorides

Nitrogen forms a trichloride as an explosive yellow oil when ammonia reacts with excess chlorine:

$$NH_3(g) + 3Cl_2(g) \rightarrow NCl_3(l) + 3HCl(g)$$

Phosphorus, on the other hand, forms a tri- and pentachloride by direct combination, ignition occurring on contact of the elements:

$$2P(s) + 3Cl_2(g) \rightarrow 2PCl_3(l)$$
$$PCl_3(l) + Cl_2(g) \rightarrow PCl_5(s)$$

The trichloride is a colourless liquid and the pentachloride a pale yellow solid which is covalent in the vapour state but is made up of $[PCl_4]^+$ and $[PCl_6]^-$ ions in the solid state. Both chlorides fume in moist air and are instantly hydrolysed by cold water:

$$PCl_3(l) + 3H_2O(l) \rightarrow H_3PO_3(aq) + 3HCl(g)$$
Phosphonic acid (phosphorous acid)

$$PCl_5(s) + H_2O(l) \rightarrow POCl_3(l) + 2HCl(g)$$
Phosphorus trichloride oxide (phosphorus oxychloride)

then
$$POCl_3(l) + 3H_2O(l) \rightarrow H_3PO_4(aq) + 3HCl(g)$$
Phosphoric (V) acid (ortho-phosphoric acid)

Both compounds are used in organic chemistry to replace hydroxyl groups by chlorine (pages 312 and 333) whilst the evolution of hydrogen chloride fumes on treatment with phosphorus pentachloride is used as a test for hydroxyl compounds.

GROUP VI – OXYGEN AND SOME SULPHUR COMPOUNDS

Oxygen, with atomic number 8, is the first element of Group VI and it has the electronic structure $1s^2 2s^2 2p^4$. It can achieve the noble gas structure by gaining two electrons and forming the O^{2-} ion as in Na_2O, by sharing the two unpaired $2p$ electrons and forming two covalent bonds as in water, or by acting as an electron pair acceptor in a co-ordinate bond. If ionic oxides are soluble in water, e.g. Na_2O, hydrolysis occurs and the oxide ion is converted into the stable hydroxide ion which is intermediate between O^{2-} and H_2O, i.e.

$$O^{2-}(s) + H_2O(l) \rightarrow 2HO^-(aq)$$

Oxygen has no available d orbitals and so it is restricted to a covalency maximum of four. A covalency of four is in fact extremely rare, but one of three is not unusual, e.g. H_3O^+, the oxonium ion. As with nitrogen, oxygen has a considerable tendency to form multiple bonds, e.g. $C=O$, $N=O$, $S=O$, etc.

Oxygen may be prepared by thermal decomposition of various oxides and oxy-salts, e.g. HgO, Pb_3O_4, BaO_2, $KClO_3$ (catalysed by MnO_2), and $KMnO_4$.

$$2KClO_3(s) \rightarrow 2KCl(s) + 3O_2(g)$$
$$2KMnO_4(s) \rightarrow MnO_2(s) + K_2MnO_4(s) + O_2(g)$$

It is manufactured by fractional distillation of liquid air. Oxygen is diatomic and paramagnetic even though it contains an even number of electrons. This suggests some unpairing of the π electrons with the formation of a diradical, $\ddot{O}-\dot{O}$: Like the rest of the group (S, Se, Te, and Po), it exhibits allotropy, the two forms being dioxygen, O_2 and trioxygen (ozone), O_3. Trioxygen is prepared by passing a silent electric discharge through air or oxygen. It is diamagnetic and has an angular structure:

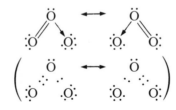

Oxygen combines directly with most elements, although heating may be required to start the reaction. It is able to stabilise high valency states as a result of its small size and strongly electronegative character, e.g. PbO_2, SO_3, Cl_2O_7, and $K_2Cr_2O_7$.

Hydrides of oxygen and sulphur

a. Water

This possesses remarkable physical properties. It has no smell or taste (in contrast to the hydrides of the rest of the group), high enthalpies (latent heats) of fusion and vaporisation, it is a liquid at room temperature (the hydrides of the remainder of the group are gases), and it has a high dielectric constant and a maximum density at 4 °C.

The high enthalpies (latent heats) of fusion and vaporisation and the high boiling point may be attributed to hydrogen bonding. The increased density at 4 °C results from the water molecules being packed together closer than they are in ice, in fact, water at low temperature has a *quasi*-crystalline structure. Its great solvent power for ionic compounds is due to its lone pairs and it being polarised so that it can solvate ions. Many covalent compounds are also soluble in it because it can form hydrogen bonds with them, e.g.

$$\underset{CH_3}{\overset{\delta-}{O}}\underset{}{\overset{\delta+}{-H}}----\overset{\delta-}{:}\underset{H^{\delta+}}{\overset{H^{\delta+}}{O}:}$$

Water is slightly dissociated:

$$2H_2O(l) \rightleftharpoons H_3O^+(aq) + HO^-(aq)$$

and at 25 °C the concentration of the oxonium and the hydroxide ions is 1×10^{-7} mol litre^{-1}.

It can react directly with many substances as a result of its lone pairs; this process is known as hydrolysis, e.g.

$$H_2\ddot{O} + \overset{\delta+}{CH_3}-\overset{\delta-}{I} \rightarrow \left[\underset{H}{\overset{H}{:}}\overset{\delta+}{O}:----CH_3----\overset{\delta-}{I} \right] \rightarrow \underset{H}{\overset{H}{:}}\overset{+}{O}-CH_3 + I^-$$

$$CH_3\ddot{O}H + HI$$

Acidity can result from co-ordinated water, e.g.

$$[Fe(H_2O)_6]^{3+} + H_2\ddot{O} \rightleftharpoons [Fe(H_2O)_5(OH)]^{2+} + H_3O^+$$

The small highly charged cation weakens the O—H bonds (see also page 192).

b. Hydrogen sulphide

This is often prepared by adding dilute hydrochloric acid to iron (II) sulphide:

$$FeS(s) + 2HCl(aq) \rightarrow FeCl_2(aq) + H_2S(g)$$

However, iron(II) sulphide usually contains some free iron and so hydrogen is an impurity. Purer hydrogen sulphide is formed by the action of cold water on aluminium sulphide:

$$Al_2S_3(s) + 6H_2O(l) \rightarrow 2Al(OH)_3(s) + 3H_2S(g)$$

It is a very poisonous, colourless gas, boiling point $-61\ °C$. The low boiling point compared with water is explained by the absence of hydrogen bonding; the sulphur atom, being larger and less electronegative than oxygen, has a very low charge density. This also explains the low solubility in water (at $20\ °C$, approximately 3 volumes of hydrogen sulphide dissolve in 1 volume of water).

Water is strongly exothermic ($\Delta H_f^{\circ} = -285.9\ kJ\,mol^{-1}$) but hydrogen sulphide is only weakly so ($\Delta H_f^{\circ} = -20.6\ kJ\,mol^{-1}$) and so, whilst water is thermally stable to above $1200\ °C$, hydrogen sulphide is decomposed into its elements above $400\ °C$. Hydrogen sulphide burns with a blue flame:

$$2H_2S(g) + 3O_2(g) \rightarrow 2H_2O(l) + 2SO_2(g)$$

In solution, it is a weak dibasic acid and gives rise to normal sulphides and hydrogensulphides:

$$2HO^-(aq) + H_2S(aq) \rightleftharpoons S^{2-}(aq) + 2H_2O(l)$$
$$HO^-(aq) + H_2S(aq) \rightleftharpoons HS^-(aq) + H_2O(l)$$

The alkali metal sulphides are soluble in water and give alkaline solutions due to hydrolysis:

$$S^{2-}(s) + H_2O(l) \rightleftharpoons HS^-(aq) + HO^-(aq)$$

Passing hydrogen sulphide through aqueous solutions of metal salts generally results in the precipitation of the metal sulphide, e.g.

$$Cu^{2+}(aq) + H_2S(g) \rightarrow CuS(s) + 2H^+(aq)$$

The concentration of S^{2-} ions in hydrogen sulphide solution can be controlled by the acidity or alkalinity of the solution. The equilibrium operating is:

$$H_2S(aq) \rightleftharpoons 2H^+(aq) + S^{2-}(aq)$$

and so in acid solution the concentration of S^{2-} will be considerably reduced. In mildly acidic solution, there is still sufficient S^{2-} to exceed the solubility products of Cu, Cd, Pb, Hg, Bi, Sn, Sb, and As sulphides whereas the higher S^{2-} concentration in alkaline solution (ammonia solution is added) is required to precipitate Co, Ni, Zn, and Mn sulphides. The facts are utilised to separate the metal ions in inorganic qualitative analysis.

Hydrogen sulphide is a powerful reducing agent:

$$S^{2-} \rightarrow S + 2e^-$$

Examples are:

$$H_2S(g) + Cl_2(g) \rightarrow S(s) + 2HCl(g)$$
$$2H_2S(g) + SO_2(g) \xrightarrow{\text{trace of moisture}} 3S(s) + 2H_2O(l)$$
$$2FeCl_3(aq) + H_2S(g) \rightarrow S(s) + 2FeCl_2(aq) + 2HCl(aq)$$
$$H_2SO_4(conc) + H_2S(g) \rightarrow S(s) + 2H_2O(l) + SO_2(g)$$
$$K_2Cr_2O_7(aq) + 4H_2SO_4(aq) + 3H_2S(g) \rightarrow 3S(s) + K_2SO_4(aq) + Cr_2(SO_4)_3(aq)$$
$$+ 7H_2O(l)$$

$$2KMnO_4(aq) + 3H_2SO_4(aq) + 5H_2S(g) \rightarrow 5S(s) + K_2SO_4(aq) + 2MnSO_4(aq) + 8H_2O(l)$$

These reactions may be considered as combinations of two half-reactions as described in Chapter 7.

Properties of sulphuric(VI) acid and its use in the laboratory

Sulphuric(VI) acid is manufactured by the contact process as discussed on page 235. The pure acid is a colourless oily liquid with a density of 1.83 g cm^{-3} and a boiling point of 330 °C. Both the viscosity and the relatively high boiling point are attributable to extensive hydrogen bonding. Concentrated sulphuric(VI) acid contains 98% of the acid and 2% water.

The laboratory use of sulphuric(VI) acid falls into four main categories.

1. *As an acid.* Sulphuric(VI) acid is a strong, dibasic acid:

$$H_2SO_4(conc.) + H_2O(l) \rightleftharpoons H_3O^+(aq) + HSO_4^-(aq)$$

$$\Big\updownarrow H_2O$$

$$H_3O^+(aq) + SO_4^{2-}(aq)$$

It will therefore react with bases to give sulphates(VI) and hydrogen-sulphates(VI).

Sulphuric(VI) acid displaces weaker acids, and the strong acids more volatile than itself, from their salts, e.g.

$$2C_6H_5 \cdot COONa(aq) + H_2SO_4(aq) \rightarrow 2C_6H_5 \cdot COOH(s) + Na_2SO_4(aq)$$
$$\text{Benzenecarboxylic (benzoic) acid}$$
$$NaCl(s) + H_2SO_4(conc.) \rightarrow NaHSO_4(s) + HCl(g)$$

2. *As a dehydrating agent.* Concentrated(VI) sulphuric acid has a great affinity for water and is used to dry various gases such as hydrogen, nitrogen, and oxygen. It has the ability to dehydrate a number of organic compounds and this is utilised in some laboratory preparations. Thus, carbon monoxide and ethene are prepared by dehydration of methanoic acid and ethanol respectively:

$$HCOOH(l) \xrightarrow[\text{room temp.}]{H_2SO_4} H_2O(l) + CO(g)$$

$$CH_3 \cdot CH_2OH(l) \xrightarrow[\text{180 °C}]{H_2SO_4} H_2O(l) + CH_2{=}CH_2(g)$$

3. *As an oxidising agent.* Hot concentrated sulphuric(VI) acid is a powerful oxidising agent (page 138). For example, it oxidises carbon and sulphur to their dioxides. The reaction with copper can be used for the laboratory preparation of sulphur dioxide:

$$Cu(s) + 2H_2SO_4(conc.) \rightarrow CuSO_4(s) + 2H_2O(l) + SO_2(g)$$

4. *As a sulphonating agent.* Aromatic hydrocarbons can be sulphonated by sulphuric(VI) acid. For example, benzene reacts with the concentrated acid at 80 °C to give benzenesulphonic acid (see also page 309):

$$C_6H_6(l) + H_2SO_4(conc.) \rightleftharpoons C_6H_5SO_3H(aq) + H_2O(l)$$

GROUP VII — THE HALOGENS

There are five halogens as shown below. However, astatine is rare, intensely radioactive, and beyond the scope of this text.

Element	Symbol	Boiling point	Appearance	Electronic structure
Fluorine	F	−188 °C	greenish yellow	2.7
Chlorine	Cl	−34 °C	green yellow	2.8.7
Bromine	Br	58 °C	red brown	2.8.18.7
Iodine	I	m.p. 114 °C	grey black	2.8.18.18.7
Astatine	At	m.p. 302 °C		2.8.18.32.18.7

Some relevant data are listed below

	F	Cl	Br	I
Covalent radius/nm	0.064	0.099	0.111	0.128
Ionic radius/nm	0.133	0.181	0.196	0.219
Electron affinity/kJ mol^{-1}	−322.6	−364	−342	−295.4
1st Ionisation energy/kJ mol^{-1}	1680	1260	1140	1010
$E^{\ominus}$ halogen/halide ion/volts	+2.87	+1.36	+1.09	+0.54
Enthalpy (heat) of atomisation/kJ mol^{-1}	79.1	121.1	112.0	106.6

Group trends

The halogens complete their octet either by gaining an electron to give the halide ion, X^- or by sharing their unpaired p electron to form a covalent bond. Ionic halides are formed with the highly electropositive elements, i.e. the alkali metals and alkaline earths whereas covalent bonds are formed with weakly electropositive metals and non-metals. In the case of fluorine, covalency is restricted to one since only its unpaired p orbital is available for overlap. However, chlorine, bromine, and iodine have d orbitals available in their outer shell and so promotion of some of their electrons can result in higher valencies, e.g. with iodine:

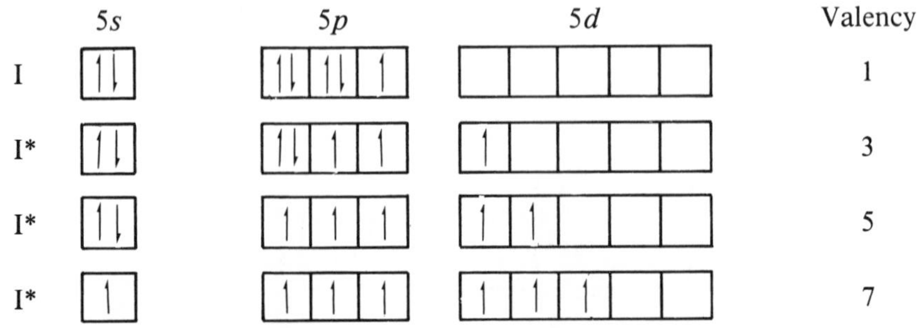

*Denotes an excited state.

Chlorine can have valencies of 1 or 3 and bromine 1, 3, or 5.

The tendency to form covalent bonds increases down the group. However, the halogen–halogen bond strengths decrease from chlorine to iodine as the bonding electrons become further from the nucleus. The enthalpies of atomisation therefore decrease steadily from chlorine to iodine. The position of fluorine is anomalous; it has an abnormally low enthalpy of atomisation because the atoms in the molecules are close together and so there are considerable repulsive forces between their non-bonding electrons.

The electron affinities indicate the ease of ion formation, but the position of fluorine is again somewhat anomalous in that it has a lower value than chlorine. This is because electron affinity is the sum of the electrostatic energy required to force the electron into the region of the outer shell of the atom and the energy evolved when X^- is formed from X and an electron. The former energy in the case of fluorine is large because the outer electron shell is very compact.

Fluorine is a stronger oxidising agent than chlorine because, even though its electron affinity is lower, its dissociation energy is smaller. Also, fluorides have larger lattice energies and hydration energies than chlorides. The standard electrode potentials show the relative tendencies to form ions in solution. It is apparent from these that a halogen of lower atomic number will oxidise ions of one of higher atomic number, e.g.

$$Cl_2(g) + 2I^-(aq) \rightarrow 2Cl^-(aq) + I_2(aq)$$

Fluorine is, in fact, the most electronegative element as a result of its small size and electronic structure. On combination with other elements it causes them to exhibit their maximum possible covalency, e.g.

$$PF_5, SF_6, IF_7$$

Metallic character increases down every group and this is illustrated here by the formation of some compounds where I^+ is present, e.g. I^+CNO^-. The I^+ ion also occurs as a stabilised complex cation, e.g. $[I(pyridine)_2]^+NO_3^-$. Similar compounds of bromine may exist.

Properties of the elements

All the halogens exist as stable diatomic molecules, the stability decreasing down the group from chlorine, but, even so, iodine does not dissociate below 600°C. They combine directly, often very vigorously, with most metals and non-metals. Interhalogen compounds can also be formed, e.g. ClF_3, ICl, BrF_5, etc.

Fluorine vigorously oxidises water in a complex reaction which gives mainly hydrogen fluoride and oxygen, together with some trioxygen (ozone) and traces of hydrogen peroxide and oxygen difluoride (fluorine monoxide), F_2O:

$$2H_2O(l) + 2F_2(g) \rightarrow 4HF(aq) + O_2(g)$$

Chlorine undergoes a much milder reaction with water. Some oxidation occurs, especially in sunlight:

$$2H_2O(l) + 2Cl_2(g) \rightarrow 4HCl(aq) + O_2(g)$$

but the main reaction is hydrolysis to chloric(I) acid (hypochlorous acid) and hydrochloric acid:

$$H_2O(l) + Cl_2(g) \rightarrow HOCl(aq) + HCl(aq)$$

The latter reaction is an example of *disproportionation*, i.e. the chlorine undergoes simultaneous oxidation and reduction. The oxidation number of chlorine changes from zero to $+1$ in the chloric(I) acid and to -1 in the hydrochloric acid (oxidation and reduction respectively).

Bromine is only slightly soluble in water and in the saturated solution (approximately 5%) it is present mainly as molecular bromine but hydrolysis increases as the solution is diluted:

$$H_2O(l) + Br_2(l) \rightarrow HOBr(aq) + HBr(aq)$$

Moist chlorine and bromine possess bleaching properties because the resultant halic(I) acid (hypohalous acid), HOX, oxidises the dye. Iodine is very sparingly soluble in water. Its appearance in organic solvents is, however, interesting. Brown solutions are obtained with oxygen containing solvents, e.g. ethers and alcohols (the oxygen acting as a lone pair donor to the iodine molecule) whilst solvents such as hydrocarbons and tetrachloromethane (carbon tetrachloride) give violet solutions.

All the halogens react with aqueous alkali, the reactivity decreasing down the group.

Cold dilute: $2KOH(aq) + 2F_2(g) \rightarrow 2KF(aq) + H_2O(l) + F_2O(g)$
Hot concentrated: $4KOH(aq) + 2F_2(g) \rightarrow 4KF(aq) + 2H_2O(l) + O_2(g)$

With cold dilute and hot concentrated potassium hydroxide solution, chlorine gives potassium chlorate(I) (potassium hypochlorite) and potassium chlorate(V) respectively.

Cold dilute: $2KOH(aq) + Cl_2(g) \rightarrow KCl(aq) + KOCl(aq) + H_2O(l)$
Hot concentrated: $6KOH(aq) + 3Cl_2(g) \rightarrow 5KCl(aq) + KClO_3(aq) + 3H_2O(l)$

Bromine undergoes similar reactions to chlorine as does iodine except that the reaction with cold alkali is reversible:

$$2KOH(aq) + I_2(s) \rightleftharpoons KI(aq) + KOI(aq) + H_2O(l)$$

and the iodate(I) (hypoiodite) changes into iodate(V) even at room temperature:

$$3KOI(aq) \rightarrow KIO_3(aq) + 2KI(aq)$$

Fluorine has a great affinity for hydrogen and reacts vigorously with hydrogen sulphide, ammonia, and organic compounds, many of which inflame spontaneously in the process.

$$H_2S(g) + 4F_2(g) \rightarrow 2HF(g) + SF_6(g)$$
$$2NH_3(g) + 3F_2(g) \rightarrow 6HF(g) + N_2(g)$$
$$CH_4(g) + 4F_2(g) \rightarrow 4HF(g) + CF_4(g)$$

Chlorine, bromine, and iodine oxidise Sn(II) in acid solution to Sn(IV), hydrogen sulphide to sulphur, etc, e.g.

$$SnCl_2(s) + Cl_2(g) \rightarrow SnCl_4(l)$$
$$H_2S(g) + Br_2(l) \rightarrow S(s) + 2HBr(g)$$

Hydrogen halides

All the hydrogen halides can be prepared by direct combination, the reaction decreasing in vigour down the group. The reaction with fluorine is explosive even in the absence of light, whereas chlorine/hydrogen mixtures are explosive only in the presence of light or on heating to 250 °C. The reaction between bromine and hydrogen is rapid at 200 °C in the presence of platinised asbestos as catalyst but the similar reaction with iodine is slow, reversible and incomplete.

Hydrogen chloride, hydrogen bromide, and hydrogen iodide are colourless gases at room temperature, whilst hydrogen fluoride is a colourless fuming liquid:

$$\text{HF} \quad \text{b.p.} \quad 20 \,°\text{C}, \quad \text{HCl} \quad \text{b.p.} \,-85 \,°\text{C},$$
$$\text{HBr} \quad \text{b.p.} \,-69 \,°\text{C} \quad \text{HI} \quad \text{b.p.} \,-35 \,°\text{C}$$

The anomalous boiling point of hydrogen fluoride is a consequence of hydrogen bonding which is extensive both in the liquid and in the vapour just above the boiling point.

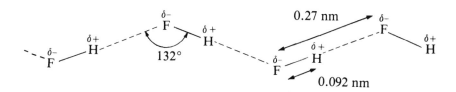

Above 90 °C, only monomer is present.

Liquid hydrogen fluoride is weakly conducting:

$$2\text{HF} \rightleftharpoons \text{H}_2\text{F}^+ + \text{F}^-$$

but the hydrogen halides are predominantly covalent. They are all very soluble in water giving acidic solutions which have a maximum boiling point.

Acid	Boiling point/°C.	Composition%
HF	111	35.6% HF
HCl	108.6	20.2% HCl
HBr	126	47.5% HBr
HI	127	57% HI

The H—X bond strength decreases from hydrogen fluoride to hydrogen iodide and this outweighs the reduced hydration energy from F^- to I^-. The acid strength therefore increases down the series, hydrofluoric acid being a relatively weak acid whilst hydriodic acid is a very strong acid.

$$\text{H}_2\ddot{\text{O}}(l) + \text{HX}(g) \rightleftharpoons \text{H}_3\text{O}^+(aq) + \text{X}^-(aq)$$

In fairly concentrated hydrofluoric acid, a second equilibrium becomes important:

$$\text{HF}(aq) + \text{F}^-(aq) \rightarrow \text{HF}_2^-(aq)$$

This is due to hydrogen bonding between F^- and HF, i.e. $F^- \overset{\delta+}{\text{-----}} \overset{\delta-}{H} - F$. Hydrogendifluorides can be prepared by evaporating solutions of fluorides in hydrofluoric acid, e.g.

$$KF(aq) + HF(aq) \rightleftharpoons KHF_2(aq)$$

The solid hydrogendifluoride decomposes on heating.

As well as undergoing the typical reactions of acids, hydrofluoric acid dissolves silicon(IV) oxide (silica) and silicates to form hydrogen hexafluorosilicate(IV) (fluorosilicic acid) and so it has to be stored in poly(ethene) and not glass bottles.

$$SiO_2(s) + 6HF(aq) \rightarrow H_2SiF_6(aq) + 2H_2O(l) \xrightarrow{\text{hydrolysis}} \text{hydrated } SiO_2(s)$$

Silicon tetrafluoride is also formed:

$$SiO_2(s) + 4HF(aq) \rightarrow 2H_2O(l) + SiF_4(g)$$

The hydrogen halides become stronger reducing agents on going from hydrogen fluoride to hydrogen iodide. Thus, hydrogen fluoride and chloride may be prepared by heating calcium fluoride and sodium chloride respectively with concentrated sulphuric(VI) acid:

$$CaF_2(s) + H_2SO_4(\text{conc.}) \rightarrow CaSO_4(s) \quad + 2HF(g)$$
$$NaCl(s) + H_2SO_4(\text{conc.}) \rightarrow NaHSO_4(s) + HCl(g)$$

Hydrogen bromide and iodide reduce the acid, e.g.

$$2KBr(s) + 3H_2SO_4(\text{conc}) \rightarrow 2KHSO_4(aq) + 2H_2O(l) + SO_2(g) + Br_2(g)$$
$$(2HBr + H_2SO_4 \rightarrow 2H_2O + SO_2 + Br_2)$$

and so have to be prepared in the laboratory by hydrolysis of the corresponding phosphorus trihalide:

$$PBr_3(l) + 3H_2O(l) \rightarrow H_3PO_3(aq) + 3HBr(g)$$
$$PI_3(s) + 3H_2O(l) \rightarrow H_3PO_3(aq) + 3HI(g)$$

Alternatively, they can be made by heating sodium or potassium bromide or iodide with concentrated phosphoric(V) acid. The reactions with phosphoric(V) acid are successful because this acid, unlike concentrated sulphuric(VI) acid, is not an oxidising agent.

Hydrogen iodide is a very powerful reducing agent and has found much application in organic chemistry for this purpose. In this connection it is usually used with red phosphorus which reconverts the iodine to hydrogen iodide.

The halides

As stated above, ionic and covalent halides are possible. Metal halides tend to be ionic, or predominantly so, but this tendency decreases on going from F^- to I^- since large ions are more readily distorted (Fajan's rules), e.g. lithium fluoride is ionic, but lithium iodide has considerable covalent character. When more than one halide is possible, the higher one tends to be covalent and the lower one ionic; such is the case with $SnCl_4$ and $SnCl_2$.

Chlorides, bromides, and iodides tend to have similar solubilities in water, but fluorides are often anomalous, e.g. silver chloride, bromide, and iodide are insoluble but the fluoride is soluble; similarly calcium chloride, bromide, and iodide are soluble whilst the fluoride is insoluble. These anomalies occur because the very small fluoride ion has large hydration and lattice energies. If, therefore, a fluoride has a very high lattice energy, low solubility will tend to result, whereas if the lattice energy is not very high, as in the case of silver fluoride, high solubility occurs.

Solutions of ionic chlorides, bromides, and iodides give white, cream, and yellow precipitates respectively with silver nitrate(V) solution. These precipitates dissolve in potassium cyanide solution forming the dicyanoargentate(I) ion, $[Ag(CN)_2]^-$. Similarly, the precipitates react with ammonia solution to give the diamminesilver(I) ion, $[Ag(NH_3)_2]^+$; the chloride is readily soluble, the bromide less readily, and the iodide only slightly so (see page 266).

Covalent halides of non-metals, which do not have their covalency maximum, are readily hydrolysed since co-ordinate bonds are possible with water, e.g.

Some metallic halides also are hydrolysed in solution, e.g. iron(III) chloride, and this tendency increases the greater the charge on the cation.

The high hydration energy of the fluoride ion results in some fluorides having water of crystallisation, whilst the other halides are anhydrous, e.g. $KF \cdot 2H_2O$ or $KF \cdot 4H_2O$, KCl, KBr, and KI.

The fluoride ion shows a distinct tendency to donate a lone pair of electrons and act as a ligand in the formation of complex ions owing to its small size and high charge density, e.g. $Na_3[AlF_6]$. Chloride, bromide, and iodide ions naturally exhibit a decreasing tendency to do this.

Many polyhalides are known, e.g. KI_3. In the case of KI_3 it is thought that the I_3^- ion results from attraction of I_2 by I^- but in other instances the central atom utilises *d* orbitals to form covalent bonds with the other halogens.

The chlorate(I) (hypochlorite) anion in solution

Halic(I) acids (hypohalous acids) and halates(I) (hypohalites) are formed when chlorine, bromine, or iodine react with water or cold dilute alkali respectively:

$$H_2O(l) + X_2 \rightarrow HOX(aq) + HX(aq)$$
$$2NaOH(aq) + X_2 \rightarrow NaOX(aq) + NaX(aq) + H_2O(l)$$

The halic (I) acids (hypohalous acids), $H-\overset{..}{\underset{..}{O}}-\overset{..}{\underset{..}{X}}:$, are weak acids, chloric(I) acid (hypochlorous acid) being the strongest and most important

$$HOCl(aq) + H_2\overset{..}{\underset{..}{O}}(l) \rightleftharpoons H_3\overset{..}{O}^+(aq) + ClO^-(aq)$$

The chlorate(I) (hypochlorite) anion is considerably hydrolysed in solution:

$$ClO^-(aq) + H_2O(l) \rightleftharpoons HO^-(aq) + HOCl(aq)$$

Aqueous solutions of chlorates(I) (hypochlorites) slowly decompose on standing:

$$3ClO^-(aq) \rightarrow 2Cl^-(aq) + ClO_3^-(aq) \quad \text{chlorate(V)}$$

and $\qquad\qquad 2ClO^-(aq) \rightarrow 2Cl^-(aq) + O_2(g)$

The latter reaction is catalysed by Co^{2+}, Fe^{3+}, Ni^{2+}, etc.

Addition of concentrated hydrochloric acid to a chlorate(I) (hypochlorite) results in evolution of chlorine:

$$ClO^-(aq) + 2H^+(aq) + Cl^-(aq) \quad \rightarrow \ H_2O(l) \ + Cl_2(g)$$

e.g. $\qquad\qquad NaOCl(aq) + 2HCl(conc) \rightarrow NaCl(aq) \ + H_2O(l) + Cl_2(g)$

Chloric(I) acid (hypochlorous acid) and chlorates(I) (hypochlorites) are fairly strong oxidising agents especially in acid solution:

$$ClO^-(aq) + 2H^+(aq) + 2e^- \rightarrow Cl^-(aq) + H_2O(l)$$

and will oxidise Fe^{2+} to Fe^{3+}, etc. Alkaline solutions will oxidise Cr^{3+} to CrO_4^{2-}, chromate(VI), and Mn^{2+} to MnO_4^-, manganate(VII) (permanganate):

$$ClO^-(aq) + H_2O(l) + 2e^- \rightarrow Cl^-(aq) + 2HO^-$$

The oxidising properties of chlorates(I) (hypochlorites) account for their use as bleaching agents and germicides. Thus, bleaching powder contains calcium chlorate(I) (calcium hypochlorite) and *Domestos, Milton* etc. contain sodium chlorate(I) (sodium hypochlorite).

The properties of the halogens and their compounds are summarised in Table 10.7.

GROUP O — THE NOBLE OR INERT GASES

Element	Symbol	Electronic configuration	Per cent by volume in air	b.p. / °C	I_1 /kJ mol^{-1}
Helium	He	2	0.00052	−269	2370
Neon	Ne	2.8	0.0018	−246	2080
Argon	Ar	2.8.8	0.94	−186	1520
Krypton	Kr	2.8.18.8	0.00011	−153	1350
Xenon	Xe	2.8.18.18.8	0.000008	−108	1170
Radon	Rn	2.8.18.32.18.8	—	−62	1040

All the noble gases are monatomic and sparingly soluble in water, the solubility increasing down the group owing to the formation of hydrates (see below). They occur in the atmosphere to some extent with the exception of radon which is formed by loss of an alpha-particle from radium:

$$^{226}_{88}Ra \rightarrow {}^{222}_{86}Rn + {}^4_2He$$

Table 10.7. Summary of the properties of the halogens and their compounds

	Fluorine	Chlorine	Bromine	Iodine
Element	Greenish-yellow gas	Greenish-yellow gas	Red-brown liquid	Violet-black solid
Electronegativity and reactivity		Decreasing →		Some tendency to form positive ions
Valencies	1	1, 3	1, 3, 5	1, 3, 5, 7
Products with water	HF, O_2, some F_2O, H_2O_2	Mainly HCl, $HOCl$	HBr, $HOBr$	Very sparingly soluble
Product with cold dilute KOH	KF, F_2O	KCl, $KOCl$	KBr, $KOBr$	(KIO), KIO_3
Product with hot conc. KOH	KF, O_2	KCl, $KClO_3$	KBr, $KBrO_3$	KI, KIO_3
Oxidation of X^- ions	Displaces Cl^-, Br^-, I^-	Displaces Br^-, I^-	Displaces I^-	—
Reaction with hydrogen	Explosive even at low temperatures	Explosive in light or at 200 °C	Rapid at 200 °C with Pt catalyst	Slow, reversible even with heat and catalyst
Halogen hydride	Liquid	Gas	Gas	Gas
Halogen acid strength	Weak	Strong	Strong	Very strong
Formation of acid salt KHX_2	Yes	No	No	No
AgX solubility in water	Soluble	Insoluble	Insoluble	Insoluble
CaX_2 solubility in water	Insoluble	Deliquescent	Deliquescent	Deliquescent

and by decay of other radioactive isotopes. Radon is itself radioactive, all its
isotopes initially producing polonium, e.g.

$$^{222}_{86}\text{Rn} \rightarrow {}^{218}_{84}\text{Po} + {}^{4}_{2}\text{He} \qquad t_{\frac{1}{2}} = 3.825 \text{ days}$$

These elements were called inert gases in view of their apparent reluctance to
form compounds. However, in 1962, Bartlett found that PtF_6 was sufficiently
electron attracting to ionise oxygen, and form the ionic compound $[PtF_6]^-O_2^+$.
He therefore thought that a similar reaction might be possible with xenon since its
first ionisation energy is similar to that of oxygen and, in fact, he obtained
$Xe^+[PtF_6]^-$ which is stable at room temperature.

Even though the higher atomic number noble gases are the most likely to form
compounds (because of their lower ionisation energies), little work has been done
with radon because it is radioactive and fairly rare. However, a considerable
number of xenon compounds are known, e.g. XeF_2, XeF_4, XeF_6, $XeOF_4$,
$Xe(OH)_6$, etc. Compounds of krypton, argon and neon have also been prepared,
examples being KrF, KrO, KrF_2, KrH, ArCl, ArO_2, ArO_4, NeI_2 and NeH.

Xenon tetrafluoride, XeF_4, may be prepared by heating xenon and fluorine at
400 °C under a pressure of about 13 atmospheres:

$$Xe(g) + 2F_2(g) \rightarrow XeF_4(s)$$

It is a colourless crystalline solid (melting point 114 °C) which is readily
hydrolysed giving xenon(VI) oxide:

$$6XeF_4(s) + 12H_2O(l) \rightarrow 2XeO_3(s) + 4Xe(g) + 24HF(g) + 3O_2(g)$$

Xenon tetrafluoride can also be prepared by passing a high voltage discharge
through xenon and fluorine at low temperature. Its structure is based on an
octahedron, the Xe and F atoms having a square planar configuration:

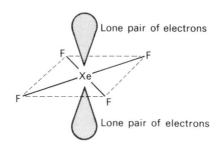

The bonding in xenon compounds is open to question. The ionisation energies
are too large for the bonds to be ionic and the bonds are too strong to be
attributable to van der Waals forces (page 36). In some compounds co-ordination
may be a possibility, but in others it is likely that some electrons are promoted
and hybridisation (page 287) occurs between the outer *s, p,* and *d* orbitals.

Hydrates are formed by argon and the higher atomic number noble gases and
these are probably formed by dipole–dipole attraction (page 36), — large atoms
are more readily polarised; xenon forms $Xe \cdot 6H_2O$.

Clathrate compounds are formed when benzene-1, 4-diol (hydroquinone),

$$HO\langle\bigcirc\rangle OH$$

is crystallised from aqueous or benzene solutions in the presence of argon, krypton, or xenon under pressure. These are not proper compounds, the noble gas is simply trapped in the benzene-1, 4-diol (hydroquinone) lattice.

Questions

10.1 (a) Describe in detail the bonding which occurs in the compounds formed between hydrogen and
   (i)   sodium (in sodium hydride),
   (ii) carbon (in methane),
   (iii) nitrogen (in ammonia).

   (b) Describe the reactions, if any, which take place between water and the hydrides of the elements in (a).

   (c) Comment upon the significance of the relative values of the following boiling points of the halogen hydrides.

HF	HCl	HBr	HI
19.5	–85	–67	–36 (°C)

[J.M.B.]

10.2   The heaviest member of Group I of the Periodic Table is francium, an element which is radioactive and whose longest lived isotope has a half-life of only 21 minutes. Consequently there has been little study of the chemistry of this element but those reactions which have been investigated yielded results which could have been predicted from the trend in properties of the Group I elements and their compounds. Discuss the chemistry you would expect for francium, paying particular attention to the halides, oxides, hydroxide, and hydride as well as to the physical properties of the element itself.   [J.M.B.]

10.3   Briefly describe the chemical properties of the simplest hydride of each of the elements sodium, carbon, nitrogen, oxygen, and fluorine.
   Comment upon the type of valence bonds occurring in each case, and show how it is related to the normal physical state of the hydride.   [J.M.B.]

10.4   A table of the following type is sometimes found in textbooks of inorganic chemistry.

	Be	Mg	Ca	Sr	Ba
Atomic number	4	12	20	38	56
Configuration	$2s^2$	$3s^2$	$4s^2$	$5s^2$	$6s^2$
Atomic radius/nm	0.089	0.136	0.174	0.191	0.198
Ionic radius/nm	0.030	0.065	0.094	0.110	0.129
Standard electrode potential/volts	–1.70	–2.34	–2.87	–2.89	–2.90

   (a) Discuss the importance of the terms given in the above table.
   (b) Show how the numerical values quoted may be used to interpret the chemical properties of this group of elements. Include in your discussion amongst other matters, some reference to the solubility of the salts of these elements, and the reactions of the metals with water.   [J.M.B.]

10.5   'The properties of the first member of a group of elements in the Periodic Table are not typical of the group as a whole'. Discuss this with reference to the chemistry of the elements of Groups I (Li–Cs) and II (Be–Ba). You should include in your answer specific properties which differentiate lithium and beryllium from other members of their respective groups as well as the reasons for the differentiation.   [J.M.B.]

10.6 (a) What is meant by 'the electronegativity of an element'? Explain briefly how the electronegativity varies with the position of the element in the periodic table.
   Account for the diagonal relationship, shown by some first and second row elements. Illustrate the existence of this relationship by stating five examples of

appropriate properties of the elements beryllium and aluminium or of their compounds.

(b) State two pieces of evidence which support the existence of the hydrogen bond in hydrogen fluoride. Why is hydrogen chloride different in this respect?

(c) Chlorine and nitrogen have similar electronegativities. Explain the difference between the reactions of (i) hydrogen chloride, (ii) ammonia, when added to water.   [J.M.B.]

10.7   Discuss briefly

(a) The anomalies of lithium and its compounds compared with the rest of the Group I elements and their compounds.

(b) The difference in reactivity of the tetrachlorides of carbon and silicon.

(c) The reasons why water is a liquid, but hydrogen sulphide is a gas.

10.8   Write a general account of the elements in Group IV B (C, Si, Ge, Sn, Pb), their oxides, hydrides, and chlorides. Refer especially to their similarities, and also to the trends shown as the atomic number increases.

[University of London]

10.9   Survey the chemistry of the Group IV elements (C–Pb) by giving

(a) a summary of the physical and chemical properties of the elements,

(b) brief descriptions of preparative routes to the chlorides and oxides,

(c) a discussion of group trends in valencies and bond-types of the chlorides and oxides,

(d) a discussion of the special properties of carbon and the ways in which its chemistry differs from the other members of the group.

[J.M.B.]

10.10   'Whereas the study of the alkali metals reveals the similarities that exist between elements of a particular group of the Periodic Table, the study of the elements C to Pb brings out the differences and dissimilarities between them'. Discuss and justify this statement.

[University of London]

10.11 (a) State two physical and two chemical properties which clearly illustrate the differences between a typical metal and a typical non-metal.

(b) 'For any given group in the periodic table, the metallic character of the element increases with increase of relative atomic mass of the element'. Discuss this statement as it applies to the Group IV elements, C, Si, Ge, Sn, Pb, indicating any properties of carbon which appear anomalous. Illustrate your answer by considering
(i)   the physical properties of the elements,
(ii) the reactions of the oxides with sodium hydroxide,
(iii) the reaction of the chlorides with water,
(iv) the stability of the hydrides to heat,
(v)   the changes in the stability of oxidation state IV with increase in relative atomic mass of the element.   [J.M.B.]

10.12   Find the element antimony Sb (atomic number 51) in the Periodic Table.

(a) Which two oxidation states would you expect this element to exhibit in its chlorides?

(b) Which of these chlorides would you expect to have the higher melting point?

(c) Explain your answer to (b).

(d) How does the thermal stability of SbH_3 compare with that of NH_3 and PH_3?

(e) How would the acidity of antimony(V) oxide compare with that of phosphorus(V) oxide?   [J.M.B.]

10.13   Using the hydrides, oxides, and chlorides as examples, describe and account for the variation in chemical properties across the first row elements Li, Be, B, C, N, O, and F.   [J.M.B.]

10.14   Compare and contrast the physical and chemical properties of the simple hydrides of each of the following elements:
sodium, calcium, carbon, nitrogen, oxygen, and fluorine.
In your answer attempt to relate these properties to the type of bonding encountered in these hydrides. [University of London]

10.15   Give explanations, with equations where appropriate, for the following statements.

(a)   Aluminium(III) chloride, when heated at atmospheric pressure, sublimes at 180 °C and melts at 193 °C under a pressure of 2.25 atm. Aluminium(III) fluoride, however, does not melt until 1290 °C, at atmospheric pressure.

(b)   Lead(II) chloride dissolves in concentrated hydrochloric acid to give a solution which, after treatment with chlorine in the cold, gives a yellow precipitate when ammonium chloride is added. A yellow oil is obtained when the yellow precipitate is added to cold, concentrated sulphuric(VI) acid.

(c)   Carbon dioxide and silicon(IV) oxide (silicon dioxide) are both covalent oxides, but carbon dioxide is a gas, whilst silicon(IV) oxide is a solid with an extremely high melting point.

(d)   In the gaseous state fluorine is extremely reactive whilst nitrogen is unreactive. [J.M.B.]

10.16   Discuss the types of bond which fluorine can form in inorganic compounds, giving one example of each type.
Compare and contrast the chemistry of fluorine and chlorine, giving reasons for any differences observed, by considering

(a)   the electrolyses of aqueous solutions of sodium fluoride and sodium chloride,

(b)   the reactions of chlorine with aqueous solutions of (i) sodium fluoride and (ii) sodium bromide,

(c)   the reactions of fluorine and chlorine with water,

(d)   the reactions of hydrogen fluoride gas with sodium fluoride and of hydrogen chloride gas with sodium chloride, both at room temperature. [J.M.B.]

10.17   Astatine (At, atomic number 85), the fifth member of the halogen group, exists only as short-lived radioactive iotopes and is not found in nature.
Use your knowledge of the properties of the halogens, and the trends in those properties, to predict the principal features of the chemistry of astatine and its compounds if a stable isotope were discovered. [University of London]

10.18   'Covalent substances tend to be volatile, of low melting point, soluble in non-aqueous solvents, but not in water, and non-electrolytes; whereas electro-valent substances tend to have the opposite properties'.
Discuss this statement critically, referring particularly to silicon(IV) oxide (SiO_2), hydrogen chloride (HCl), sodium chloride (NaCl), trichloromethane ($CHCl_3$), and ammonium chloride (NH_4Cl). [University of London]

10.19   Hydrogen resembles both the alkali metals and the halogens in its chemical behaviour. Survey the evidence which leads to this conclusion.
Use the compounds which you have described to illustrate the changes in bond-type of the hydrides of elements in a period of the Periodic Table. [J.M.B.]

10.20 (a) Describe and suggest reasons for the trends across and down the Periodic Table in (i) atomic radius, (ii) atomic volume and (iii) electronegativity.

(b)   Discuss the bonding in (i) sodium chloride and (ii) hydrogen chloride.

(c)   Describe and account for the bonding you would expect to be present in (i) rubidium chloride and (ii) iodine chloride (ICl). [J.M.B.]

10.21   Suggest with explanations the identities of A, B, C, D, and E from the following observations.

(a)   An element A reacts with hydrogen sulphide giving a deposit of sulphur. A also turns a solution of potassium iodide brown, but does not react with concentrated sulphuric (VI) acid.

(b)   A solution of a salt of a metal B deposits a white solid on boiling. When B is burnt in air and water is added to the residue, some ammonia is evolved.

(c)   A gaseous compound C forms a brown addition product with iron (II) sulphate (VI) and decolourises an acidified solution of potassium manganate (VII) (potassium permanganate).

(d)   An element D forms a solid chloride which is hydrolysed by water. The chloride readily forms co-ordination compounds with donor molecules such as ammonia.

(e)   An orange-red solid E leaves a purple-brown residue on treatment with dilute nitric (V) acid. A yellow green gas is evolved when the residue is heated with concentrated hydrochloric acid.

11 Some Elements of Industrial Importance

ALUMINIUM

Aluminium comprises about 8% of the Earth's crust and it is the most abundant metal. It occurs as the silicate(IV) in many clays and rocks but these sources are of little commercial value due to the dificulty in removing the silicon(IV) oxide (silica). The main ore is the hydrated oxide, $Al_2O_3 \cdot 2H_2O$, known as *bauxite*.

Extraction

The extraction is not as simple as may have been expected because the chloride is covalent (page 192), the oxide is unaffected by hydrogen, and the carbide is formed on heating the oxide with coke. It is obtained by electrolysis of the oxide dissolved in fused cryolite, Na_3AlF_6, a little fluorspar, CaF_2, being added to lower the operating temperature. Careful purification of the oxide is necessary since the aluminium itself cannot be purified.

The crushed ore is digested in hot sodium hydroxide solution, whereupon the aluminium oxide (alumina) dissolves as sodium aluminate(III) leaving impurities such as titanium(IV) and iron(III) compounds to be filtered off:

$$Al_2O_3(s) + 2NaOH(aq) + 3H_2O(l) \rightarrow 2NaAl(OH)_4(aq)$$

The filtrate is cooled and the aluminate(III) hydrolysed by diluting with water and adding a little freshly prepared aluminium hydroxide:

$$[Al(OH)_4]^-(aq) + 3H_2O(l) \rightleftharpoons Al(OH)_3(H_2O)_3(s) + HO^-(aq)$$

The aluminium hydroxide is filtered, washed, dried, and heated to give the oxide, whilst the sodium hydroxide is recirculated

$$2Al(OH)_3(s) \rightarrow Al_2O_3(s) + 3H_2O(g)$$

The alumina, cryolite, and fluorspar are then electrolysed at about 1000 °C using graphite electrodes, the temperature being maintained by the passage of the current. The reactions occurring may be represented as:

$$Al^{3+} + 3e^- \rightarrow Al \quad \text{at the cathode and}$$
$$2O^{2-} \rightarrow O_2 + 4e^- \quad \text{at the anode}$$

but this is a simplification of a complex process.

The extraction is generally limited to places where cheap electricity, i.e. hydroelectric power, is available. The cost of the process is increased by the liberated oxygen fairly rapidly oxidising the graphite anode to carbon dioxide.

Properties

Aluminium is a low-density, hard metal which is malleable and ductile. Its electronic structure is 2.8.3. and its successive ionisation energies are I_1 580, I_2 1800, and I_3 2700 kJ mol^{-1}. Since the energy required to form Al^{3+} is very high and this ion is very small, the compounds of aluminium with the exception of the oxide, fluoride, and sulphate(VI), are predominantly covalent.

Aluminium has a standard electrode potential of -1.66 V and this suggests that it would react readily with dilute acids, evolving hydrogen. However, it rapidly forms a protective layer of oxide on exposure to moist air and this inhibits attack by water and dilute sulphuric(VI) acid. It does dissolve in warm dilute hydrochloric acid, in concentrated hydrochloric acid, and in hot concentrated sulphuric(VI) acid.

$$2Al(s) + 6HCl(aq) \rightarrow 2AlCl_3(aq) + 3H_2(g)$$
$$2Al(s) + 6H_2SO_4(conc) \rightarrow Al_2(SO_4)_3(aq) + 6H_2O(l) + 3SO_2(g)$$

Nitric(V) acid oxidises the surface of the metal to the oxide and so renders it passive (cf. iron).

Aluminium dissolves in warm sodium hydroxide solution giving the aluminate(III) and hydrogen:

$$2Al(s) + 2NaOH(aq) + 6H_2O(l) \rightarrow 2NaAl(OH)_4(aq) + 3H_2(g)$$

Aluminium reacts with the halogens, oxygen, sulphur, etc. on heating. Its strong attraction for oxygen is utilised in the *thermite process*. Thus, a mixture of powdered iron(III) oxide and aluminium is ignited with a magnesium flare, whereupon the iron oxide is reduced:

$$Fe_2O_3(s) + 2Al(s) \rightarrow 2Fe(s) + Al_2O_3(s)$$

The heat of the reaction melts the iron and so the process is used for welding steel *in situ*. Manganese, chromium, and molybdenum are obtained from their oxides by similar reactions.

Uses

The strength and lightness of the metal results in it having great utility. Aluminium and its compounds are non-toxic and so it is used for milk bottle tops, wrapping foodstuffs, and for making cooking utensils. It and its alloys are used in constructional work, and to make car engines and many aircraft parts. Aluminium is used for making transmission lines because these are lighter than copper ones, and so fewer pylons are required. It is also employed in making non-tarnishing mirrors and anti-corrosion paints.

The corrosion resistance of aluminium may be increased by a process known as *anodising*. The aluminium item is made the anode of a cell in which a solution of sodium phosphate(V), chromic(VI) acid, or sulphuric(VI) acid is the electrolyte. The surface of the aluminium is oxidised by the liberated oxygen, and so the thickness of the protective film of the oxide is increased.

IRON

Iron is the second most abundant metal and occurs chiefly as different forms of its oxides, e.g. haematite, Fe_2O_3, magnetite or loadstone, Fe_3O_4, and limonite, $Fe_2O_3 \cdot H_2O$. Other common ores are iron pyrites, FeS_2 and siderite, $FeCO_3$.

Extraction

Iron is extracted by reduction of its oxide in a blast furnace. This *smelting* process yields the crude metal which is subsequently refined. Before carbonate and sulphide ores can be smelted, they must be converted to the oxide by roasting in air:

$$4FeCO_3(s) + O_2(g) \rightarrow 2Fe_2O_3(s) + 4CO_2(g)$$
$$4FeS_2(s) + 11O_2(g) \rightarrow 2Fe_2O_3(s) + 8SO_2(g)$$

Iron ore, coke, and limestone are admitted in alternate layers through the top of the furnace and hot air is forced in at the bottom. Combustion of the coke at the bottom raises the temperature to about 1500 °C, but this falls to about 200 °C at the top. The major reaction in the hottest parts of the furnace is reduction of the ore by carbon:

$$Fe_2O_3(s) + 3C(s) \rightarrow 2Fe(s) + 3CO(g)$$
$$Fe_3O_4(s) + 4C(s) \rightarrow 3Fe(s) + 4CO(g)$$

but carbon monoxide is the reducing agent in the cooler parts:

$$Fe_2O_3(s) + 3CO(g) \rightarrow 2Fe(s) + 3CO_2(g)$$
$$Fe_3O_4(s) + 4CO(g) \rightarrow 3Fe(s) + 4CO_2(g)$$

Molten iron sinks to the bottom of the furnace and is run off at intervals.

The limestone decomposes to yield calcium oxide and carbon dioxide:

$$CaCO_3(s) \rightarrow CaO(s) + CO_2(g)$$

The carbon dioxide is subsequently reduced to the monoxide by carbon while the calcium oxide reacts with silicon(IV) oxide (silica) and other acidic oxides to form a slag:

$$CaO(s) + SiO_2(s) \rightarrow CaSiO_3(s)$$

The slag floats on the molten iron and prevents it from being oxidised.

The gas escaping at the top of the furnace contains almost 30% carbon monoxide and so it is burnt and used to heat the air for the air blast.

The iron which is run off is allowed to solidify in moulds, and is known as *pig iron*. Pig iron is about 95% pure, the main impurity being carbon, but smaller amounts of phosphorus, silicon, and manganese are also present.

Conversion of iron into steel

Several processes are in use for converting iron into steel, but they all have the same basic principle, which is, to remove all the impurities from molten pig iron by oxidation, and then add whatever elements are needed to give steel of the required specification.

In the Bessemer process the converter is lined according to the phosphorus content of the pig iron. A silicon(IV) oxide (silica) lining (acid lining) is used for pig iron with a very low phosphorus content whilst a basic lining of calcined dolomite ($MgCO_3 \cdot CaCO_3$) is used when the phosphorus content is appreciable. In the 'acid process' molten, low phosphorus content pig iron is introduced into the converter and oxygen is blown through. The manganese and silicon are converted into their oxides and form a slag on the molten metal. The converter does not need to be heated because the oxidations are sufficiently exothermic to keep the iron molten even though the melting point of the iron increases as its purity increases (see depression of the freezing point, page 92). The carbon present is oxidised to the monoxide which burns as it leaves the converter. When the oxidations are near completion, a Fe—C—Mn alloy is added to de-oxidise the steel and to bring the carbon content to the desired level. The slag is skimmed off and the steel poured into moulds. In conversions with the basic lining, the oxygen blast converts silicon and phosphorus to silicon(IV) oxide (silicon dioxide) and phosphorus(V) oxide (phosphorus pentoxide) respectively. Calcium oxide is added to react with these acidic oxides and a basic slag of calcium silicate(IV) and calcium phosphate(V) is produced. The slag is used as a fertiliser.

Types of iron and steel

Some pig iron is remelted along with scrap steel and then cooled in moulds. The product is cast iron; it is used where cheapness is more important than strength, e.g. drain pipes. Cast iron is brittle because it contains up to about 3.5% of carbon in the form of graphite between the crystals.

If almost all the impurities are removed from pig iron, the product is wrought iron. It is soft, but can be easily forged and is used for decorative work.

Steels are alloys of iron and carbon — there is no free graphite present. The amount of carbon present varies, between 0.1 and 1.5%, and it determines the hardness of the steel. Mild steel contains about 0.1% carbon and is ductile and rather similar to wrought iron, whereas high carbon steel contains up to about 1.5%, and it can be hardened and used for making tools, etc. The hardening is done by plunging the red hot steel into cold water.

The elements in the first transition series all have similar atomic volumes (see page 172) and so they readily alloy with one another. The properties of steel can be modified appreciably by alloy formation. Thus, stainless steel contains up to about 18% chromium or chromium and nickel, and is very resistant to corrosion. Vanadium and tungsten make the steel very hard and suitable for high-speed drills, etc. whilst a little manganese imparts elasticity and high tensile strength.

Physico-chemical principles of the blast furnace process

It was seen in Chapter 4 that in many reactions there is

(a)   an increase in the degree of disorder, i.e. ΔS is positive, and

(b)   an evolution of heat, i.e. ΔH is negative.

Now, these factors may be linked by the equation

$$\Delta G = \Delta H - T\Delta S \qquad (11.1)$$

where ΔG is the change in *free energy*. The temperature term has to be introduced with ΔS because ΔS is measured in units of $J\,K^{-1}$ whereas the units for ΔG and ΔH are J. Thus, multiplying ΔS by T gives all the quantities in joules.

In general, reactions proceed if ΔH is negative and ΔS is positive and so it is seen from equation (11.1) that this would require ΔG to be negative. Note that a negative value of ΔG does not necessarily mean that the reaction is practicable — the rate may be extremely slow.

Considering the reaction

$$Fe(s) + \tfrac{1}{2}O_2(g) \rightarrow FeO(s)$$

$\Delta G^{\ominus} = -244\ kJ\,mol^{-1}$. This means that to separate one mole of iron(II) oxide at 25 °C into its elements would require 244 $kJ\,mol^{-1}$ of free energy and so to obtain iron from iron(II) oxide, the reducing agent must be capable of supplying this.

Using carbon as the reducing agent, the corresponding free energy change is

$$C(s) + \tfrac{1}{2}O_2(g) \rightarrow CO(g); \qquad \Delta G^{\ominus} = -137\ kJ\,mol^{-1}$$

and so the reduction is not possible under standard conditions. However, the free energy changes alter with the temperature. Thus, at 900 °C

$$C(s) + \tfrac{1}{2}O_2(g) \rightarrow CO(g); \qquad \Delta G = -217\ kJ\,mol^{-1} \qquad (11.2)$$
$$Fe(s) + \tfrac{1}{2}O_2(g) \rightarrow FeO(s); \qquad \Delta G = -197\ kJ\,mol^{-1} \qquad (11.3)$$

Subtracting equation (11.3) from (11.2) gives

$$C(s) + FeO(s) \rightarrow Fe(s) + CO(g); \qquad \Delta G = -217 - (-197)$$
$$= -20\ kJ\,mol^{-1}$$

Therefore, provided that the activation energy (page 156) is not so high as to make the rate very slow, reduction of iron(II) oxide by carbon is feasible at 900 °C.

The variation of free energy changes with temperature has great relevance to extraction processes and the graphical representation of these changes is known as an Ellingham diagram. The Ellingham diagram for carbon monoxide, carbon dioxide, and iron(II) oxide is shown in Figure 11.1. It is seen that ΔG for carbon becomes more negative as the temperature is increased and so its efficiency as a reducing agent increases at higher temperatures. Carbon will reduce iron(II) oxide if the free energy of formation of carbon monoxide is more negative than the free energy of formation of the iron(II) oxide.

Figure 11.1 shows that at temperatures of up to 650 °C (the point where the

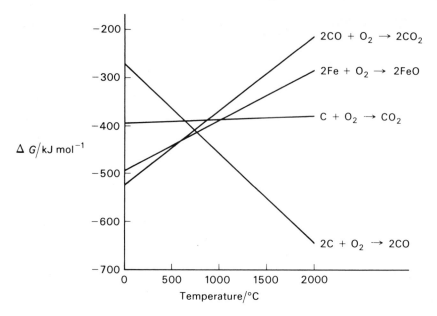

Figure 11.1   Ellingham diagram of some oxides

line for ΔG_f of carbon dioxide from carbon monoxide cuts the line for ΔG_f of iron(II) oxide) carbon monoxide will reduce iron(II) oxide, whilst above 770 °C, carbon will effect the reduction, the reduction being

$$FeO(s) + C(s) \rightarrow Fe(s) + CO(g)$$

The reduction of iron oxides in the blast furnace is a complex process. However, it is likely that stepwise reduction occurs with the reduction of iron(II) oxide being the final stage. The foregoing discussion illustrates the factors which affect these reductions.

Corrosion of iron

Iron is slowly attacked by moist air, the product being rust or hydrated iron(III) oxide, $Fe_2O_3 \cdot xH_2O$. Both moisture and oxygen are essential for rust to form and the process is catalysed by carbon dioxide.

Iron dissolves in dilute and concentrated hydrochloric acid and in dilute sulphuric(VI) acid giving the corresponding iron(II) salt and hydrogen. With dilute nitric(V) acid the reaction is:

$$4Fe(s) + 10HNO_3(aq) \rightarrow 4Fe(NO_3)_2(aq) + NH_4NO_3(aq) + 3H_2O(l)$$

Hot concentrated sulphuric(VI) acid undergoes reduction with iron:

$$2Fe(s) + 6H_2SO_4(conc) \rightarrow Fe_2(SO_4)_3(aq) + 6H_2O(l) + 3SO_2(g)$$

Concentrated nitric(V) acid renders iron 'passive' due to the formation of an impervious film of iron(II) diiron(III) oxide (ferrosoferric oxide), Fe_3O_4. Phosphoric(V) acid also renders iron 'passive'. Other ways of preventing corrosion are alloying the iron with other transition metals (see stainless steel above), galvanising with zinc (page 143) or tin plating.

SULPHUR

It was noted on page 189 that the first element in each group tends to exhibit some properties different to the rest of the group because the atoms are very small. These differences are particularly evident in Group VI between oxygen and sulphur. Thus, oxygen exists as diatomic molecules and is a gas whereas sulphur exists as S_8 molecules and is a solid. Oxygen has very little tendency to form O—O bonds and, even when it does in trioxygen (ozone) and peroxides, they are readily broken by heat. Sulphur, on the other hand, readily forms chains of S—S bonds and these are stable even at fairly high temperatures. The relative stabilities of O—O and S—S bonds are indicated by the bond enthalpies (energies) which are 146 (in hydrogen peroxide) and 250 kJ mol^{-1} respectively.

Oxygen exists as dioxygen, O_2 and trioxygen (ozone), O_3, but there are four allotropes of sulphur: rhombic, monoclinic, plastic, and amorphous sulphur. Both oxygen and sulphur can form covalent and ionic bonds. However, oxygen is almost invariably divalent (although it is trivalent in the oxonium ion, H_3O^+) whilst sulphur can have other valencies because it has available d orbitals. An oxidation chart for sulphur is shown in Table 11.1.

Table 11.1. Oxidation chart for sulphur

Oxidation number	Examples
+7	Peroxodisulphuric(VI) acid (perdisulphuric acid), $H_2S_2O_8$, and its salts
+6	SF_6, H_2SO_4, $SO_4{}^{2-}$
+4	SO_2, H_2SO_3, $SO_3{}^{2-}$
+2	SCl_2, thiosulphates(VI), i.e. $S_2O_3{}^{2-}$
0	S_8
−2	H_2S, S^{2-}

Before sulphur can exhibit a valency of four, it is necessary for one of the $3p$ electrons to be promoted to the $3d$ level. For example, sulphur dioxide is formed as follows:

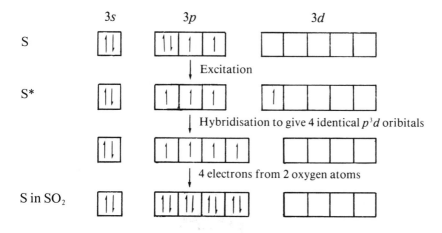

Similarly, the valency of 6 is a result of promotion of an *s* as well as the *p* electron to the *d* level:

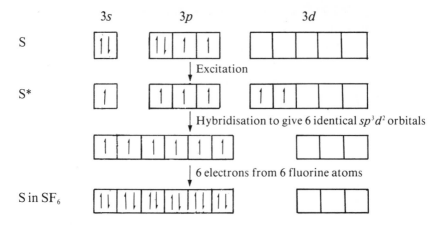

The interconversions between some of the oxidation states are illustrated by the reactions below.

Hydrogen sulphide is readily oxidised by MnO_4^-, $Cr_2O_7^{2-}$, H_2O_2, etc., the oxidation number changing from –2 to zero in the process:

$$2MnO_4^-(aq) + 6H^+(aq) + 5H_2S(g) \rightarrow 5S(s) + 2Mn^{2+}(aq) + 8H_2O(l)$$

Sulphates(IV) (sulphites) are oxidised by similar reagents:

$$5SO_3^{2-}(aq) + 2MnO_4^-(aq) + 6H^+(aq) \rightarrow 5SO_4^{2-}(aq) + 2Mn^{2+}(aq) + 3H_2O(l)$$

These salts may be titrated with iodine as well as with the above reagents, the oxidation number again changing from $+4$ to $+6$:

$$SO_3^{2-}(aq) + I_2(aq) + H_2O(l) \rightarrow SO_4^{2-}(aq) + 2H^+(aq) + 2I^-(aq)$$

With cerium(IV) salts, however, the reaction is rather different since both sulphate(VI) and dithionate, $S_2O_6^{2-}$ are formed. This may be the result of the following sequence of reactions:

$$SO_3^{2-}(aq) + H^+(aq) \rightleftharpoons HSO_3^-(aq)$$
$$HSO_3^-(aq) + Ce^{4+}(aq) \rightarrow HSO_3(aq) + Ce^{3+}(aq)$$

$$2HSO_3(aq) \underset{\text{disproportionation}}{\Big\langle} \begin{array}{l} \longrightarrow 2H^+(aq) + S_2O_6^{2-}(aq) \\ \\ \longrightarrow SO_2(g) + SO_4^{2-}(aq) + 2H^+(aq) \end{array}$$

(*Disproportionation* means that simultaneous oxidation and reduction of the species occurs.) In the case of the iodine reaction, no dithionate is found in the final products — it is easily oxidised by iodine.

Boiling solutions of sulphates(IV) (sulphites) are oxidised by sulphur, the product being a thiosulphate(VI):

$$S(s) + SO_3^{2-}(aq) \rightarrow S_2O_3^{2-}(aq)$$

Thiosulphates(VI) are oxidised by iodine to tetrathionates:

$$2S_2O_3^{2-}(aq) + I_2(aq) \rightarrow 2I^-(aq) + S_4O_6^{2-}(aq)$$

the change in oxidation number being from $+2$ to $+2.5$.

Acidification of a thiosulphate(VI) leads to disproportionation:

$$S_2O_3^{2-}(aq) + 2H^+(aq) \rightarrow S(s) + H_2O(l) + SO_2(aq)$$

Hot concentrated sulphuric(VI) acid oxidises carbon, sulphur, copper, etc. and is itself reduced to sulphur dioxide:

$$C(s) + 2H_2SO_4(conc) \rightarrow 2H_2O(l) \quad + CO_2(g) \quad + 2SO_2(g)$$
$$S(s) + 2H_2SO_4(conc) \rightarrow 2H_2O(l) \quad + 3SO_2(g)$$
$$Cu(s) + 2H_2SO_4(conc) \rightarrow CuSO_4(aq) + 2H_2O(l) \quad + SO_2(g)$$

The acid reacts with alkali metal fluorides and chlorides giving the corresponding halogen acid, but it is reduced by bromides and iodides:

$$SO_4^{2-} + 2Br^- + 4H^+ \rightarrow 2H_2O + Br_2 + SO_2$$

Electrolysis of fairly concentrated sulphuric(VI) acid at $0\,°C$ gives peroxodisulphuric(VI) acid (perdisulphuric acid):

$$H-\overset{\overset{\displaystyle O}{\|}}{\underset{\underset{\displaystyle O}{\|}}{\ddot{O}-S}}-\ddot{O}-\ddot{O}-\overset{\overset{\displaystyle O}{\|}}{\underset{\underset{\displaystyle O}{\|}}{S}}-\ddot{O}-H$$

The peroxodisulphates(VI) (perdisulphates or persulphates) are used as oxidising agents. They oxidise Mn^{2+} ions to manganese(IV) oxide:

$$Mn^{2+}(aq) + S_2O_8^{2-}(aq) + 2H_2O(l) \rightarrow MnO_2(s) + 2SO_4^{2-}(aq) + 4H^+(aq)$$

Using a trace of Ag^+ as catalyst, however, the Mn^{2+} is oxidised to a manganate(VII) (permanganate):

$$2Mn^{2+}(aq) + 5S_2O_8^{2-}(aq) + 8H_2O(l) \rightarrow 2MnO_4^-(aq) + 16H^+(aq) + 10SO_4^{2-}(aq)$$

In both of these reactions, the $+7$ oxidation number of the sulphur in the $S_2O_8^{2-}$ ion falls to $+6$ in the SO_4^{2-} ion:

$$S_2O_8^{2-} + 2e^- \rightarrow 2SO_4^{2-}$$

Manufacture and uses of sulphuric(VI) acid

Sulphur dioxide is made by burning sulphur in air:

$$S(s) + O_2(g) \rightarrow SO_2(g)$$

or by roasting various metal sulphides (zinc blende, galena, and iron pyrites) in air as part of their extraction processes:

$$2ZnS(s) + 3O_2(g) \rightarrow 2ZnO(s) + 2SO_2(g)$$
$$2PbS(s) + 3O_2(g) \rightarrow 2PbO(s) + 2SO_2(g)$$
$$4FeS_2(s) + 11O_2(g) \rightarrow 2Fe_2O_3(s) + 8SO_2(g)$$

The sulphur dioxide is then purified and converted into the trioxide by the contact process:

$$2SO_2(g) + O_2(g) \xrightarrow[\text{slight pressure}]{\text{Pt or } V_2O_5/450\,°C} 2SO_3(g)$$

The physico-chemical principles of the process are discussed on page 117. The yield of sulphur(VI) oxide is in excess of 95%; this could be raised by increasing the pressure, but the higher yields do not justify the large increase in cost of building high pressure plant. The reaction between sulphur(VI) oxide and water is very vigorous and it is found more convenient to absorb the gas in concentrated sulphuric(VI) acid and then dilute this with water to the required concentration.

$$H_2SO_4(conc) + SO_3(g) \rightarrow H_2S_2O_7(l)$$
$$H_2S_2O_7(l) + H_2O(l) \rightarrow 2H_2SO_4(conc)$$

A large amount of sulphuric(VI) acid is used in the fertilizer industry. Thus, 'superphosphate' is made by adding the concentrated acid to calcium phosphate(V):

$$Ca_3(PO_4)_2 + 2H_2SO_4 \rightarrow \underbrace{Ca(H_2PO_4)_2 + 2CaSO_4}_{\text{'superphosphate'}}$$

while reaction with ammonia gives ammonium sulphate(VI):

$$2NH_3(aq) + H_2SO_4(aq) \rightarrow (NH_4)_2SO_4(aq)$$

Other large scale uses include the manufacture of detergents, dyestuffs, drugs, explosives and rayon (artificial silk). It is also used in accumulators and for cleaning up metal before galvanising, tin plating, etc.

NITROGEN

The chemistry of the element was discussed in the previous chapter but two important nitrogen compounds, i.e. nitric(V) acid and ammonia, will be considered here.

The structure of the acid and the shape of the ammonia molecule are discussed on pages 65 and 68.

Redox reactions involving nitric(V) acid

The acid is a powerful oxidising agent and in this type of reaction it may be reduced to a variety of products depending upon the pH and the nature of the substance being oxidised. The reduction products may be nitrogen dioxide, nitrogen oxide (nitric oxide), dinitrogen oxide (nitrous oxide), nitrogen, hydroxylamine (NH_2OH), or ammonia.

Addition of the concentrated acid to copper leads to the evolution of nitrogen dioxide:

$$2NO_3^- + 2e^- + 4H^+ \rightarrow 2H_2O + 2NO_2$$
$$(Cu(s) + 4HNO_3(conc) \rightarrow Cu(NO_3)_2(aq) + 2H_2O(l) + 2NO_2(g))$$

but with 50% nitric(V) acid, it is nitrogen oxide (nitric oxide) which is produced:

$$2NO_3^- + 6e^- + 8H^+ \rightarrow 4H_2O + 2NO$$
$$(3Cu(s) + 8HNO_3(aq) \rightarrow 3Cu(NO_3)_2(aq) + 4H_2O(l) + 2NO(g))$$

Further dilution of the acid leads to the formation of some dinitrogen oxide (nitrous oxide).

Reaction with zinc gives a variety of products depending upon the temperature and concentration of the acid. With cold dilute acid, the main reduction product is ammonium nitrate(V)

$$NO_3^- + 8e^- + 10H^+ \rightarrow NH_4^+ + 3H_2O$$
$$(4Zn(s) + 10HNO_3(aq) \rightarrow 4Zn(NO_3)_2(aq) + 3H_2O(l) + NH_4NO_3(aq))$$

If a very dilute (less than 2%) nitric(V) acid is added to magnesium, hydrogen is formed:

$$Mg(s) + 2HNO_3(aq) \rightarrow Mg(NO_3)_2(aq) + H_2(g)$$

and so it is the H^+ not the NO_3^- ions which are reduced here.

Heating a nitrate(V) with sodium hydroxide solution and Devarda's alloy (roughly equal amounts of aluminium and copper with a little zinc) gives a quantitative yield of ammonia:

$$8Al(s) + 3NO_3^-(aq) + 5HO^-(aq) + 18H_2O(l) \rightarrow 8Al(OH)_4^-(aq) + 3NH_3(g)$$

(see page 192 re the formula of the aluminate(III) ion). Hydrogen also is produced owing to reaction between the aluminium and alkali:

$$2Al(s) + 2HO^-(aq) + 6H_2O(l) \rightarrow 2Al(OH)_4^-(aq) + 3H_2(g)$$

Nitric(V) acid oxidises hydrogen sulphide to sulphur:

$$2HNO_3(conc) + H_2S(g) \rightarrow S(s) + 2H_2O(l) + 2NO_2(g)$$

If the resultant mixture is boiled, the sulphur is slowly oxidised to sulphuric(VI) acid:

$$S(s) + 6HNO_3(conc) \rightarrow H_2SO_4(aq) + 2H_2O(l) + 6NO_2(g)$$

The acid oxidises iodides to iodine and is itself reduced to nitric(III) acid (nitrous acid):

$$HNO_3(conc) + 2HI(aq) \rightarrow I_2(s) + HNO_2(aq) + H_2O(l)$$

Nitric(V) acid oxidises the metal in some compounds to higher valency states, e.g.

$$3Fe^{2+} + NO_3^- + 4H^+ \rightarrow 3Fe^{3+} + 2H_2O + NO$$
$$(6FeSO_4(aq) + 3H_2SO_4(aq) + 2HNO_3(conc) \rightarrow 3Fe_2(SO_4)_3(aq) + 4H_2O(l) + 2NO(g))$$

The manufacture of ammonia

Ammonia is manufactured by the Haber process; this utilises nitrogen from the atmosphere and hydrogen from the petroleum industry. The gases react at 550 °C, and pressures of up to 1000 atmospheres, over a finely divided iron catalyst:

$$N_2(g) + 3H_2(g) \rightleftharpoons 2NH_3(g)$$

The yield of ammonia is in the region of 15% and this is condensed out and the unchanged gases recycled. The physico-chemical principles of the process are discussed on page 117. Note that this is another process where economics override the physical requirements for optimum yield.

Huge quantities of hydrogen are produced as by-product from cracking processes, and so it is now usual to manufacture ammonia at or near oil refineries to save on transport costs.

Nitrogen compounds in nutrition, agriculture, and industry

Nitrogen is an essential constituent of all living matter, both animal and vegetable; it is present in the form of proteins (page 357). Soluble nitrogen compounds are therefore important fertilizers, and so vast quantities of ammonium salts such as the sulphate(VI), nitrate(V) and phosphate(V) are made. Not all fertilizers are nitrogen compounds, however, and substances such as 'superphosphate' (page 236) and 'potash' are necessary to give a balanced fertilizer. ('Potash' or potassium carbonate is manufactured by passing carbon dioxide through potassium hydroxide solution, evaporating the resulting solution, and then heating the residue:

$$KOH(aq) + CO_2(g) \rightarrow KHCO_3(aq)$$
$$2KHCO_3(s) \rightarrow K_2CO_3(s) + H_2O(g) + CO_2(g)$$

As well as being used to make fertilizers, a large amount of ammonia is used to manufacture nitric(V) acid by the Ostwald process. The ammonia is oxidised by air by rapid passage over a platinum-rhodium catalyst at 900 °C.

$$4NH_3(g) + 5O_2(g) \rightarrow 6H_2O(g) + 4NO(g)$$

The gases are cooled, mixed with more air to convert the nitrogen oxide (nitric oxide) to nitrogen dioxide, and then passed into water:

$$2NO(g) + O_2(g) \rightarrow 2NO_2(g)$$
$$H_2O(l) + 3NO_2(g) \rightarrow 2HNO_3(aq) + NO(g)$$

The nitrogen oxide produced subsequently combines with more oxygen and water to give more acid.

Nitric(V) acid is used to nitrate aromatic hydrocarbons (page 308); the resultant nitro compounds are reduced to amines for use in the dye industry (see coupling reactions of diazonium salts, page 344). It is also used for making fertilizers, explosives, and as an oxidising agent to prepare the starting materials for making nylon and terylene (pages 361 and 362).

Questions

11.1   Describe how sulphur dioxide is obtained industrially from either anhydrite (calcium sulphate(VI)) or iron pyrites and discuss the essential reactions involved in the conversion of sulphur dioxide into sulphuric(VI) acid by the Contact process.

Describe and explain the reaction between sulphuric(VI) acid and

(a) sodium bromide,

(b) copper,

(c) sulphur. [University of London]

11.2 Give an account of the chemical principles involved in the manufacture of iron in a blast furnace and its subsequent conversion into steel. Structures made of steel are often protected from rusting by 'galvanising' with zinc. Explain the principles underlying this method of rust prevention.

[University of London]

11.3 'Nitric(V) acid can behave in various ways according to the conditions under which it is used'. Illustrate this statement by classifying the reactions of nitric(V) acid and taking examples from all branches of chemistry.

Give the electronic structure of nitric(V) acid and comment on it.

[University of London]

11.4 With reference to either sulphuric(VI) acid or ammonia:

(a) outline the major physico-chemical principles underlying the industrial manufacture of the chosen compound from suitable starting materials;

(b) discuss the important uses of the substance chosen in industry and related areas. [University of London]

11.5 For EITHER aluminium OR iron:

Give a brief account of the occurrence of the metal or its compounds in nature, and of the method by which the metal is extracted from its ores.

Summarise the advantages and disadvantages of the metal in everyday use. Describe with essential practical details how you would convert the metal into the chloride in which it has oxidation number $+3$. Certain physical properties of the metal chloride are set out at the end of the question. What conclusions can you draw from them about the chloride of the metal you have chosen?

	Aluminium chloride	*Iron(III) chloride*
Melting	190 °C	290 °C
Boiling point	420 °C	Not given in tables
	Soluble in benzene	Soluble in benzene
Relative molecular mass	267 in benzene, and also in vapour just above b.p.	324 in benzene 167 in ethanol
pH of aqueous solution	less than 7	less than 7

[University of London]

11.6 EITHER

The principal oxidation states of sulphur are -2, $+2$, $+4$, and $+6$. Give the names and formulae of four compounds containing sulphur, one in each of these oxidation states. Describe briefly how three of the compounds you mention can be obtained from elemental sulphur.

Outline how a named compound with sulphur in its lowest oxidation state may be converted into one with sulphur in its highest oxidation state.

OR

Four common oxidation states of nitrogen are -3, $+3$, $+4$, and $+5$. Give the names and formulae of four compounds containing nitrogen, one in each of these oxidation states.

Describe briefly how three of the compounds you mention could be conveniently prepared in the laboratory.

Describe how you would obtain a specimen of nitrogen from one of the compounds you have discussed. [University of London]

11.7   EITHER

Describe the Contact process for the manufacture of sulphuric acid from sulphur, paying particular attention to the physico-chemical principles involved.

Mention briefly three industrial processes in which sulphuric acid plays a part, indicating in each case the particular role which the acid plays.

OR

Describe the Haber process for the manufacture of ammonia, paying particular attention to the sources of raw materials and to the physico-chemical principles involved.

Mention briefly three important uses of nitrogen-containing compounds, indicating in each case why they are important.

[University of London]

12 The First Row Transition Elements

The periodic table is built up by electrons gradually filling orbitals of increasing energy. Thus, from hydrogen to calcium, the orbitals are progressively filled in the order 1s, 2s, 2p, 3s, 3p, 4s so that calcium has the electronic structure Ca, $1s^22s^22p^63s^23p^64s^2$. Now the next lowest energy orbital is the 3d (see page 16) and so scandium has the electronic configuration $1s^22s^22p^63s^23p^63d^14s^2$. The 3d orbital is complete with zinc, i.e. Zn, $1s^22s^22p^63s^23p^63d^{10}4s^2$ and the 4p orbital then starts to fill, i.e. Ga, $1s^22s^22p^63s^23p^63d^{10}4s^24p^1$. In the series scandium to zinc an inner shell (the 3d) is being filled up after an outer one (the 4s) has been started and the elements involved are known as *transitional elements*. The first row transition elements and their outer electronic configurations are given below.

Element	Symbol	Atomic number (Z)	Outer electronic structure
Scandium	Sc	21	$3d^1\ 4s^2$
Titanium	Ti	22	$3d^2\ 4s^2$
Vanadium	V	23	$3d^3\ 4s^2$
Chromium	Cr	24	$3d^5\ 4s^1$
Manganese	Mn	25	$3d^5\ 4s^2$
Iron	Fe	26	$3d^6\ 4s^2$
Cobalt	Co	27	$3d^7\ 4s^2$
Nickel	Ni	28	$3d^8\ 4s^2$
Copper	Cu	29	$3d^{10}4s^1$
Zinc	Zn	30	$3d^{10}4s^2$

The irregularities in- the electronic structures of chromium and copper occur because full and half-full orbitals are more stable, i.e. have a little less energy, than other arrangements.

The elements in this transition series are known as *d*-block elements, because *d* electrons are involved in the formation of some of their compounds.

Relevant physical properties are:

	Sc	Ti	V	Cr	Mn	Fe	Co	Ni	Cu	Zn
M.p./°C	1400	1677	1917	1903	1244	1539	1495	1455	1083	420
B.p./°C	2477	3277	3377	2642	2041	2887	2877	2837	2582	908
Covalent radii/nm	0.144	0.132	0.122	0.117	0.117	0.116	0.116	0.115	0.117	0.125
Density g cm^{-3}	3.0	4.5	6.1	7.2	7.4	7.9	8.9	8.9	8.9	7.1
I_1/kJ mol^{-1}	630	660	650	650	720	760	760	740	750	910
M^{2+} ionic radii/nm	—	0.090	0.088	0.088	0.080	0.076	0.074	0.072	0.069	0.074
M^{3+} ionic radii/nm	0.081	0.076	0.074	0.063	0.066	0.064	0.063	0.062	—	—
$E^{\ominus}M^{2+}/M$ /volts	*-2.08	-1.63	-1.2	-0.56	-1.19	-0.44	-0.28	-0.25	+0.34	-0.76
*Sc^{3+}/Sc										

With the exception of copper, the metals in the first transition series are silvery white. Their covalent radii decrease from scandium to nickel because the nuclear charge increases but the extra electrons enter the same shell. Hence the outer electrons are progressively pulled in closer to the nucleus. The small increase in the radius of the copper atom, and the larger increase for zinc, reflect the increased shielding of the outer s electrons from the nucleus as the d orbitals are completed. The changes in atomic radius and hence atomic volume account for the variation in density. The high melting points suggest that the $3d$ electrons as well as the $4s$ are delocalised in the crystal lattice. Zinc's melting point is much lower because its d orbitals are full and these electrons are not involved in bond formation. Alloy formation is common (e.g. brass is an alloy of copper and zinc) since the atoms are similar in size and character.

The $E^{\ominus}$ values show that all the metals except copper (positive $E^{\ominus}$) are more easily oxidised than hydrogen and so they should displace hydrogen from hydrochloric and dilute sulphuric(VI) acids. Furthermore, zinc should reduce Fe^{2+}, Co^{2+}, Ni^{2+}, Cr^{2+} and Cu^{2+} to the metal but not Sc^{3+}, Ti^{2+}, V^{2+}, or Mn^{2+}.

Their oxides tend to be amphoteric and are, in fact, sometimes entirely acidic, e.g. CrO_3. Bonding is often predominantly covalent, e.g. iron(III) chloride, Fe_2Cl_6, and tetracarbonylnickel(0), $Ni(CO)_4$, are covalent.

Transition metals have a number of characteristic properties as outlined below.

1 Variable valency or oxidation states

The variety of oxidation number or states can be illustrated by the oxides:

Element	Oxides	Oxidation numbers
Sc	Sc_2O_3	$+3$
Ti	TiO, Ti_2O_3, TiO_2	$+2$, $+3$, $+4$
V	VO, V_2O_3, VO_2, V_2O_5	$+2$, $+3$, $+4$, $+5$
Cr	CrO, Cr_2O_3, CrO_3	$+2$, $+3$, $+6$
Mn	MnO, Mn_2O_3, MnO_2, K_2MnO_4, $KMnO_4$	$+2$, $+3$, $+4$, $+6$, $+7$
Fe	FeO, Fe_2O_3	$+2$, $+3$
Co	CoO, Co_2O_3	$+2$, $+3$
Ni	NiO, Ni_2O_3, NiO_2	$+2$, $+3$, $+4$
Cu	Cu_2O, CuO	$+1$, $+2$
Zn	ZnO	$+2$

It is seen that the maximum oxidation number rises to a peak on going from scandium to manganese, and then falls. The reason for the variable valency and oxidation numbers is that these elements have $4s$ and $3d$ electrons all of similar energy and available for bonding purposes. The maximum oxidation number is never greater than the total number of $3d$ and $4s$ electrons, since further electrons would have to come from the argon core and this would require very high energy. In general, covalent character increases as the oxidation number increases. The higher oxidation numbers are formed with more reluctance on traversing the row because the ionisation energies of the d and s electrons increase with atomic number. (The electrons are in the same shells, but the nuclear charge becomes greater.)

The transition elements have high catalytic activity and this is a result of their ability to exhibit different valencies. Thus Co^{2+} catalyses the decomposition of the chlorate(I) (hypochlorite) anion because of its ready interconversion with Co^{3+}:

$$4Co(OH)_2(s) + 2ClO^-(aq) + 2H_2O(l) \rightarrow 4Co(OH)_3(s) + 2Cl^-(aq)$$
$$4Co(OH)_3(s) \rightarrow 4Co(OH)_2(s) + 2H_2O(l) + O_2(g)$$

The transition metals and their compounds are used as catalysts in many important industrial processes. Examples are iron in the Haber process (page 117), nickel in hydrogenations (page 160), vanadium(V) oxide in the Contact process (page 117), and titanium(IV) chloride in the polymerisation of alkenes using Ziegler catalysts (page 360).

2 Paramagnetism and coloured ions

All the ions other than those with completely filled or empty d orbitals possess unpaired electrons and so are paramagnetic (see page 59). All the hydrated ions possessing unpaired d electrons are coloured; this may be illustrated by. the examples below.

Number of unpaired electrons	Colour of ion
0	Sc^{3+}, Ti^{4+}, Cu^+, Zn^{2+} all colourless
1	Ti^{3+} purple, V^{4+} blue, Cu^{2+} blue
2	V^{3+} green, Ni^{2+} green
3	V^{2+} violet, Cr^{3+} green, Co^{2+} pink
4	Cr^{2+} blue, Mn^{3+} violet, Fe^{2+} green
5	Mn^{2+} pink, Fe^{3+} yellow

The colour is a consequence of absorption of light promoting the unpaired d electrons to higher energy levels (colour is mainly an absorption rather than an emission effect). $CuSO_4 \cdot 5H_2O$ is blue, because it absorbs the red and yellow elements of white light leaving only the green and blue to be transmitted. The colour may well change when different ligands are present (see below).

3 Complex ion formation

This is very prevalent and occurs by other molecules or ions, known as ligands, becoming bonded to the central cation. For example, dissolving nickel(II) sulphate(VI) in water gives $[Ni(H_2O)_6]^{2+} SO_4^{2-}$, an oxygen in each of the six water molecules using a lone pair of electrons to form a co-ordinate bond with the Ni^{2+} ion. The Ni^{2+} ion is said to have a co-ordination number of six because it has six nearest neighbours, i.e. the $6H_2O$. These water molecules can be replaced by other ligands more willing to donate a lone pair, e.g.

$$[Ni(H_2O)_6]^{2+} + 6NH_3 \rightarrow [Ni(NH_3)_6]^{2+} + 6H_2O$$
Green   Blue-violet
Hexaaquanickel(II) ion   Hexaamminenickel(II) ion

$$[Fe(H_2O)_6]^{2+} + 6CN^- \rightleftharpoons [Fe(CN)_6]^{4-} + 6H_2O$$
Green   Brown
Hexaaquairon(II) ion   Hexacyanoferrate(II)
(ferrocyanide) ion

$$[Co(H_2O)_6]^{2+} + 4Cl^- \rightleftharpoons [CoCl_4]^{2-} + 6H_2O$$
Pink   Blue
Hexaaquacobalt(II) ion   Tetrachlorocobaltate(II) ion

The above reactions are readily demonstrated. Note that the names of anionic complexes end in '-ate'.

The tendency to form complex ions increases along the series as the charge density of the metal ion increases (the ionic radius decreases as the atomic number increases).

Geometry of complex ions

Co-ordination numbers of 2, 4, and 6 are encountered with transition metals. Co-ordination number 2 is not very common, but examples are $[Cu(CN)_2]^-$ and $[Cu(NH_3)_2]^+$. These examples are linear, i.e. $[NH_3{\rightarrow}Cu{\leftarrow}NH_3]^+$.

The tetrahedral configuration with co-ordination number 4 is fairly common. It generally occurs when the element is in its maximum oxidation state and no d electrons are available, e.g. $[Zn(CN)_4]^{2-}$, $[Zn(NH_3)_4]^{2+}$, and MnO_4^-. The formation of the tetracyanozinacate(II) ion is illustrated below.

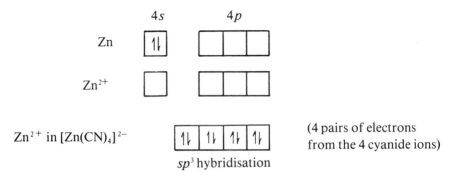

If the transition metals have oxidation numbers less than their maximum, they have partially filled d orbitals and so form few tetrahedral complexes, but $[CoCl_4]^{2-}$ and $[FeCl_4]^-$ are examples. The formation of the tetrachloro-cobaltate(II) ion may be represented as follows:

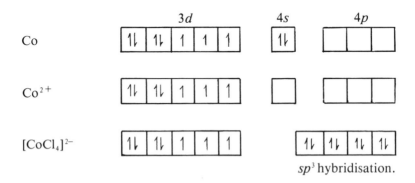

Square planar complexes are also known, but they are not very common. They tend to occur with metal ions having a d^8 configuration, e.g. Ni^{2+} in $[Ni(CN)_4]^{2-}$:

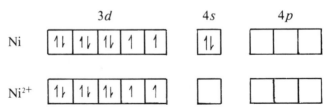

The *d* electrons then rearrange to accept incoming electrons from ligands

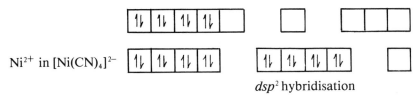

Ni²⁺ in [Ni(CN)₄]²⁻

*dsp*² hybridisation

A further example is the co-ordination of butanedione dioxime (dimethyl-glymoxime) with Ni²⁺ ions:

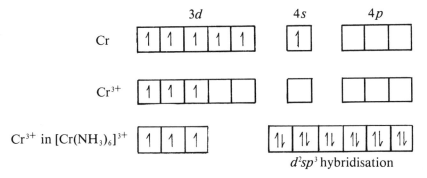

$$Ni^{2+} + 2 \quad \substack{CH_3-C=NOH \\ | \\ CH_3-C=NOH} \quad \rightarrow \quad \begin{matrix} HO & O^- \\ | & | \\ CH_3-C=N & N^+=C-CH_3 \\ \diagdown & \diagup \\ & Ni \\ \diagup & \diagdown \\ CH_3-C=N^+ & N=C-CH_3 \\ | & | \\ O^- & OH \end{matrix} \quad + 2H^+$$

This reaction gives a pink precipitate of the complex and it is used to detect Ni²⁺ ions. Note that the butanedione dioxime occupies two positions simultaneously round the central cation; such groups are known as *bidentate* groups as opposed to groups such as CN⁻ which are said to be *monodentate*.

Co-ordination number 6 is very common and these complexes are all octahedral, e.g. [Cr(NH₃)₆]³⁺, [Co(NH₃)₆]³⁺, [Fe(CN)₆]³⁻, and [Cu(H₂O)₆]²⁺. The hexaamminechromium(III) ion is formed as follows:

	3*d*	4*s*	4*p*

Cr

Cr³⁺

Cr³⁺ in [Cr(NH₃)₆]³⁺

*d*²*sp*³ hybridisation

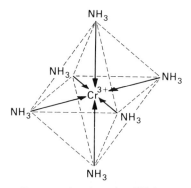

Hexaamminechromium(III) ion

Complex ion equilibria and stability constants

Complex ion formation is an equilibrium process which takes place in stages and the equilibrium constants for the various stages can often be measured. For example, $[Cu(NH_3)_4(H_2O)_2]^{2+}$ is formed in four stages, each stage consisting of replacement of one water ligand by an ammonia molecule. Each stage has its own equilibrium constant:

(a) $[Cu(H_2O)_6]^{2+} + NH_3 \rightleftharpoons [Cu(H_2O)_5NH_3]^{2+} + H_2O$

$$k_1 = \frac{[[Cu(H_2O)_5(NH_3)]^{2+}]\,[H_2O]}{[[Cu(H_2O)_6]^{2+}]\,[NH_3]} = 1.41 \times 10^4 \quad \text{at } 20\,°C$$

(b) $[Cu(H_2O)_5NH_3]^{2+} + NH_3 \rightleftharpoons [Cu(H_2O)_4(NH_3)_2]^{2+} + H_2O$

$$k_2 = \frac{[[Cu(H_2O)_4(NH_3)_2]^{2+}]\,[H_2O]}{[[Cu(H_2O)_5NH_3]^{2+}]\,[NH_3]} = 3.16 \times 10^3$$

(c) $[Cu(H_2O)_4(NH_3)_2]^{2+} + NH_3 \rightleftharpoons [Cu(H_2O)_3(NH_3)_3]^{2+} + H_2O$

$$k_3 = \frac{[[Cu(H_2O)_3(NH_3)_3]^{2+}]\,[H_2O]}{[[Cu(H_2O)_4(NH_3)_2]^{2+}]\,[NH_3]} = 7.77 \times 10^2$$

(d) $[Cu(H_2O)_3(NH_3)_3]^{2+} + NH_3 \rightleftharpoons [Cu(H_2O)_2(NH_3)_4]^{2+} + H_2O$

$$k_4 = \frac{[[Cu(H_2O)_2(NH_3)_4]^{2+}]\,[H_2O]}{[[Cu(H_2O)_3(NH_3)_3]^{2+}]\,[NH_3]} = 1.35 \times 10^2$$

These four equilibria exist in ammoniacal solutions of Cu^{2+}, the actual concentration of each complex being determined by the concentration of free ammonia. The concentration of the final complex rises sharply as the free ammonia concentration is increased.

Adding the above four equations together gives:

$$[Cu(H_2O)_6]^{2+} + 4NH_3 \rightleftharpoons [Cu(H_2O)_2(NH_3)_4]^{2+} + 4H_2O$$

for which

$$\frac{[[Cu(H_2O)_2(NH_3)_4]^{2+}]\,[H_2O]^4}{[[Cu(H_2O)_6]^{2+}]\,[NH_3]^4} = k_1 \times k_2 \times k_3 \times k_4 = 4.67 \times 10^{12}$$

This value is known as the stability constant, K_{st}, for the $[Cu(H_2O)_2(NH_3)_4]^{2+}$ ion. Stability constants show the extent to which complex ion formation occurs between the cation and the ligands and so gives an indication of the stability of the complex in solution. The greater the size of K_{st} the more stable the complex.

Generally, the ligands in a complex ion can be replaced by adding other ligands more willing to donate a lone pair of electrons. The K_{st} values will indicate how effective such reactions will be.

If the complex ions are coloured, their concentrations can be readily determined by colorimetric methods.

Types and properties of complexes

Complexes can be cationic, e.g. $[Co(NH_3)_6]^{3+}$, neutral, e.g. $Ni(CO)_4$, or anionic, e.g. $[Fe(CN)_6]^{3-}$. The central ion in a complex generally tries to attain the noble gas structure although it does not always succeed.

In the $[Co(NH_3)_6]^{3+}$ ion the Co^{3+} has 24 electrons and it is supplied with a further 12 electrons from the 6 ammonia molecules giving a total of 36. In the case of $[Co(NH_3)_6]^{2+}$, however, the Co^{2+} has a total of 37 electrons. Hence the former complex, in which the cobalt has the electronic structure of krypton, is the most stable.

Some of the properties of complexes are illustrated by the sequence below:

$$[Co^{III}(NH_3)_6]^{3+} \xleftarrow[\text{Oxidation by air}]{NH_3} [Co^{II}(H_2O)_6]^{2+} \xrightarrow{Cl^-} [Co^{II}Cl_4]^{2-}$$

Hexaamminecobalt(III) ion   Hexaaquacobalt(II) ion   Tetrachlorocobaltate(II) ion
Orange-brown   Pink   Blue

The $[Co^{II}(H_2O)_6]^{2+}$ complex is readily oxidised because the Co^{III} complexes are more stable (see above), in contrast to the simple ions. The simple Co^{3+} ion is unstable because d orbital electrons are reluctant to enter into bonding when the d level is more than half full (Co $1s^2 2s^2 2p^6 3s^2 3p^6 3d^7 4s^2$). The Co^{3+} ion is, in fact, a very strong oxidising agent and oxidises water to oxygen:

$$4Co^{3+}(aq) + 2H_2O(l) \rightarrow 4Co^{2+}(aq) + 4H^+(aq) + O_2(g)$$

the electrons being taken from the hydroxide ions:

$$HO^- \rightarrow HO + e^-$$

$$4HO \rightarrow 2H_2O + O_2$$

Chloride ions will donate a pair of electrons more readily than H_2O and so $[Co^{II}(H_2O)_6]^{2+}$ has its water replaced by Cl^- ions, the co-ordination falling to four in the process (Cl^- ions are larger than water molecules). Similar reduction in the co-ordination number occurs when excess chloride ions are added to the hexaaquairon(III), nickel(II), and copper(II) ions.

Some complexes are so stable that they do not give the reactions of the constituent ions, e.g. potassium hexacyanoferrate(II) (ferrocyanide), $K_4[Fe(CN)_6]$, does not give the reactions of Fe^{2+} or CN^-.

Titanium

Titanium is very resistant to corrosion due to the formation of a thin protective layer of titanium(IV) oxide, TiO_2. It is not attacked by water although it decomposes steam on strong heating:

$$Ti(s) + 2H_2O(g) \rightarrow TiO_2(s) + 2H_2(g)$$

The metal is attacked by hydrofluoric acid and dilute sulphuric(VI) acid:

$$Ti(s) + 3H^+(aq) \rightarrow Ti^{3+}(aq) + 1\tfrac{1}{2}H_2(g)$$

Titanium is unaffected by cold, dilute or concentrated, hydrochloric acid but with the hot concentrated acid, a violet solution of titanium(III) chloride is obtained. Similarly, titanium does not react with cold nitric(V) acid but with the hot concentrated acid, a white precipitate of hydrated titanium(IV) oxide, $TiO_2 \cdot xH_2O$, is formed. (In this reaction titanium resembles tin.) There is no reaction between titanium and alkalis.

Heating titanium strongly in air gives a mixture of the oxide, TiO_2 and nitride, TiN. It reacts with chlorine at 300 °C:

$$Ti(s) + 2Cl_2(g) \rightarrow TiCl_4(l)$$

The outer electronic configuration of titanium is $3d^2 4s^2$ and oxidation states of 2, 3, and 4 are encountered:

Oxidation state	Bonding	d-configuration	Comments
Ti(II)	Ionic	d^2	Not very important
Ti(III)	Ionic	d^1	Powerful reducing agents
Ti(IV)	Ionic or covalent	d^0	Most stable state

The hydrated Ti(IV) ion is acidic because the high charge density on the titanium ion weakens the O—H bonds in the water molecules (cf. $[Al(H_2O)_6]^{3+}$, page 192).

$$[Ti(H_2O)_6]^{4+}(aq) \rightarrow [Ti(OH)_2(H_2O)_4]^{2+}(aq) + 2H^+(aq)$$

The Ti(IV) compounds are generally covalent and are readily reduced to Ti(III) in aqueous solution, e.g.

$$TiCl_4(l) \xrightarrow{\text{Zn/dilute HCl}} [Ti(H_2O)_6]^{3+}(aq)$$
Colourless \qquad\qquad Violet

The chemistry of Ti(IV) is similar to that of Sn(IV) as a result of their similar sizes; this is very apparent in the case of the M(IV) oxides. There is some tendency to form the titanyl(IV) radical, $=Ti=O$, e.g. the reaction between hot concentrated sulphuric(VI) acid and titanium gives titanium(IV) oxide sulphate(VI) (titanyl sulphate), $[TiO][SO_4]$.

Titanium(IV) oxide is amphoteric

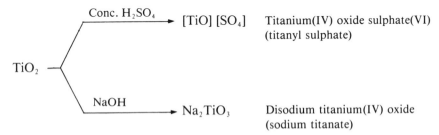

It is important commercially as a white pigment for paint and as a filler for paper.

Titanium(IV) chloride is a colourless liquid which fumes in moist air (this reaction is used to make smokescreens):

$$TiCl_4(l) + 2H_2O(l) \rightarrow TiO_2(s) + 4HCl(aq)$$

It dissolves in concentrated hydrochloric acid giving hydrogen hexachlorotitanate(IV):

$$TiCl_4(l) + 2HCl(conc) \rightarrow H_2TiCl_6(aq)$$

The hydrated Ti^{3+} ion, $[Ti(H_2O)_6]^{3+}$, in contrast to $[Ti(H_2O)_6]^{4+}$, does not lose protons and addition of alkali results in a purple precipitate of hydrated oxide, $Ti_2O_3 \cdot nH_2O$. The salts of Ti^{3+} are coloured and paramagnetic owing to their unpaired electron.

Titanium(III) chloride may be prepared by reduction of the tetrachloride with hydrogen. It is used as a reducing agent in volumetric analysis:

$$Ti^{3+} - e^- \rightarrow Ti^{4+}$$

Since the Ti^{3+} is very readily oxidised, air must be excluded. Disproportionation occurs when the trichloride is heated:

$$2TiCl_3(s) \rightarrow TiCl_2(s) + TiCl_4(l)$$

Titanium(II) compounds are oxidised by dissolution in water or dilute acid, hydrogen being liberated in the process:

$$Ti^{2+}(s) + H_3O^+(aq) \rightarrow Ti^{3+}(aq) + H_2O(l) + \tfrac{1}{2}H_2(g)$$

They are coloured owing to their unpaired electrons.

Addition of hydrogen peroxide to titanium salts gives a golden yellow colour due to the formation of a peroxo ion of uncertain structure.

Vanadium

Vanadium is a very hard metal which is unaffected at room temperature by air, water, solutions of alkalis, or non-oxidising acids. It dissolves in oxidising acids, e.g. nitric(V) acid and sulphuric(VI) acid.

Heating vanadium with oxygen and fluorine gives vanadium(V) oxide and fluoride respectively. Chlorine, however, oxidises the metal only to the $+4$ oxidation state.

Vanadium has the outer electronic structure $3d^3 4s^2$ and it exhibits valencies of 2, 3, 4, and 5:

Oxidation state	d-configuration	Comments
V(II)	d^3	Forms $[V(H_2O)_6]^{2+}$ in solution, powerful reducing agents, basic oxide
V(III)	d^2	Forms $[V(H_2O)_6]^{3+}$ in solution, reducing agents, basic oxide
V(IV)	d^1	Stable covalent compounds which in solution give the $[VO(H_2O)_5]^{2+}$ ion, amphoteric oxide
V(V)	d^0	Stable state, tends to give the hydrated vanadate(V) ion $[VO_2(OH)_4]^{3-}$ in solution, amphoteric oxide

Vanadium(II) may be prepared by reducing a vanadium(IV) oxide salt (vanadyl salt), e.g. $VOCl_2$ with zinc and hydrochloric acid. The hexaaquavanadium(II) ion, $[V(H_2O)_6]^{2+}$ is violet. Vanadium(II) readily forms complexes, e.g. $K_4[V(CN)_6]3H_2O$.

Vanadium(III) compounds may be prepared by electrolytic reduction of vanadium(IV) or (V) salts in acid solution. The hexaaquavanadium(III) ion $[V(H_2O)_6]^{3+}$ is green. Vanadium(III) compounds are oxidised to vanadium(IV) compounds by air.

Vanadium(IV) chloride is covalent and is made by heating the elements. The

+4 oxidation state is, however, generally encountered as the VO^{2+} ion, e.g. vanadium(IV) oxide sulphate(VI) (vanadyl sulphate), $VOSO_4$.

Vanadium(V) fluoride is made by direct combination of the elements on heating or by disproportionation of vanadium(IV) fluoride:

$$VCl_4(l) \xrightarrow{\text{HF}} VF_4(s)$$
$$2VF_4(s) \longrightarrow VF_3(s) + VF_5(l)$$

It is the only oxygen-free halide of vanadium(V). Vanadium(V) oxide halides (vanadyl oxytrihalides), VOX_3, with the exception of the iodide, are obtained by action of the hydrogen halide on vanadium(V) oxide:

$$V_2O_5(s) + 6HCl(g) \rightarrow 2VOCl_3(l) + 3H_2O(l)$$

Vanadium(V) oxide may be prepared by direct combination or by heating ammonium vanadate(V) which, for simplicity, may be represented as NH_4VO_3:

$$2NH_4VO_3(s) \rightarrow V_2O_5(s) + H_2O(l) + 2NH_3(g)$$

Vanadium(V) oxide is orange despite the fact that it has no unpaired electrons. This is due to transitions of electrons between the orbitals of the oxygen and vanadium atoms; the phenomenon is known as a *charge transfer spectrum*. The oxide dissolves in alkali to form vanadates(V), for example, in concentrated sodium hydroxide solution, sodium vanadate(V) (sodium orthovanadate) is formed:

$$V_2O_5(s) + 6NaOH(aq) \rightarrow 2Na_3VO_4(aq) + 3H_2O(l)$$

Addition of H^+ to this produces condensed oxy-anions, e.g. $V_2O_7{}^{4-}$, $V_3O_9{}^{3-}$.

Acidified solutions of vanadium(V) salts turn orange on addition of hydrogen peroxide due to the formation of complex peroxo ions (cf. titanium).

The following standard electrode potentials indicate that zinc should be able to bring about the reduction of vanadium(V) to vanadium(II):

$$Zn^{2+}(aq)\,|\,Zn(s) \qquad E^{\ominus} = -0.76 \text{ V}$$
$$VO_2{}^+(aq)\,|\,VO^{2+}(aq) \qquad E^{\ominus} = +1.0 \text{ V}$$
$$VO^{2+}(aq)\,|\,V^{3+}(aq) \qquad E^{\ominus} = +0.36 \text{ V}$$
$$V^{3+}(aq)\,|\,V^{2+}(aq) \qquad E^{\ominus} = -0.25 \text{ V}$$
$$V^{2+}(aq)\,|\,V(s) \qquad E^{\ominus} = -1.2 \text{ V}$$

The colour due to each oxidation state is in fact observed when a solution of ammonium vanadate(V) is treated with zinc and dilute hydrochloric acid:

Change	$VO_2{}^+(aq)$	$\rightarrow$	$VO^{2+}(aq)$	$\rightarrow$	$[V(H_2O)_6]^{3+}(aq)$	$\rightarrow$	$[V(H_2O)_6]^{2+}(aq)$
Oxidation state	$+5$		$+4$		$+3$		$+2$
Colour	yellow		blue		green		violet

Chromium

Chromium has the outer electronic structure $3d^5 4s^1$. The most important valencies are 2, 3, and 6 and these are illustrated by the oxides, etc. below

Valency	Oxide	Properties	Cation	Properties	Anion	Properties
2	CrO	Basic	Cr^{2+}	Readily oxidised	—	—
3	Cr_2O_3	Amphoteric	Cr^{3+}	Stable	$[Cr(HO)_4]^-$	Stable
6	CrO_3	Acidic	—	—	CrO_4^{2-}	Stable
					$Cr_2O_7^{2-}$	Readily reduced

The ions $[Cr(OH)_4]^-$, CrO_4^{2-}, and $Cr_2O_7^{2-}$ are known as chromate(III) (chromite), chromate(VI) and dichromate(VI) ions respectively.

The most stable and common chromium compounds involve chromium(III) and if these contain water of crystallisation they are often violet coloured. Dissolving chromium(III) (chromic) salts in water gives $[Cr(H_2O)_6]^{3+}$ and this complex ion readily loses protons to give chromium(III) hydroxide. Thus, when alkali is added to a solution of $[Cr(H_2O)_6]^{3+}$, a pale green precipitate of hydrated chromium(III) hydroxide is formed via the following sequence of steps:

$$[Cr(H_2O)_6]^{3+}(aq) + HO^-(aq) \rightleftharpoons [Cr(H_2O)_5OH]^{2+}(aq) + H_2O(l)$$
$$[Cr(H_2O)_5OH]^{2+}(aq) + HO^-(aq) \rightleftharpoons [Cr(H_2O)_4(OH)_2]^+(aq) + H_2O(l)$$
$$[Cr(H_2O)_4(OH)_2]^+(aq) + HO^- \rightleftharpoons [Cr(H_2O)_3(OH)_3](s) + H_2O(l)$$

The hydrated hydroxide will, in fact, dissolve in excess alkali to give a chromate(III) (chromite), e.g. $NaCr(OH)_4$.

Chromium(III) compounds readily form complexes. For example, $[Cr(NH_3)_6]^{3+}$, $[Cr(CN)_6]^{3-}$, and $[CrCl_6]^{3-}$ are formed when aqueous solutions of a chromium(III) salt are treated with excess ammonia solution, cyanide ion, or chloride ion respectively. Some of its complexes exhibit structural isomerism. Thus, the existence of the isomers $[Cr(H_2O)_4Cl_2]^+$ $Cl^- \cdot 2H_2O$, $[Cr(H_2O)_5Cl]^{2+}$ $(Cl^-)_2 \cdot H_2O$, and $[Cr(H_2O)_6]^{3+}(Cl^-)_3$ may be illustrated by the fact that addition of silver nitrate(V) solution gives one, two, and three moles of silver chloride respectively.

Chromium(III) salts are oxidised to chromium(VI) by, for example, fusing with sodium peroxide or simply by adding sodium peroxide to a chromium(III) salt solution or freshly prepared chromium(III) hydroxide:

$$2Cr(OH)_3(s) + 3Na_2O_2(s) \rightarrow 2Na_2CrO_4(aq) + 2NaOH(aq) + 2H_2O(l)$$

The use of sodium peroxide as oxidising agent ensures alkaline conditions and the formation of a chromate(VI). Chromates(VI) are yellow and are converted into dichromates(VI) which are orange, in acid solution:

$$2CrO_4^{2-}(aq) + 2H^+(aq) \underset{\text{Alkali}}{\overset{\text{Acid}}{\rightleftharpoons}} Cr_2O_7^{2-}(aq) + H_2O(l)$$

Note that the chromium is in the $+6$ oxidation state in both ions:

Careful addition of concentrated sulphuric(VI) acid to a concentrated chromate(VI) or dichromate(VI) solution gives bright red needle-shaped crystals of chromium(VI) oxide, e.g.

$$Na_2CrO_4(aq) + 2H_2SO_4(conc) \rightarrow CrO_3(s) + 2NaHSO_4(aq) + H_2O(l)$$
$$Na_2Cr_2O_7(aq) + 2H_2SO_4(conc) \rightarrow 2CrO_3(s) + 2NaHSO_4(aq) + H_2O(l)$$

The mixture has powerful oxidising properties and is used for cleaning glassware. The dichromate(VI) ion in acid solution is a powerful oxidising agent:

$$Cr_2O_7{}^{2-}(aq) + 14H^+(aq) + 6e^- \rightarrow 2Cr^{3+}(aq) + 7H_2O(l)$$

and potassium dichromate(VI) finds considerable use in volumetric oxidations. Unlike potassium manganate(VII) (potassium permanganate), it does not oxidise aqueous solutions of chloride ions and so the presence of hydrochloric acid is no problem.

Both hydrogen sulphide and sulphur dioxide reduce acidified solutions of potassium dichromate(VI) to green Cr^{3+} and this reaction is used as a test for them:

$$K_2Cr_2O_7(aq) + 4H_2SO_4(aq) + 3H_2S(g) \rightarrow Cr_2(SO_4)_3(aq) + K_2SO_4(aq) + 7H_2O(l) + 3S(s)$$

$$K_2Cr_2O_7(aq) + H_2SO_4(aq) + 3SO_2(g) \rightarrow Cr_2(SO_4)_3(aq) + K_2SO_4(aq) + H_2O(l)$$

Crystallisation of the solution from the latter reaction gives purple crystals of chromium(III) potassium sulphate(VI)-12-water (chrome alum), $KCr(SO_4)_2 \cdot 12H_2O$.

Potassium dichromate(VI) reacts with chlorides and hot concentrated sulphuric(VI) acid to give red fumes of chromium(VI) dichloride dioxide (chromyl chloride), e.g.

$$K_2Cr_2O_7(s) + 4KCl(s) + 6H_2SO_4(conc) \rightarrow 6KHSO_4(aq) + 3H_2O(l) + 2CrO_2Cl_2(l)$$

It can be seen from the following standard electrode potentials that zinc will reduce dichromate(VI) to chromium(II).

$$Zn^{2+}(aq)|Zn(s) \qquad E^{\ominus} = -0.76 \text{ V}$$
$$Cr_2O_7{}^{2-}(aq)|Cr^{3+}(aq) \qquad E^{\ominus} = +1.33 \text{ V}$$
$$Cr^{3+}(aq)|Cr^{2+}(aq) \qquad E^{\ominus} = -0.41 \text{ V}$$
$$Cr^{3+}(aq)|Cr(s) \qquad E^{\ominus} = -0.74 \text{ V}$$

The following changes are observed when zinc is added to an acidified solution of potassium dichromate(VI):

Change	$Cr_2O_7{}^{2-}(aq)$	$\rightarrow$	$[Cr(H_2O)_6]^{3+}(aq)$	$\rightarrow$	$[Cr(H_2O)_6]^{2+}(aq)$
Oxidation state	+6		+3		+2
Colour	orange		violet or green		blue

Manganese

The outer electronic structure of manganese is $3d^5 4s^2$ and it exhibits valencies of 2, 3, 4, 6, and 7. Details are:

Valency				
2	Mn^{2+}	$3d^5 4s^0$	—	most stable valency (d shell exactly half full)
3	Mn^{3+}	$3d^4 4s^0$	—	unstable
4	Covalent		—	stable

6   Covalent — manganates (VI)   unstable

7   Covalent — manganates(VII)   strong oxidising agents.
      (permanganates)

The variable valency can be illustrated by the sequence:

$$Mn^{2+}(aq) \rightleftharpoons Mn^{IV}O_2 \rightleftharpoons Mn^{VI}O_4{}^{2-} \rightleftharpoons Mn^{VII}O_4{}^-$$

$$\text{Manganates(VI)} \quad \text{Manganates(VII)}$$
$$\text{(Permanganates)}$$

Hydrated Mn^{2+} ions are pale pink and they can be oxidised in acidic solution by potassium chlorate(V), etc. or in alkaline solution by, for example, sodium peroxide or sodium chlorate(I) (sodium hypochlorite) to give manganese(IV) oxide which is brown-black:

$$Mn^{2+}(aq) + 2H_2O(l) \rightarrow MnO_2(s) + 4H^+(aq) + 2e^- \text{ (in acid solution)}$$
$$Mn^{2+}(aq) + 4HO^-(aq) \rightarrow MnO_2(s) + 2H_2O(l) + 2e^- \text{ (in alkaline solution)}$$

Fusing manganese(IV) oxide (manganese dioxide) with an alkali and an oxidising agent such as potassium chlorate(V) or potassium nitrate(V) gives a manganate(VI),

$$\left[\begin{array}{c} :O: \\ \| \\ :\ddot{O}-Mn-\ddot{O}: \\ \| \\ :O: \end{array} \right]^{2-}$$

e.g.

$$3MnO_2(s) + KClO_3(s) + 6KOH(s) \rightarrow 3K_2MnO_4(s) + KCl(s) + 3H_2O(l)$$

These are green salts which disproportionate (i.e. undergo simultaneous oxidation and reduction) in neutral or acid solution to give manganese(IV) oxide and a purple manganate(VII) (permanganate),

$$\left[\begin{array}{c} :O: \\ \| \\ :\ddot{O}-Mn=\ddot{O} \\ \| \\ :O: \end{array} \right]^{-}$$

e.g. $3MnO_4{}^{2-}(s) + 2H_2O(l) \rightarrow 2MnO_4{}^-(aq) + MnO_2(s) + 4HO^-(aq)$

i.e. $Mn^{VI} \rightarrow Mn^{VII} + Mn^{IV}$

Manganates(VI) can, therefore, exist only in alkaline solution.

Manganates(VII) (permanganates) react slowly with concentrated solutions of alkali giving a manganate(VI):

$$4MnO_4{}^-(aq) + 4HO^-(aq) \rightarrow 4MnO_4{}^{2-}(aq) + 2H_2O(l) + O_2(g)$$

This change can also be brought about by heating a manganate(VII) to about 200 °C:

$$2KMnO_4(s) \rightarrow K_2MnO_4(s) + MnO_2(s) + O_2(g)$$

Manganese(IV) oxide is reduced to Mn^{2+} on heating with concentrated hydrochloric acid:

$$MnO_2(s) + 4HCl(conc) \rightarrow MnCl_2(aq) + 2H_2O(l) + Cl_2(g)$$

The above interconversions between Mn^{IV}, Mn^{VI}, and Mn^{VII} can be predicted on the basis of the relevant $E^{\ominus}$ values. For example, a likely way of producing Mn^{VI} is from a reaction between a compound containing Mn^{IV} and one containing Mn^{VII}. Under standard acid conditions, however, the $E^{\ominus}$ values are so far apart that the reaction is not feasible even if the acidity is adjusted:

$$Mn^{VI}O_4{}^{2-} + 4H^+ + 2e^- \rightarrow Mn^{IV}O_2 + 2H_2O, \quad E^{\ominus} = +2.26 \text{ V}$$
$$2Mn^{VII}O_4{}^- + 2e^- \rightarrow 2Mn^{VI}O_4{}^{2-}, \qquad\qquad E^{\ominus} = +0.56 \text{ V}$$

On the other hand, the $E^{\ominus}$ values are sufficiently close under standard alkaline conditions to suggest that altering the conditions could alter the polarity of the electrodes:

$$Mn^{VI}O_4{}^{2-} + 2H_2O + 2e^- \rightarrow Mn^{IV}O_2 + 4HO^-, \quad E^{\ominus} = +0.60 \text{ V}$$
$$2Mn^{VII}O_4{}^- + 2e^- \rightarrow 2Mn^{VI}O_4{}^{2-}, \qquad\qquad E^{\ominus} = +0.56 \text{ V}$$

It follows that increasing the alkalinity should be effective whereas making the solution acidic would reverse the desired reaction.

Manganese(IV) oxide and potassium manganate(VII) (potassium permanganate) are used as oxidising agents. Manganese(IV) oxide is used in the laboratory preparation of chlorine as in the equation above and as a depolariser in Leclanché cells and dry batteries — it oxidises hydrogen atoms to water and so prevents the formation of hydrogen bubbles on the cathode:

$$2MnO_2 + 2H \rightarrow Mn_2O_3 + H_2O$$
$$\text{Manganese(III) oxide}$$

Potassium manganate(VII) (permanganate) can be used as an oxidising agent under acid or alkaline conditions:

$$MnO_4{}^- + 8H^+ + 5e^- \rightarrow Mn^{2+} + 4H_2O$$
$$MnO_4{}^- + 2H_2O + 3e^- \rightarrow MnO_2 + 4HO^-$$

but generally acidic conditions are preferred so as to avoid precipitation of manganese(IV) oxide. It is used extensively in volumetric analysis, e.g. $Fe^{2+} \rightarrow Fe^{3+}$, $C_2O_4{}^{2-}$, ethanedioates (oxalates) $\rightarrow CO_2$, $NO_2{}^- \rightarrow NO_3{}^-$, and $SO_3{}^{2-} \rightarrow SO_4{}^{2-}$. The colour change is virtually purple to colourless because the hydrated Mn^{2+} concentration is too low to colour the solution pink. Potassium manganate(VII) (permanganate) also finds considerable use in organic chemistry, eg. oxidation of alcohols, aldehydes, alkenes, etc.

Iron

Iron has the outer electronic structure $3d^6 4s^2$ and has two common oxidation states: Fe^{2+}, $3d^6 4s^0$ and Fe^{3+}, $3d^5 4s^0$. The $+3$ oxidation state is the most stable because the $3d$ orbitals are exactly half full.

Iron has a greater tendency to form complex ions than manganese since its atomic number is greater and its ionic radius smaller. The Fe^{2+} ion is often hydrated by the co-ordination of six water molecules — that is, it forms the hexa-aquairon(II) ion, $[Fe(H_2O)_6]^{2+}$, which is pale green — e.g. $FeSO_4 \cdot 7H_2O(6H_2O$ being co-ordinated to the Fe^{2+} ion and one to the $SO_4{}^{2-}$ ion), $FeCl_2 \cdot 6H_2O$, $Fe(NO_3)_2 \cdot 6H_2O$, and $Fe_3(PO_4)_2 \cdot 8H_2O$ ($6H_2O$ being co-ordinated to the Fe^{2+} ion and two to the $PO_4{}^{3-}$ ions).

Fe^{2+} ions are fairly stable in acid solution, but are oxidised by strong oxidising agents such as potassium manganate(VII) (permanganate), potassium dichromate(VI), nitric(V) acid, and hydrogen peroxide, etc. In neutral or alkaline solution, however, they are readily oxidised even by atmospheric oxygen. Thus, adding sodium hydroxide solution to iron(II) sulphate(VI) solution leads to the precipitation of greenish iron(II) hydroxide (white if all air is excluded), but on exposure to air this darkens and turns brown as oxidation to iron(III) hydroxide occurs:

$$Fe(OH)_2(s) + HO^-(aq) \rightarrow Fe(OH)_3(s) + e^-$$

The complete equation is:

$$4Fe(OH)_2(s) + 2H_2O(l) + O_2(g) \rightarrow 4Fe(OH)_3(s)$$

the electrons from the Fe^{2+} being utilised by the oxygen and water to form hydroxide ions:

$$2H_2O(l) + O_2(g) + 4e^- \rightarrow 4HO^-(aq)$$

The reverse process, i.e.

$$Fe^{3+}(aq) + e^- \rightarrow Fe^{2+}(aq),$$

is accomplished by hydrogen sulphide, sulphur dioxide, Sn^{2+} in hydrochloric acid, etc.

Addition of cyanide ion to an iron(II) salt solution results in the replacement of the water molecules round the Fe^{2+} (CN^- is more electron-donating than $H_2\overset{..}{O}$), the hexacyanoferrate(II) ion (ferrocyanide) being produced:

$$[Fe(H_2O)_6]^{2+}(aq) + 6CN^-(aq) \rightleftharpoons [Fe(CN)_6]^{4-}(aq) + 6H_2O(l)$$

The potassium salt is yellow. It is a stable complex, its solution does not give the reactions of Fe^{2+} or CN^- ions, and it is non-poisonous. Altering the ligands has therefore considerably affected the stability of the Fe^{2+} ion. The hexacyanoferrate(II) ion (ferrocyanide) can be oxidised to hexacyanoferrate(III) (ferricyanide) by chlorine:

$$2[Fe(CN)_6]^{4-}(aq) + Cl_2(g) \rightarrow 2[Fe(CN)_6]^{3-}(aq) + 2Cl^-(g)$$

Potassium hexacyanoferrate(III) (ferricyanide) exists as dark red crystals. It is an oxidising agent and readily reverts to hexacyanoferrate(II). Fe^{2+} complex ions have the noble gas structure.

Solutions of Fe^{2+} and Fe^{3+} ions may be distinguished by adding potassium or ammonium thiocyanate; Fe^{3+} forms a soluble blood red complex of uncertain composition whereas no change is observed with Fe^{2+} ions. A simplified equation for the reaction is:

$$[Fe(H_2O)_6]^{3+}(aq) + SCN^-(aq) \rightarrow [Fe(H_2O)_5SCN]^{2+}(aq) + H_2O(l)$$

Other aspects of iron chemistry are discussed in the previous chapter.

Copper

The outer electronic structure of copper is $3d^{10}4s^1$ and there are two important oxidation states: Cu^+, $3d^{10}4s^0$ which is diamagnetic (page 59) and Cu^{2+},

$3d^9 4s^0$ which has an unpaired electron and so is paramagnetic (page 59).

It may have been expected from the electronic structures that Cu^{2+} would not be very stable compared to Cu^+. However, it is energetically favourable for copper to form Cu^{2+} even though this necessitates its taking an electron from the full d orbitals of the Cu^+ ion. Thus, compensation for the energy needed to disrupt the stable d configuration is obtained, in the solid state, by the increased lattice energy of copper(II) compounds and, in aqueous solution, by the increased hydration energy. In fact, the following standard electrode potentials indicate that, in aqueous solution, Cu^+ is unstable with respect to Cu^{2+} ions and copper:

$$Cu^+(aq) + e^- \rightarrow Cu(s), \qquad E^\ominus = +0.52 \text{ V}$$
$$Cu^{2+}(aq) + e^- \rightarrow Cu^+(aq), \qquad E^\ominus = +0.15 \text{ V}$$

Hence, as is seen below, Cu^+ disproportionates in aqueous solution into Cu^{2+} and copper. Copper(I) is, however, the most stable state at high temperatures, e.g. copper(II) oxide gives copper(I) oxide on strong heating:

$$4CuO(s) \rightarrow 2Cu_2O(s) + O_2(g)$$

Copper(II) compounds are prepared by the standard methods. For example, copper(II) oxide is prepared by thermal decomposition of the hydroxide, nitrate(V), or carbonate, e.g.

$$2Cu(NO_3)_2(s) \rightarrow 2CuO(s) + 4NO_2(g) + O_2(g)$$

Copper(II) sulphate(VI) is obtained from the reaction between the oxide and hot dilute sulphuric(VI) acid:

$$CuO(s) + H_2SO_4(aq) \rightarrow CuSO_4(aq) + H_2O(l)$$

Crystallisation yields the pentahydrate. The similar reaction with concentrated hydrochloric acid gives blue copper(II) chloride dihydrate (green if moist). Brown anhydrous copper(II) chloride is obtained by heating copper in chlorine.

Copper(I) compounds are formed by a variety of methods. Copper(I) chloride is made by the reduction of the copper(II) salt. Thus, boiling a solution of copper(II) chloride in concentrated hydrochloric acid with copper results in reduction of the hydrogen tetrachlorocuprate(II) to hydrogen dichlorocuprate(I). Addition of this brown solution to a large excess of water gives a white precipitate of copper(I) chloride.

$$CuCl_2(s) + 2HCl(conc) \rightarrow H_2[CuCl_4]$$
$$H_2[CuCl_4] + Cu(s) \rightarrow 2H[CuCl_2]$$
$$H[CuCl_2] \xrightarrow{H_2O} CuCl(s) + HCl(aq)$$

Copper(I) chloride may also be prepared by reducing copper(II) chloride solution with sulphur dioxide:

$$2CuCl_2(aq) + 2H_2O(l) + SO_2(g) \rightarrow 2CuCl(s) + 2HCl(aq) + H_2SO_4(aq)$$

Copper(I) oxide may be prepared by reduction of copper(II) oxide, in the form of a complex salt, by warming it with an aldehyde. This reaction is used as a test for aliphatic aldehydes — see Fehling's solution, page 328).

Copper(I) iodide is obtained as a white precipitate when potassium iodide is added to a solution of a copper(II) salt, the reason being that some of the iodide ions reduce the Cu^{2+} to Cu^+ and are themselves oxidised to iodine:

$$2Cu^{2+}(aq) + 4I^-(aq) \rightarrow 2CuI(s) + I_2(aq)$$

Similarly, addition of cyanide ion to a solution of a copper(II) salt results in the formation of copper(I) cyanide and cyanogen:

$$2Cu^{2+}(aq) + 4CN^-(aq) \rightarrow 2CuCN(s) + (CN)_2(g)$$

Both the copper(I) cyanide and iodide are the final products in these reactions only because they are insoluble in water. As mentioned above, Cu^+ ions disproportionate in aqueous solution. Thus, in water, copper(I) sulphate(VI) yields copper and copper(II) sulphate(VI):

$$Cu_2SO_4(aq) \rightarrow Cu(s) + CuSO_4(aq)$$

i.e. $$2Cu^+(aq) \rightarrow Cu(s) + Cu^{2+}(aq)$$

Copper(I) sulphate(VI) is prepared by heating copper(I) oxide with dimethyl sulphate(VI):

$$Cu_2O(s) + (CH_3)_2SO_4(l) \rightarrow Cu_2SO_4(s) + CH_3 \cdot O \cdot CH_3(l)$$

Complex ion formation is very common with copper compounds. Aqueous solutions of copper(II) compounds are blue due to the hexaaquacopper(II) ion, $[Cu(H_2O)_6]^{2+}$. In the solid state, however, the Cu^{2+}, does not have more than four water molecules co-ordinated to it. Copper(II) sulphate(VI) pentahydrate, for example, has the fifth water molecule held by hydrogen bonds to the sulphate(VI) ion and the other water molecules.

Addition of ammonia solution to aqueous solutions of copper(II) salts initially results in a pale blue precipitate of copper(II) hydroxide but this dissolves in excess of the reagent to give a dark blue solution of diaquatetraamminecopper(II) hydroxide:

$$Cu(OH)_2(s) + 4NH_3(aq) + 2H_2O(l) \rightarrow [Cu(NH_3)_4(H_2O)_2]^{2+} (OH^-)_2(aq)$$

Copper(I) and copper(II) chlorides dissolve in concentrated hydrochloric acid giving hydrogen dichlorocuprate(I), $H[CuCl_2]$, and hydrogen tetrachlorocuprate(II), $H_2[CuCl_4]$, respectively. Complex ion formation also occurs when excess cyanide ion is added to copper(I) cyanide, $[Cu(CN)_4]^{3-}$ being formed.

Catalysis by transition metals in industrial reactions

As has already been stated, the ability of transition elements to exhibit different valencies conveys on them catalytic activity and this has been utilised in many industrial processes. A few examples are given below.

The Contact Process $\qquad 2SO_2(g) + O_2(g) \xrightarrow[\text{slight pressure}]{450\,°C,\ V_2O_5\,\text{catalyst,}} 2SO_3(g)$

The Haber Process $\qquad N_2(g) + 3H_2(g) \xrightarrow[\text{Fe catalyst}]{500\,°C,\ \text{up to}\ 1000\ \text{atm,}} 2NH_3(g)$

The Ostwald process for the manufacture of nitric(V) acid involves catalytic oxidation of ammonia as the first stage:

$$4NH_3(g) + 5O_2(g) \xrightarrow{\text{900 °C, Pt–Rh catalyst}} 4NO(g) + 6H_2O(g)$$

Catalytic hydrogenation of oils

$$\begin{array}{c} \diagdown \quad \diagup \\ C = C \\ \diagup \quad \diagdown \end{array} + H_2 \xrightarrow{\text{Ni, 200–300 °C}} \begin{array}{c} \diagdown \quad \diagup \\ CH - CH \\ \diagup \quad \diagdown \end{array}$$

Preparation of aldehydes by dehydrogenation of primary alcohols, e.g.

$$CH_3 \cdot CH_2OH(l) \xrightarrow{\text{Cu, 300 °C}} CH_3 \cdot CHO(l) + H_2(g)$$

The Ziegler method of polymerising alkenes often involves the use of titanium(IV) chloride as catalyst.

Catalytic action of transition metals in biological systems

A number of the elements present in very small quantities in plants and animals are essential for the enzymes to function. (An *enzyme* is a biological catalyst which brings about chemical reactions in living cells.) A particularly important trace element is copper and it catalyses several processes, e.g. energy is produced by the stepwise oxidation of food and the final stage of this oxidation for animals and many plants is catalysed by an enzyme which cannot function in the absence of copper. Without this final stage the whole process stops. The workings of these trace elements are very complex and often not fully understood. Nevertheless, it is known that several transition elements, e.g. iron, manganese, cobalt, zinc, and molybdenum, are involved in various processes.

Questions

12.1   Identify the oxidation number of the transition element in each of the following compounds:

$$Cr_2(SO_4)_3 \cdot 18H_2O \qquad MnO_3F$$
$$Mn(CH_3CO_2)_2 \cdot 4H_2O \qquad K_2Ca[Fe(CN)_6]$$
$$K_2MnO_4 \qquad [Co(NH_3)_5Cl]Cl_2$$
$$CrO_2Cl_2$$

One of the characteristic properties of the transition elements is their ability to form compounds in which the metal has variable oxidation states. Explain, by reference to vanadium (atomic number 23), manganese (atomic number 25), and iron (atomic number 26), how the various oxidation states of these elements may be accounted for in terms of their electronic structures.
                                                                    [University of London]

12.2   Survey the chemistry and structures of complexes formed by the first row transition elements. Your answer should include discussion of complexes of three different co-ordination numbers and deal mainly with the ligands water, ammonia, and chloride ions.   [J.M.B.]

12.3   Discuss (a) the acidity and (b) the substitution reactions of metal hexaaqua cations, $[M(H_2O)_6]^{n+}$ (where $n = 2$ or 3), giving two examples of each type of reaction.

Discuss the effect upon the stabilities of the $+2$ and $+3$ oxidation states of

(a)   increasing the pH in iron chemistry, and

(b)   complex formation (with ligands other than water) in cobalt chemistry.

[J.M.B.]

12.4   An excess of a black compound of manganese, A, was stirred into molten potassium hydroxide containing an oxidising agent. The melt was cooled, shaken with water and filtered to give a green solution of compound B. When this solution was acidified, a black powder identical with A was obtained, and when this was filtered off a purple solution of compound C remained. Solutions of B and C each oxidised iron(II) in acid solution to iron(III).

(a)   Write the name and formula of A, B, and C.

(b)   Write an ionic half equation for the conversion of A to B.

(c)   Give full ionic equations for
        (i)   the conversion of B to A plus C,
        (ii)   the action of C on iron(II) in acid solution.   [J.M.B.]

12.5   When cobalt(II) chloride was dissolved in water, a pink solution A was formed. The addition of concentrated hydrochloric acid to A gave a blue solution B. If solution A was treated with concentrated ammonia solution a blue-green precipitate was formed; upon addition of further ammonia solution followed by the passage of air through the mixture, an orange-red solution C was produced.

(a)   Write down the formulae of the species containing cobalt which are present in each of A, B, and C.

(b)   How are the ligands arranged spatially around the cobalt in A and B?

(c)   Name two other metallic ions, giving their appropriate oxidation states, which undergo reactions similar to that of cobalt(II) when aqueous solutions of their salts are treated with concentrated hydrochloric acid.   [J.M.B.]

12.6   Define the term transition element. State, with examples, three characteristic properties of transition metals. Account for what happens when:

(a)   potassium iodide solution is added to copper(II) sulphate(VI) solution;

(b)   ammonia solution is added to a cobalt(II) solution;

(c)   potassium manganate(VII) (permanganate) is added to a concentrated solution of alkali;

(d)   sodium chromate(VI) is added to dilute hydrochloric acid.

12.7   What are transition elements?

Discuss briefly, with examples, why zinc differs from the remainder of the first transition series?

Explain the following:

(a)   addition of copper(I) sulphate(VI) to water gives a coloured solution and an insoluble residue;

(b)   dissolving titanium(III) and titanium(IV) salts in dilute sulphuric(VI) acid gives coloured and colourless solutions respectively;

(c)   addition of acidified potassium dichromate(VI) solution to sodium sulphate(IV) (sodium sulphite) results in a colour change from orange to green;

(d)   a precipitate is formed when ammonia solution is added to an iron(II) salt solution unless excess sodium cyanide is added first.

12.8   A green crystalline compound Z has the formula $K_xFe_y(C_2O_4)_z \cdot 3H_2O$, where x, y, and z are whole numbers, and $C_2O_4^{2-}$ is the ethanedioate (oxalate) ion. The crystals were analysed quantitatively as follows.

(i)   In order to determine the ethanedioate, a weighed quantity was dissolved in $2 M\ H_2SO_4$, warmed to $60\ °C$, and titrated with $0.02 M$ potassium manganate(VII) (permanganate). It was found that $p\ cm^3$ were required. (In the titration, the iron was not affected.)

$$5C_2O_4^{2-} + 2MnO_4^- + 16H^+ \rightarrow 2Mn^{2+} + 10CO_2 + 8H_2O$$

(ii) To determine the iron, the solution left after titration with the manganate(VII) in (i) was reduced by shaking with zinc amalgam. The reduced solution was again titrated with 0.02 M $KMnO_4$ requiring $p/6$ cm^3

$$5Fe^{2+} + MnO_4^- + 8H^+ \rightarrow 5Fe^{3+} + 4H_2O + Mn^{2+}$$

(a) Deduce the ratio of z to y.

(b) The potassium can be determined 'by difference', i.e. by knowing the weights of iron, ethanedioate (oxalate), and water in a given mass of crystals and subtracting. It is found that the ratio of x to y is 3 : 1.
   (i) What value does this give for the oxidation number of the iron in compound Z?
   (ii) Is this value consistent with what you would expect from the information given earlier in the question? Explain.

(c) The compound Z is said to contain a 'complex ion'. Explain what is meant by the term 'complex ion'.

(d) What explanation can you give of the fact that a solution of Z gives a red colouration with ammonium thiocyanate solution, but only after acidification.

(e) Explain what is meant by the 'stability constant' with particular reference to the complex ion in Z.   [University of London]

12.9 In their compounds, the early members of the first transition series are usually in the highest possible oxidation states, whereas this is not true for the later numbers of the series. Discuss this statement, including in your answer reference to the following.

(a) The common oxidation states of titanium and vanadium.

(b) The common oxidation states of cobalt and nickel, including examples of actual compounds. Suggest why cobalt has two stable oxidation states.

(c) The reasons for the differences between (a) and (b).

(d) The effect of change of ligand on oxidation-reduction behaviour.   [J.M.B.]

12.10 (a) In aqueous solution, the colours of certain vanadium ions are as follows: VO_2^+ [i.e., V(V)] yellow, V^{4+} blue, V^{3+} green. When sulphur dioxide is passed through an acidic solution of ammonium vanadate (NH_4VO_3), the solution turns green. How would you find out experimentally whether the green colour was due to V^{3+}, or to a mixture of vanadium in oxidation states IV and V?

(b) The reaction between peroxodisulphate ions, $S_2O_8^{2-}$, and iodide ions is catalysed by some transition metal ions. Describe how you would compare the catalytic effect of Cr^{3+} and Fe^{3+} ions on this reaction.

(c) In fact, Fe^{3+} is found to be a catalyst, but not Cr^{3+}. Suggest a possible mechanism, in view of the following standard electrode potentials:

$$\begin{array}{ll} Fe^{3+}(aq), Fe^{2+}(aq) & +0.77 \text{ V} \\ S_2O_8^{2-}(aq), 2SO_4^{2-}(aq) & +2.01 \text{ V} \\ I_2(aq), 2I^-(aq) & +0.54 \text{ V} \end{array}$$

What prediction can be made about the standard electrode potential of $Cr^{3+}(aq)$, $Cr^{2+}(aq)$ ions in view of the observation that Cr^{3+} ions do not catalyse the reaction?   [University of London]

12.11 List the commonly occurring oxidation states in aqueous solution for each of the first row transition elements from Ti to Cu. Indicate any changes in stability of these states across the row.

Give two examples (using two different transition elements) to illustrate how a change of ligand can influence the oxidation–reduction behaviour of particular oxidation states.

Compare and contrast the chemistry of copper(II)(aq) and cobalt(II)(aq) by discussing the reactions of aqueous solutions of their sulphates(VI) with
(i) concentrated hydrochloric acid,
(ii) concentrated aqueous ammonia in the presence of air.   [J.M.B.]

12.12 2.135 g of a pure copper(II) sulphate(VI) hydrate were dissolved in water and the solution made up to 250 cm³. A 25 cm³ portion of this solution was treated with an excess of potassium iodide. The iodine liberated in the reaction was found to react with exactly 20 cm³ of 0.05 M sodium thiosulphate(VI) solution.

(a) Write an equation for the reaction between copper(II) sulphate(VI) solution and potassium iodide.

(b) Write an equation for the reaction of iodine with sodium thiosulphate(VI) solution.

(c) State an indicator suitable for the titration of iodine with sodium thiosulphate(VI).

(d) Give the name and empirical formula of the white precipitate remaining at the end of the titration.

(e) Calculate the mass of anhydrous copper(II) sulphate(VI) in the 2.135 g sample of the above hydrate.

(f) Calculate the mole ratio $CuSO_4 : H_2O$ in the sample and hence determine the formula of the hydrate. [J.M.B.]

13 Selected Chemistry of Some Cations and Anions

The first part of this chapter is concerned with the reactions of a number of aqua-cations with HO^-, CO_3^{2-}, Cl^-, Br^-, and I^- ions. This is followed by the identification of several anions from their reactions with silver nitrate(V), barium chloride, and sulphuric(VI) acid.

REACTIONS OF THE CATIONS Ag^+, Pb^{2+}, Fe^{2+}, Co^{2+}, Ni^{2+}, Cu^{2+}, Al^{3+}, Cr^{3+}, and Fe^{3+}

1. With sodium hydroxide

Precipitation occurs in each case when sodium hydroxide is added to solutions containing the above cations. However, a number of the hydroxides are amphoteric and so they redissolve if excess of the reagent is added.

The reaction between solutions of hydroxide and silver ions results in a transient white precipitate of silver hydroxide but this is unstable and immediately decomposes to give a dirty brown residue of silver oxide:

$$Ag^+(aq) + HO^-(aq) \rightarrow [AgOH](s)$$
$$2[AgOH](s) \rightarrow Ag_2O(s) + H_2O(l)$$

Addition of sodium hydroxide to an aqueous solution of Fe^{2+} ions gives the expected precipitate of iron(II) hydroxide:

$$Fe^{2+}(aq) + 2HO^-(aq) \rightarrow Fe(OH)_2(s)$$

If all oxygen is excluded, the precipitate is white but in the presence of air it is dirty green and this rapidly darkens as oxidation occurs. After a short period of time, any precipitate exposed to the atmosphere has the characteristic red-brown colour of iron(III) hydroxide.

Nickel(II), copper(II), and iron(III) hydroxides are obtained as pale green, blue, and red-brown precipitates respectively when hydroxide ions are added to solutions of the corresponding salts.

Lead(II) hydroxide is a white solid but this dissolves in excess sodium hydroxide giving a colourless solution of the plumbate(II). The plumbate(II) ion may have the structure $[Pb(OH)_4]^{2-}$ or $[Pb(OH)_6]^{4-}$.

Addition of hydroxide ion to a cobalt(II) salt solution gives a royal blue precipitate of cobalt(II) hydroxide which dissolves in excess of the reagent giving a blue solution containing the tetrahydroxocobaltate(II) ion, $[Co(OH)_4]^{2-}$. Solutions of aluminium salts react with sodium hydroxide yielding initially a white precipitate of aluminium hydroxide, but this dissolves in the presence of excess hydroxide ion to give a colourless solution of sodium aluminate(III), $NaAl(OH)_4$ (see page 192). Amphoteric behaviour also occurs with chromium(III) hydroxide, the greenish solid dissolving in excess alkali to give a green solution of a chromate(III):

$$[Cr(H_2O)_6]^{3+}(aq) \xrightarrow{\ HO^-\ } Cr(OH)_3(H_2O)_3(s) \xrightarrow{\ HO^-\ } [Cr(OH)_4]^-(aq)$$

2. With sodium carbonate

Carbonates are precipitated when aqueous sodium carbonate is added to solutions containing Ag^+, Pb^{2+}, Fe^{2+}, and Co^{2+} ions. These carbonates are coloured pale yellow, white, off-white, and mauve respectively.

As noted previously (page 181), the carbonate ion undergoes some hydrolysis in aqueous solution, i.e.

$$CO_3^{2-}(aq) + H_2O(l) \rightleftharpoons HO^-(aq) + HCO_3^-(aq)$$

Co-precipitation of the carbonate and hydroxide therefore occurs when solutions of some cations are treated with sodium carbonate, i.e. a basic carbonate is formed. Nickel(II) and copper(II) salts yield pale green and blue basic carbonates respectively from these reactions.

The addition of sodium carbonate to solutions of aluminium, chromium(III), and iron(III) salts results in the precipitation of the metal hydroxide, not the carbonate. This may be explained by the fact that the Al^{3+}, Cr^{3+}, and Fe^{3+} ions are all very small and highly charged and so are intensely polarising. Thus, their aqueous solutions are acidic owing to reactions of the type:

$$[Fe(H_2O)_6]^{3+} + H_2\ddot{O} \rightleftharpoons [Fe(H_2O)_5(OH)]^{2+} + H_3\overset{..}{O}{}^+$$

Any strongly basic anion such as CO_3^{2-} or S^{2-} will abstract the protons and so the insoluble hydroxides will be formed.

3. With chloride ions

Of the metals under discussion, only silver and lead form insoluble chlorides. A white precipitate of silver chloride is produced when silver nitrate(V) is added to a solution of a chloride. The precipitate dissolves in the presence of excess Cl^- giving $[AgCl_2]^-$, dichloroargentate(I) and is soluble in ammonia solution due to the formation of the diamminesilver(I) ion:

$$Ag^+(aq) + Cl^-(aq) \rightarrow AgCl(s) \xrightarrow{\ NH_3(aq)\ } [Ag(NH_3)_2]^+(aq) + Cl^-(aq)$$

Lead(II) chloride is obtained as a white precipitate when aqueous solutions of Pb^{2+} and Cl^- are mixed. The precipitate dissolves in the presence of excess Cl^-. For example, addition of concentrated hydrochloric acid results in the formation of a colourless solution of hydrogen tetrachloroplumbate(II).

$$Pb^{2+}(aq) + 2Cl^-(aq) \rightarrow PbCl_2(s)$$
$$PbCl_2(s) + 2HCl(conc) \rightarrow H_2[PbCl_4](aq)$$

Dilution of this solution causes re-precipitation of the lead(II) chloride.

A number of the other cations form complex ions in the presence of excess chloride ion. For example, pink solutions of cobalt(II) salts turn blue when excess concentrated hydrochloric acid is added (see page 243):

$$[Co(H_2O)_6]^{2+}(aq) \; \underset{H_2O}{\overset{HCl(conc)}{\rightleftharpoons}} \; [CoCl_4]^{2-}(aq)$$

In similar reactions, pale green $[Ni(H_2O)_6]^{2+}$ gives green $[NiCl_4]^{2-}$, and blue $[Cu(H_2O)_6]^{2+}$ gives green $[CuCl_4]^{2-}$. Addition of concentrated hydrochloric acid to iron(III) solutions results in the formation of yellow-brown $[FeCl_4]^-$ and $[FeCl_6]^{3-}$.

4. With bromide ions

Again precipitation occurs only with silver and lead.

$$Ag^+(aq) + Br^-(aq) \rightarrow AgBr(s) \quad (cream)$$
$$Pb^{2+}(aq) + 2Br^-(aq) \rightarrow PbBr_2(s) \quad (yellow)$$

Silver bromide is only slightly soluble in ammonia solution (see page 266). Lead(II) bromide dissolves in concentrated hydrobromic acid giving hydrogen tetrabromoplumbate(II), $H_2[PbBr_4]$. Some of the other cations form complexes in the presence of excess bromide ion but there is little visual change and they are relatively unimportant.

5. With iodide ions

Precipitation occurs here with Ag^+, Pb^{2+}, and Cu^{2+}. Silver iodide is a yellow solid, insoluble in ammonia solution (see page 266). Lead(II) iodide is a golden yellow solid which, like the chloride and bromide, dissolves in hot water and may be crystallised from it. It dissolves in concentrated solutions of iodide ions but is reprecipitated on dilution:

$$Pb^{2+}(aq) + 2I^-(aq) \rightarrow PbI_2(s) \; \underset{H_2O}{\overset{I^-}{\rightleftharpoons}} \; [PbI_3]^-(aq) \quad (colourless)$$

The reaction between aqueous iodide ions and copper(II) ions gives copper(I) iodide not the expected copper(II) iodide (see page 256). The white precipitate dissolves in the presence of excess iodide ions giving a brown solution containing the $[CuI_2]^-$ ion:

$$2Cu^{2+}(aq) + 4I^-(aq) \rightarrow 2CuI(s) + I_2(aq)$$
$$CuI(s) + I^-(aq) \rightarrow [CuI_2]^-(aq)$$

A summary of the above reactions of the cations is given in Table 13.1.

Ion

Reagent	Ag^+	Pb^{2+}	Fe^{2+}	Co^{2+}	Ni^{2+}	Cu^{2+}	Al^{3+}	Cr^{3+}	Fe^{3+}
H_2O	Ag^+(aq) Colourless	$[Pb(H_2O)_6]^{2+}$ Colourless	$[Fe(H_2O)_6]^{2+}$ Pale green	$[Co(H_2O)_6]^{2+}$ Pink	$[Ni(H_2O)_6]^{2+}$ Pale green	$[Cu(H_2O)_6]^{2+}$ Blue	$[Al(H_2O)_6]^{3+}$ Colourless	$[Cr(H_2O)_6]^{3+}$ Green or blue	$[Fe(H_2O)_6]^{3+}$ Yellow
HO^-	Ag_2O(s) Brown	$Pb(OH)_2$(s) White $\xrightarrow{HO^-}$ $[Pb(OH)_4]^{2-}$(aq) Colourless	$Fe(OH)_2$(s) Dirty green (in air)	$Co(OH)_2$(s) Royal blue $\xrightarrow{HO^-}$ $[Co(OH)_4]^{2-}$(aq) Blue	$Ni(OH)_2$(s) Pale green	$Cu(OH)_2$(s) Blue	$Al(OH)_3$(s) White $\xrightarrow{HO^-}$ $[Al(OH)_4]^-$(aq) Colourless	$Cr(OH)_3$(s) Dirty Green $\xrightarrow{HO^-}$ $[Cr(OH)_4]^-$(aq) Green	$Fe(OH)_3$(s) Red-brown
CO_3^{2-}	Ag_2CO_3(s) Pale yellow	$PbCO_3$(s) White	$FeCO_3$(s) Off-white	$CoCO_3$(s) Mauve	Basic carbonate Pale green	Basic carbonate Blue	$Al(OH)_3$(s) White	$Cr(OH)_3$(s) Dirty green	$Fe(OH)_3$(s) Red-brown
Cl^-	$AgCl$(s) White $\xrightarrow{NH_3(aq)}$ $[Ag(NH_3)_2]^+$(aq)	$PbCl_2$(s) White $\xrightarrow{Cl^-}$ $[PbCl_4]^{2-}$(aq)		With conc. HCl $[CoCl_4]^{2-}$(aq) Blue	With conc. HCl $[NiCl_4]^{2-}$(aq) Little visible change	With conc. HCl $[CuCl_4]^{2-}$(aq) Green			Little visible change $[FeCl_4]^-$ and $[FeCl_6]^{3-}$ formed
Br^-	$AgBr$(s) Cream Slightly soluble in NH_3(aq)	$PbBr_2$(s) Yellow $\xrightarrow{Br^-}$ $[PbBr_4]^{2-}$(aq)							
I^-	AgI(s) Yellow Insoluble in NH_3(aq)	PbI_2(s) Golden yellow $\xrightarrow{I^-}$ $[PbI_3]^-$(aq) Colourless				$[CuI_2]$ $\xrightarrow{}$ $CuI + I_2$ White $\xrightarrow{I^-}$ $[CuI_2]^-$(aq) Brown			

SOME REACTIONS OF THE ANIONS F⁻, Cl⁻, Br⁻, I⁻, NO_2^-, NO_3^- CO_3^{2-}, SO_3^{2-} AND SO_4^{2-}

1. With silver nitrate(V) solution

The only anions from the above list which are not precipitated by silver ions are fluoride and nitrate(V). These exceptions are not unexpected. It has been explained previously (page 219) that the solubilities of fluorides are often out of step with those of the other halides, whilst practically all nitrates(V) are soluble.

Mixing solutions of chlorides, bromides, and iodides with silver nitrate(V) solution yields precipitates of the silver halides, the colours being white, cream, and yellow respectively. These precipitates differ in their behaviour towards ammonia solution. Silver chloride is slightly soluble in water and so undissolved solid is in equilibrium with its ions in solution. If ammonia solution is added, Ag^+ (aq) ions are removed as the soluble diamminesilver(I) complex and so more silver chloride dissolves:

$$Ag^+Cl^-(s) + H_2O(l) \rightleftharpoons Ag^+(aq) + Cl^-(aq)$$

$$\downarrow NH_3(aq)$$

$$[Ag(NH_3)_2]^+(aq)$$

All the silver chloride will dissolve if sufficient ammonia solution is added. Now, the formation of the diammine complex is an equilibrium process:

$$Ag^+(aq) + 2NH_3(aq) \rightleftharpoons [Ag(NH_3)_2]^+(aq)$$

and so its formation is aided by high Ag^+ or NH_3 concentrations. Since silver bromide is far less soluble than the chloride, the Ag^+ concentration in solution is very low and so the diammine complex can be formed only by using much higher concentrations of ammonia. Thus, silver bromide is only sparingly soluble in dilute ammonia solution. Silver iodide is even less soluble than the bromide and so its solubility in dilute ammonia solution is negligible. The solubility products of silver chloride, bromide, and iodide are 2×10^{-10}, 5×10^{-13}, and 8×10^{-17} mol² litre⁻² respectively.

A white precipitate of silver nitrate(III) (silver nitrite) is obtained when aqueous silver nitrate(V) is added to a concentrated nitrate(III) (nitrite) solution, e.g.

$$AgNO_3(aq) + NaNO_2(aq) \rightarrow AgNO_2(s) + NaNO_3(aq)$$

The precipitate is not obtained if an acidified nitrate(III) (nitrite) solution is used since the resultant nitric(III) acid (nitrous acid) is unstable and decomposes even at room temperature:

$$3HNO_2(aq) \rightarrow HNO_3(aq) + H_2O(l) + 2NO(g)$$
$$2NO(g) + O_2(g) \rightarrow 2NO_2(g)$$

Silver carbonate is obtained as a white precipitate when solutions of a carbonate and silver nitrate(V) are mixed. The precipitate dissolves in dilute nitric(V) acid due to destruction of the carbonate:

$$Ag_2CO_3(s) + 2HNO_3(aq) \rightarrow 2AgNO_3(aq) + H_2O(l) + CO_2(g)$$

i.e. $$CO_3^{2-}(aq) + 2H^+(aq) \rightarrow H_2O(l) + CO_2(g)$$

Similarly, mixing solutions of sulphates(IV) (sulphites) and silver nitrate(V) gives a white precipitate of silver sulphate(IV) (sulphite) which is soluble in dilute nitric(V) acid:

$$2AgNO_3(aq) + SO_3^{2-}(aq) \rightarrow Ag_2SO_3(s) + 2NO_3^-(aq)$$
$$Ag_2SO_3(s) + 2HNO_3(aq) \rightarrow 2AgNO_3(aq) + H_2O(l) + SO_2(g)$$

i.e.
$$SO_3^{2-}(aq) + 2H^+(aq) \rightarrow H_2O(l) + SO_2(g)$$

Silver sulphate(VI) is moderately soluble in water and so is precipitated only when a solution of a sulphate(VI) is added to concentrated silver nitrate(V). Silver sulphate(VI) is white and is insoluble in dilute nitric(V) acid.

In summary, it is seen that treating solutions of anions, acidified with nitric(V) acid, with dilute silver nitrate(V) solution will lead to precipitation only in the case of chlorides, bromides, and iodides. The resultant silver halides can be distinguished by their behaviour with ammonia solution.

2. With barium chloride solution

Barium chloride, bromide, iodide, nitrate(III) (nitrite), and nitrate(V) are all water-soluble. Hence, addition of aqueous barium chloride to solutions of the anions will result in precipitation only in the cases of fluoride, carbonate, sulphate(IV) (sulphite), and sulphate(VI). The precipitate is white in each case. Barium carbonate and sulphate(IV) (sulphite), like the corresponding silver salts (see above), are destroyed by acid. Thus, barium fluoride and sulphate(VI) are the only anions precipitated from solutions acidified with dilute hydrochloric acid. Fluorides can be distinguished from sulphates(VI) by their reaction with concentrated sulphuric(VI) acid (see below).

3. With sulphuric(VI) acid

Nitrates(III) (nitrites), carbonates, and sulphates(IV) (sulphites) react with dilute sulphuric(VI) acid. Treatment of a solid nitrate(III) with dilute sulphuric(VI) acid gives nitric(III) acid (nitrous acid) and, as stated above, this readily decomposes. The nitrogen oxide produced reacts with oxygen in the air giving nitrogen dioxide and so brown fumes are observed. (If a cooled solution of a nitrate(III) (nitrite) is acidified with cold dilute sulphuric(VI) acid, decomposition of the nitric(III) acid is retarded and the solution may be used to form diazonium salts — see page 343.)

Carbon dioxide and sulphur dioxide are formed when dilute sulphuric(VI) acid is added to a carbonate or sulphate(IV) (sulphite) respectively. Both gases are colourless and turn limewater milky. However, they may· be distinguished by the fact that the latter gas is a reducing agent and turns filter paper dipped in acidified potassium dichromate(VI) from orange to green:

$$Cr_2O_7^{2-}(aq) + 2H^+(aq) + 3SO_2(g) \rightarrow 2Cr^{3+}(aq) + 3SO_4^{2-}(aq) + H_2O(l)$$

The halides and nitrates(V) react with concentrated sulphuric(VI) acid slowly in the cold but more rapidly on warming. With fluorides and chlorides, the hydrogen halides are produced:

$$X^-(s) + H_2SO_4(conc) \rightarrow HSO_4^- + HX(g)$$

Both are colourless gases which fume in moist air but they may be distinguished by the fact that hydrogen fluoride attacks moist glass. Thus, a moist glass rod held in hydrogen fluoride becomes covered with a gelatinous deposit of hydrated silicon(IV) oxide.

$$SiO_2(s) + 4HF(g) \rightarrow 2H_2O(l) + SiF_4(g)$$
$$4H_2O(l) + 3SiF_4(g) \rightarrow SiO_2 \cdot 2H_2O(s) + 2H_2SiF_6(aq)$$

A red-brown vapour is produced when bromides are warmed with concentrated sulphuric(VI) acid. This is due to the redox reaction between the acid and the hydrogen bromide:

$$Br^-(s) + H_2SO_4(conc) \rightarrow HSO_4^- + HBr(g)$$
$$H_2SO_4(conc) + 2HBr(g) \rightarrow 2H_2O(l) + SO_2(g) + Br_2(g)$$

The products from the reaction fume in moist air and so it is apparent that not all the hydrogen bromide is oxidised. Similarly, violet iodine vapour is produced when an iodide is warmed with concentrated sulphuric(VI) acid. Hydrogen iodide is a powerful reducing agent and so the reaction products contain less hydrogen halide than is the case with the corresponding bromide reaction.

When a nitrate(V) is warmed with concentrated sulphuric(VI) acid, nitric(V) acid is formed. However, the vapours are brown due to the presence of nitrogen dioxide resulting from some decomposition of the latter acid.

$$NO_3^-(s) + H_2SO_4(conc) \rightarrow HSO_4^- + HNO_3(l)$$
$$4HNO_3(l) \rightarrow 2H_2O(l) + 4NO_2(g) + O_2(g)$$

Although similar results are obtained with bromides and nitrates(V) in this reaction, these anions may be distinguished by their reaction with silver nitrate(V) — see above. Nitrates(V) are also detected by the brown ring test (page 204).

The reactions of the anions are summarised in Table 13.2.

Question

13.1   You are given six aqueous solutions labelled A to F, each containing one of the following solutes (no two solutions are the same): NaOH; K_2CO_3; $BaCl_2$; $Cr_2(SO_4)_3$; $NiCl_2$; $(NH_4)_2SO_4$. Using these solutions together with a known solution of dilute hydrochloric acid and red and blue litmus paper, devise a logical series of tests which allows you to identify the solute in each case. (No other reagent may be used and test-tubes and a bunsen burner are the only apparatus permitted.)

Suggest also one confirmatory test for the cation in each solution, for which any suitable reagents and apparatus may be used.   [J.M.B.]

Table 13.2 Some reactions of anions

Reagent	Ion								
	F^-	Cl^-	Br^-	I^-	NO_2^-	NO_3^-	CO_3^{2-}	SO_3^{2-}	SO_4^{2-}
$AgNO_3$	No precipitate (AgF is soluble)	$AgCl(s)$ White Insoluble in dil. HNO_3 but soluble in $NH_3(aq)$	$AgBr(s)$ Cream Insoluble in dil. HNO_3, slightly soluble in $NH_3(aq)$	$AgI(s)$ Yellow Insoluble in dil. HNO_3 and in $NH_3(aq)$	$AgNO_2(s)$ (From conc. solns.) White	—	$Ag_2CO_3(s)$ White Soluble in dil. HNO_3	$Ag_2SO_3(s)$ White Soluble in dil. HNO_3	$Ag_2SO_4(s)$ (From conc. solns. of $AgNO_3$) White Insoluble in dil. HNO_3
$BaCl_2$	$BaF_2(s)$ White Insoluble in dilute HCl	—	—	—	—	—	$BaCO_3(s)$ White Soluble in dilute HCl	$BaSO_3(s)$ White Soluble in dilute HCl	$BaSO_4(aq)$ White Insoluble in dilute HCl
Dilute H_2SO_4	—	—	—	—	Gives HNO_2 which decomposes giving brown fumes	—	$CO_2(g)$ Turns lime-water milky but no reaction with $H^+/Cr_2O_7^{2-}$	$SO_2(g)$ Turns lime-water milky and $H^+/Cr_2O_7^{2-}$ green	—
Warm conc. H_2SO_4	$HF(g)$ Fumes in moist air, attacks wet glass	$HCl(g)$ Fumes in moist air	$Br_2(g) + HBr(g)$ Brown mixture fumes in moist air	$I_2(g) + HI(g)$ Violet vapour fumes in moist air	Very vigorous reaction	$HNO_3(g)$ Some reddish brown vapour due to NO_2	Very vigorous reaction	Very vigorous reaction	—

14 Fundamentals of Organic Chemistry

Organic chemistry refers to the compounds of carbon with the exception of a few compounds such as carbon monoxide, carbon dioxide, the metal carbonates, and the disulphide which are usually classed as inorganic compounds.

There are far more compounds of carbon than of all the other elements put together, even though a relatively narrow range of elements is involved with the carbon. This is mainly due to carbon having a valency of four and its ability to form stable bonds with itself in chains of virtually any length (see catenation of the group IV elements on page 194). There are, for example, thousands of compounds containing only carbon and hydrogen — the carbon atoms can be in an unbranched chain, e.g. butane, $CH_3 \cdot CH_2 \cdot CH_2 \cdot CH_3$ and poly(ethene), $\left(CH_2 - CH_2 \right)_n$ where n is several thousand, in branched chains, e.g. 2-methyl-propane, $CH_3 - \overset{\displaystyle |}{CH} - CH_3$, or in rings, e.g. cyclohexane,

$$
\begin{array}{ccc}
 & CH_2 & \\
 \diagup & & \diagdown \\
CH_2 & & CH_2 \\
| & & | \\
CH_2 & & CH_2 \\
 \diagdown & & \diagup \\
 & CH_2 &
\end{array}
$$

The carbon atoms can also form double or triple bonds with themselves, e.g. propene, $CH_3 \cdot CH = CH_2$, and ethyne (acetylene), $CH \equiv CH$ or with other elements, e.g. propanone (acetone),

$$
\begin{array}{c}
O \\
\| \\
CH_3 - C - CH_3
\end{array}
$$

and ethanenitrile (methyl cyanide), $CH_3 - C \equiv N$. The molecules may be very simple as in the examples above, more complex as in the case of nicotine* or extremely complex as in some living cells.

Organic compounds are extremely important since they are present in all living matter and in many commercial products, for example, petroleum products,

*The structure of nicotine is given in the universally accepted shorthand method. Thus, it is understood that at each intersection of the lines there is a carbon atom together with sufficient hydrogen atoms to make the valency of the carbon up to four.

rubber, perfumes and flavours, antibiotics, vitamins, insecticides, explosives and propellants, plastics and synthetic fibres, refrigerants, dyes, sugars, proteins, paints, etc.

PURIFICATION OF ORGANIC COMPOUNDS

Before an organic compound can be characterised, it must be separated from the reaction mixture, or from other substances if it is obtained from a naturally occurring source, and then purified. Some of the common techniques employed for this are briefly described below.

1. Solvent extraction

In organic preparations the required product is often obtained as an aqueous solution or as a suspension in water. The most effective way of separating it is to shake it with an organic solvent in which it is soluble and which is itself almost insoluble in water. Repeated extraction will eventually transfer the compound almost entirely into the organic solvent (see page 99). Ethoxyethane (diethyl ether) is usually the most suitable solvent because:
(a)   it is chemically inert and so unlikely to react with the compound,
(b)   it is very volatile (b.p. 35 °C) and so readily removed from the compound.

The separations are performed in a pear shaped separating funnel and during the shaking the pressure of the ethoxyethane (ether) caused by the heat of the hands has to be released occasionally. The two layers are then allowed to separate, the lower aqueous layer is run off and then re-extracted with more ethoxyethane (ether). The various extracts are then combined, dried, and distilled.

If the compound is very soluble in water, then many extractions would be necessary. To overcome this the aqueous solution is saturated with salt to lower the compound's solubility — this process is known as *salting out*.

Common drying agents with their practical limitations are given below.

Calcium chloride   — cannot be used for alcohols, phenols, and amines since it forms addition compounds with them, or for acidic solutions because it contains some calcium hydroxide.

Potassium hydroxide   — used for drying amines. It cannot be used with acids, phenols, or esters because it reacts with them.

Sodium sulphate(VI) (anhydrous)   — may be used with almost anything but it is slow.

Magnesium sulphate(VI) (anhydrous)   — may be used with almost anything, faster than sodium sulphate(VI).

Calcium sulphate(VI) ('drierite')   — may be used with all liquids.

Sodium wire   — used for drying ethers or hydrocarbons but not compounds which are affected by alkalis or compounds which are readily reduced $(2Na + 2H_2O \rightarrow 2NaOH + H_2)$.

2. Distillation

If the impurities in a liquid are non-volatile they can be separated by distillation, the impurities remaining in the distillation flask (Figure 14.1). The liquid is placed in the flask which should not be more than three-fifths full. A few pieces of unglazed porcelain are added to provide nuclei for the vapour bubbles to form and so ensure steady boiling — if these are not added the liquid is liable to superheat and bump violently. The thermometer bulb is level with the side-arm so that the temperature of the vapour is measured as it leaves the flask. If the liquid boils below 140 °C a water condenser is used but, if it boils above this temperature, an air condenser (a straight glass tube with no jacket) is used to eliminate the risk of cracking. The flask is heated by a water-bath for distillations of low-boiling-point flammable liquids, otherwise a bunsen or an oil-bath is used.

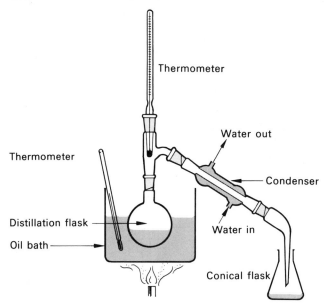

Figure 14.1   Simple distillation apparatus

If a liquid contains volatile impurities, then the distillation above would result only in partial separation because the vapours of both liquids would distil over. In such cases a fractionating column is used vertically between the flask and condenser. (For details see page 103.)

For very high boiling liquids or for liquids that decompose fairly easily, the distillation is performed under reduced pressure.

3. Steam distillation

Many liquid or solid compounds which are insoluble in water can be purified by distillation in a current of steam, provided that they are volatile under these conditions, and the impurities are not. The apparatus used is illustrated in Figure 14.2.

Water vapour and the steam volatile compound distil over. The purified compound may be separated from the water by filtration in the case of a solid or ethoxyethane (ether) extraction in the case of a liquid.

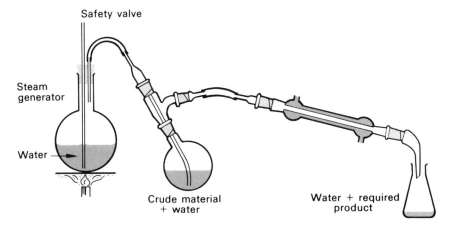

Figure 14.2   Apparatus for steam distillation

Steam distillation is based on the fact that the vapour pressure above an immiscible mixture is equal to the sum of the two separate vapour pressures. As a result of this, the water and the compound will distil over at a lower temperature than the boiling point of either (see page 105).

4. Crystallisation

The process of crystallisation involves finding a solvent which will readily dissolve the crude material when hot, but only to a small extent when cold. The crude material is then dissolved in the minimum volume of the boiling solvent, the solution filtered if necessary to remove any insoluble impurities, and then cooled.

The impurities will remain in solution and most of the required product will crystallise out. The crystals are filtered off and dried, and, if necessary, the process is repeated until the crystals are pure (as indicated by a sharp melting point, etc.).

Recrystallisation does not necessarily depend on the assumption that the impurities are more soluble than the required product. Consider a mixture containing 97% of the required compound A and 3% of impurity B. Two alternatives arise.

(a) A is less soluble than B in a given solvent. Here recrystallisation will obviously soon leave all B in the mother liquors (i.e. filtrate).

(b) A is more soluble than B in a given solvent. Suppose that a given volume of the cold solvent can dissolve 15 g of A and 5 g of B. If 100 g of the mixture is dissolved in that volume of hot solvent and the solution is then allowed to cool, then 82 g (97–15 g) of A will crystallise out whilst all B remains in solution since the solution is not saturated with B. This ignores the small mutual effect on the solubility of each substance due to the presence of the other. It should be noted that in all crystallisations, some of the required product is lost, i.e. it remains in the solvent. In the above case the loss is 15 g.

The choice of solvent is largely a matter of trial and error, tests being made on very small samples. Methylated spirits is very useful, but ethoxyethane (ether) is rather unsatisfactory because of its flammability and its tendency to creep up the walls of the containing vessel and deposit crystals by complete evaporation.

In organic chemistry, crystals are invariably removed from solvent by filtration under reduced pressure using a Buchner funnel and flask (Figure 14.3). This method is used because it is much more rapid and effective than the normal way. When no more solvent can be sucked out, the crystals may be dried by pressing them between layers of filter paper. The final traces of solvent may then be removed by drying in a steam oven, provided the substance is physically and chemically stable up to 100 °C. Alternatively the substance is placed in a vacuum desiccator containing concentrated sulphuric(VI) acid if water is the solvent, sodium hydroxide flakes if an acid solvent is used, e.g. ethanoic acid (acetic acid) or hydrochloric acid, or paraffin wax if a hydrocarbon solvent, e.g. petroleum ether, benzene, etc. is used.

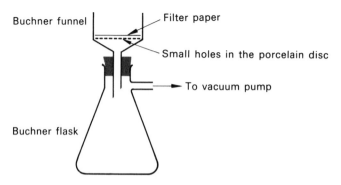

Figure 14.3   Buchner funnel and flask

5. Chromatography

Organic compounds may be separated by various chromatographic methods. For example, a solution of the impure compound can be allowed to run down a column packed with a suitable material (Figure 14.4). The various components split up into bands as they pass down the column at different rates. Solvent can then be allowed to run through until the various bands have been eluted and each collected or else the column material may be carefully blown out and cut up into sections and each section extracted with a suitable solvent.

If the components of the mixture are colourless, then their location is more difficult and various methods have to be tried; for example, seeing if they fluoresce under ultraviolet light or collecting many small samples of the eluate, evaporating off the solvent and examining any residue.

Column chromatography may be divided into two types — adsorption and partition. In adsorption chromatography, separation occurs because as the various components pass down the column they are adsorbed to different extents by the packing material (often alumina, Al_2O_3). In partition chromatography the separation is due to differential partitioning (see page 99) between the solvent and the water held by the packing material. A common partitioning material is cellulose.

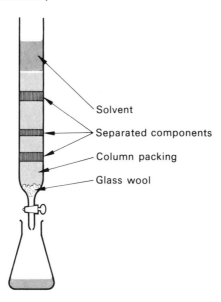

Figure 14.4   Column chromatography

Very small quantities (a few mg) of material can be separated by thin layer or paper chromatography. Here the basic principles are the same — a solution of the mixture passes over a thin layer of alumina, cellulose, etc., and the various components are adsorbed or partitioned at different rates. The ratio of the distance travelled by the substance to the distance travelled by the solvent front is known as the R_f value and is characteristic of the substance provided that the conditions are unaltered.

Gas–liquid chromatography (G.L.C.) or vapour phase chromatography (V.P.C.) also finds considerable application. The basis of this is that a gas, or the vapour of a liquid or volatile solid, passes down a column containing an inert solid impregnated with a non-volatile liquid known as the *stationary phase*. The vapour is carried along by an inert carrier gas such as nitrogen or argon, the inert solid can be anything from celite (a form of calcium carbonate) to crushed fire-brick, and the stationary phase is often a non-volatile ester or a silicone oil. The column length may vary from about one to ten metres, and it can be heated in a thermostatically controlled oven if necessary. The retention time (i.e. the time it spends in the column) depends on how strongly the substance is attracted by the stationary phase, and it is characteristic of the substance for a given set of conditions. The apparatus is illustrated in Figure 14.5.

In gas–liquid chromatography, the presence of a substance in the carrier gas, as it emerges from the column, may be detected by measuring the changes in thermal conductivity of the effluent. By using a recorder in conjunction with the detector, the results can be obtained on a chart as a peak in an otherwise straight line. If a sample gives more than one peak on gas–liquid chromatographic analysis it is obviously not pure. Chromatography is, in fact, used for testing the purity of organic compounds as well as a means of purification.

The purity of a solid is often indicated by its melting point. Thus, a pure solid usually has a very sharp melting point whilst an impure one melts over a range of temperature lower than that of the pure substance. This fact is utilised in the technique of mixed melting point determinations in which the sample under test is

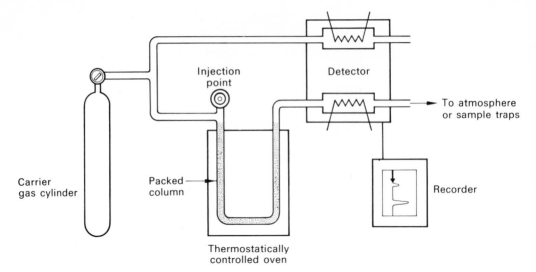

Figure 14.5   The essentials of a gas chromatograph

mixed with some of the substance which is known to be pure. If the sample is impure, the mixture will be found to have a depressed melting point.

The identity of a substance may be confirmed by mixing it with an authentic sample, and then determining the melting point of the mixture. If no depression of the melting point occurs, the two substances are the same.

DETERMINATION OF EMPIRICAL, MOLECULAR, AND STRUCTURAL FORMULAE

When a substance has been purified, attempts can be made to elucidate its structure. This involves determining in turn its empirical, molecular, and structural formulae. The *empirical formula* shows the simplest whole number ratio of the atoms of each element in the molecule. The *molecular formula* gives the actual number of atoms of each element in the molecule, whilst the *structural formula* illustrates how these atoms are arranged in the molecule. For example, ethanoic acid (acetic acid) has an empirical formula CH_2O, molecular formula $C_2H_4O_2$, and the structural formula

$$
\begin{array}{ccc}
\text{H} & & \text{O} \\
 & \diagdown & \diagup \\
\text{H}-\text{C}-\text{C} & & \\
 & \diagup & \diagdown \\
\text{H} & & \text{O}-\text{H}
\end{array}
$$

The determination of these formulae will be discussed briefly below.

Empirical formula determination

The initial step is qualitative analysis (finding which elements are present). Carbon and hydrogen are detected by heating a little of the compound with excess copper(II) oxide. Any carbon present is oxidised to carbon dioxide which is detected by its action on limewater, and hydrogen is oxidised to water which is detected by its property of turning anhydrous copper(II) sulphate(VI) blue. Nitrogen, sulphur, and the halogens are detected by the Lassaigne sodium fusion method. This involves fusing a little of the compound with sodium, whereby, any nitrogen, sulphur, chlorine, bromine, or iodine is converted to sodium cyanide, sulphide, chloride, bromide, or iodide respectively and these anions are detected by the standard inorganic tests. Metals are detected by heating some of the compound in air and examining any residue by the standard inorganic methods. Oxygen is not normally tested for.

Once the elements present are known, the compound is analysed quantitatively, i.e. to find how much of each element is present.

Carbon and hydrogen are determined by heating a known mass of the compound with copper(II) oxide in a stream of oxygen. The hydrogen is converted to water which is absorbed in weighed magnesium chlorate(VII) (magnesium perchlorate or 'anhydrone') tubes and the carbon is converted into carbon dioxide and absorbed in weighed soda-asbestos tubes (sodium hydroxide held on asbestos gauze). The percentages of carbon and hydrogen can be calculated from the masses of carbon dioxide and water produced respectively from the known masses of compound as illustrated by the example below.

Example   0.0234 g of an organic compound produced 0.0792 g carbon dioxide and 0.0162 g water on combustion analysis. Given that the relative atomic masses of carbon, hydrogen, and oxygen are 12, 1, and 16 respectively, calculate the percentage of carbon and hydrogen present in the compound.

44 g of carbon dioxide contain 12 g of carbon

∴   0.0792 g of carbon dioxide contain $\dfrac{12}{44} \times 0.0792 = 0.0216$ g carbon

∴   Percentage carbon in the compound $= \dfrac{0.0216}{0.0234} \times 100 = 92.3\%$

18 g of water contain 2 g of hydrogen

∴   0.0162 g of water contain $\dfrac{2}{18} \times 0.0162 = 0.0018$ g hydrogen

∴   Percentage hydrogen in the compound $= \dfrac{0.0018}{0.0234} \times 100 = 7.7\%$

Nitrogen is often determined by the Kjeldahl method. A known mass of the compound is boiled with concentrated sulphuric(VI) acid and a catalyst so that the nitrogen is converted into ammonium sulphate(VI). The ammonium sulphate(VI) is boiled with sodium hydroxide solution and the liberated ammonia, is absorbed in a known volume of standard acid. The amount of acid consumed is found by back titration using standard alkali, and the percentage of nitrogen in the compound is calculated from the results as in the example below.

Example   In a Kjeldahl determination using 0.250 g of an organic compound, the ammonia produced was absorbed in 20.0 cm³ of 0.5 M hydrochloric acid. The resultant solution required 16.3 cm³ of 0.5 M sodium hydroxide to neutralise it.

Calculate the percentage of nitrogen in the compound given that the relative atomic masses of hydrogen and nitrogen are 1 and 14 respectively.

From the results it is apparent that the acid used by the ammonia = $20.0 - 16.3 = 3.7 \text{ cm}^3$ of 0.5 M HCl.

Now, since, by definition, 1000 cm^3 of 0.5 M hydrochloric acid contain 0.5 mole hydrochloric acid, 3.7 cm^3 will contain $\frac{0.5}{1000} \times 3.7 = 0.00185$ mole hydrochloric acid.

However, the equation $HCl + NH_3 \rightarrow NH_4Cl$ shows that 1 mole of hydrochloric acid is equivalent to 1 mole of ammonia. Therefore, in the above determination 0.00185 mole of ammonia must have been produced. Since 1 mole of ammonia contains 14 g of nitrogen, 0.00185 mole contain 14×0.00185 g nitrogen.

$$\therefore \quad \text{The percentage nitrogen in the compound} = \frac{14 \times 0.00185}{0.250} \times 100 = 10.4\%$$

Halogens are determined by the Carius method. A known mass of the compound is heated in a sealed tube with fuming nitric(V) acid and a few crystals of silver nitrate(V). The silver halide produced is filtered, washed, dried, and weighed. The calculation involved in finding the percentage of halogen is illustrated below.

Example 0.2355 g of an organic compound known to contain bromine produced 0.2819 g of silver bromide in a Carius determination. Given that the relative atomic masses of bromine, and silver are 80 and 107.9 respectively, calculate the percentage of bromine in the compound.

The mass of bromine in the compound is the same as the mass in the silver bromide, i.e.

$$\frac{80}{187.9} \times 0.2819 = 0.1200 \text{ g bromine}$$

$\therefore$ The percentage bromine in the compound

$$= \frac{0.1200}{0.2355} \times 100 = 51.0\%$$

Sulphur is also determined by the Carius method, but in this case no silver nitrate(V) is added. The sulphur is converted into sulphuric(VI) acid and addition of barium chloride solution to this gives barium sulphate(VI) which is filtered off, washed, dried, and weighed.

Example 0.2700 g of an organic sulphur compound produced 0.3510 g of barium sulphate(VI) in a Carius determination. Calculate the percentage of sulphur in the compound. The relative atomic masses are

$$S = 32, \quad O = 16, \quad Ba = 137.4$$

The mass of sulphur in the barium sulphate(VI) $= \frac{32}{233.4} \times 0.3510 = 0.0481 \text{ g}$

$\therefore$ The percentage of sulphur in the compound $= \frac{0.0481}{0.2700} \times 100 = 17.8\%$

Oxygen is normally estimated by difference, i.e. it is taken as the remainder when all the other percentages have been subtracted from one hundred.

Now that the percentage by mass of each element in the compound is known the empirical formula may be calculated. The percentage of each element is divided by its relative atomic mass to give the fractional ratio of the moles of atoms present. All the ratios are then divided by the smallest to give the simplest whole number ratio.

Example An organic compound contains 40.0% carbon and 6.7% hydrogen; calculate the empirical formula.

Element	C	H	O
Percentage	40.0	6.7	53.3 (by difference)
Percentage/relative atomic mass	$\dfrac{40.0}{12}$	$\dfrac{6.7}{1}$	$\dfrac{53.3}{16}$
Fractional ratio of atoms	3.33	6.7	3.33
Fractional ratio/smallest number	$\dfrac{3.33}{3.33}$	$\dfrac{6.7}{3.33}$	$\dfrac{3.33}{3.33}$
Whole number ratio of atoms	1	2	1

The empirical formula is therefore CH_2O.

Small errors in the ratio will obviously occur owing to experimental errors in the quantitative analyses. However, if significant differences from whole numbers are obtained in the final line of the calculation, these figures must be multiplied up until they do give the simplest whole number ratio.

Example Calculate the empirical formula of an organic compound containing 81.8% carbon and 18.2% hydrogen.

Element	C	H
Percentage	81.8	18.2
Percentage/relative atomic mass	$\dfrac{81.8}{12}$	$\dfrac{18.2}{1}$
Fractional ratio of atoms	6.82	18.2
Fractional ratio/smallest number	$\dfrac{6.82}{6.82}$	$\dfrac{18.2}{6,82}$
'Whole number ratio'	1	2.67

The $C:H$ ratio of $1:2.67$ is too far from whole numbers to be attributed to experimental error and so the lowest multiple of these which gives a whole number ratio is the empirical formula, i.e.

$$1:2.67 \quad 2:5.34 \quad 3:8.01 \quad 4:10.68 \quad \text{etc.}$$

It is apparent from this that the empirical formula is C_3H_8.

Molecular formula determination

The molecular formula is the same as, or a multiple of, the empirical formula. It is found by multiplying the empirical formula by the number of times its relative molecular mass goes into the true relative molecular mass.

Example If the empirical formula of a compound is CH_2O and its relative molecular mass is 60, what is its molecular formula? Relative atomic masses:

$$C = 12, \qquad H = 1, \qquad \text{and} \quad O = 16.$$

The relative molecular mass of $CH_2O = 12 + (2 \times 1) + 16 = 30$

$\therefore$ The molecular formula $= (CH_2O) \times \frac{60}{30}$,   i.e. $(CH_2O) \times 2 = C_2H_4O_2$.

The relative molecular mass determination does not need to be highly accurate since the molecular formula must be a whole number multiple of the empirical formula. The following methods are commonly used and are discussed in Chapter 5.

(a) Relative density (vapour density) measurements (relative molecular mass $= 2 \times$ relative density). This method is used for vapours and volatile liquids.

(b) Depression of the freezing point or elevation of the boiling point methods. These methods are used for non-volatile liquids and solids.

It should be noted that the molecular formula of a compound does not fully characterise it. For example, both methyl methanoate (methyl formate),

$$H-C{\underset{\textstyle OCH_3}{\overset{\textstyle O}{<}}}$$

and ethanoic acid (acetic acid),

$$CH_3-C{\underset{\textstyle OH}{\overset{\textstyle O}{<}}}$$

have a molecular formula of $C_2H_4O_2$. The next stage is, therefore, the structural formula determination.

Structural formula determination

Many organic compounds contain *functional groups*, that is, groups of atoms which occur together and the group as a whole possesses its own characteristic properties, e.g. the carboxylic acid group,

$$-C{\underset{\textstyle OH}{\overset{\textstyle O}{<}}}$$

Once the molecular formula is known, the arrangement of the atoms in the molecule is deduced by studying its reactions and finding what functional groups are present. More complicated molecules are often broken down into smaller molecules of known structure. Once a provisional structure has been decided upon, the compound is synthesised via an unambiguous route to confirm that the proposed structure is correct.

There are now several important physical methods which can aid structure determination and some of these are discussed briefly below.

(a) *X-ray crystallography* (see page 39). The diffraction pattern can be used to deduce the relative positions of atoms in crystalline compounds. However, the accurate location of hydrogen atoms by the method is difficult or impossible.

(b) *Infrared spectroscopy.* Atoms joined by covalent bonds are in a state of continual vibration and each type of bond vibrates at its own characteristic frequency. If a compound is subjected to infrared radiation of gradually varying frequency, each type of bond will absorb the radiation at its own characteristic frequency. An infrared spectrum shows which frequencies are absorbed and so the various types of bond and functional groups may be detected. Figure 14.6 shows the infrared spectrum of ethyl 3-oxo-butanoate ($CH_3 \cdot CO \cdot CH_2 \cdot COOC_2H_5$).

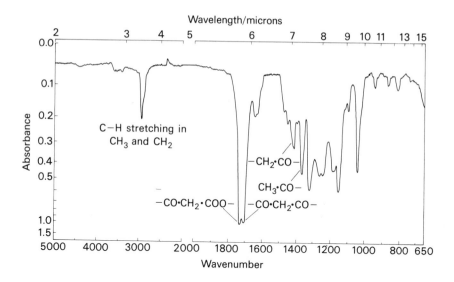

Figure 14.6   The infrared spectrum of ethyl 3-oxobutanoate

(c) *Mass spectroscopy.* Bombarding organic compounds with high energy electrons results in the formation of positively charged molecule ions, e.g.

$$C_4H_9Br + e^- \rightarrow [C_4H_9Br]^+ + 2e^-$$

Generally, the mass spectrometer enables the charge-to-mass ratio of these ions to be determined before they break down. Hence, very accurate relative molecular mass determinations are possible. However, the method also gives other extremely useful information. Thus the molecular ions produced in the mass spectrometer split up into charged fragments, and a knowledge of the relative masses and relative abundance of these fragments enables the structure of the parent compound to be deduced. It should be remembered that the mass spectrometer identifies only the positively charged fragments (page 5) and so atoms and radicals are not detected.

The mass spectrum of 1-bromobutane is illustrated in Figure 14.7. The peak for the most abundant ion, in this case $C_4H_9^+$, is known as the base peak and the ion is said to have an abundance of 100. The abundances of the other ions are then expressed relative to the base peak. The $C_4H_9^+$ ion is produced by the molecular ion losing a bromine atom. However, heterolytic fission (page 284) of the C—Br bond to give the Br^+ ion does occur to a small extent. The spectrum shows two peaks of similar size for Br^+ ions because bromine has two naturally occurring isotopes, $^{79}_{35}Br$ and $^{81}_{35}Br$, in approximately equal proportions. Obviously doublets are also given by ions containing bromine. Some of the smaller peaks in the spectrum arise from ions containing the ^2_1H and $^{13}_6C$ isotopes.

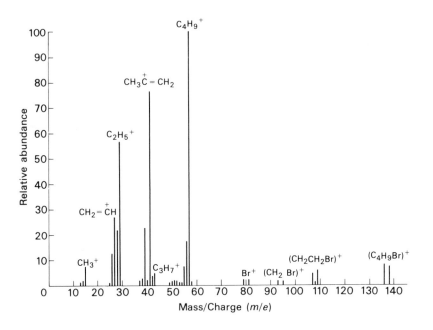

Figure 14.7   The mass spectrum of 1-bromobutane

The charge-to-mass ratios indicate that there is some tendency for each of the C—C bonds in the $[C_4H_9Br]^+$ ion to split. Thus, CH_3^+, $C_2H_5^+$, and $C_3H_7^+$ are observed in addition to $C_4H_9^+$. Carbonium ions and radicals can undergo elimination reactions to form alkenes. The peaks at 27 and 41 are due to the alkenyl cations $CH_2{=}\overset{+}{C}H$ and $CH_3{\cdot}\overset{+}{C}{=}CH_2$ ions respectively; this type of ion is a characteristic intermediate in the breakdown of alkyl chains.

A large number of mass spectra have now been studied and many empirical correlations have been obtained for the cracking patterns. For example, branched chain alkanes are known to have a very strong tendency to break at branch points since this will produce relatively stable carbonium ions or radicals. The elucidation of cracking patterns has been facilitated by the construction of correlation tables and mass spectrometry is now a major method for structure determination.

(d) *Nuclear magnetic resonance spectroscopy.* This method is based on the fact that the hydrogen atoms in a compound absorb radiant energy when placed in a strong magnetic field. Hydrogen atoms in different environments absorb radiation of slightly different wavelengths. From the wavelength and amount of energy absorbed, it is possible to determine the numbers and positions of hydrogen atoms in the molecule; for example, how many hydrogen atoms are present in CH_3 groups and in CH_2 groups, etc.

Homologous series

A *homologous series* is a series of compounds in which adjacent members differ in their formula by a CH_2 unit, e.g.

CH_4 (methane), C_2H_6 (ethane), C_3H_8 (propane), etc. — the alkane series;
CH_3OH (methanol), C_2H_5OH (ethanol), C_3H_7OH (propanol), etc. — the alcohol series.

The members of a homologous series have similar chemical properties, but exhibit a steady gradation of physical properties, for example, the $C_1 - C_4$ alkanes are gases, $C_5 - C_{16}$ are liquids, and C_{17} upwards are solids with gradually increasing melting points and boiling points.

Since each member of the series differs in composition from adjacent members by a CH_2 unit, it follows that the series will have a general formula. The general formulae of the alkane and alcohol series above are C_nH_{2n+2} and $C_nH_{2n+1}OH$ respectively where n is any whole number.

Structural or constitutional isomerism

Ethanoic acid (acetic acid) and methyl methanoate (methyl formate) have the same molecular formula but different structural formulae:

$$CH_3-C{\overset{\displaystyle O}{\underset{\displaystyle OH}{<}}} \quad \text{and} \quad H-C{\overset{\displaystyle O}{\underset{\displaystyle O-CH_3}{<}}}$$

Similarly, there are two compounds with molecular formula C_4H_{10}:

$$CH_3-CH_2-CH_2-CH_3 \quad \text{and} \quad CH_3-CH-CH_3$$
$$\mid$$
$$CH_3$$

Butane   2-Methylpropane

The phenomenon of compounds having the same molecular but different structural formulae is known as *structural isomerism* and it is very common in organic chemistry. Structural isomers have different physical and chemical properties.

BASIC THEORETICAL PRINCIPLES

1. Breaking and forming bonds

A covalent bond may be broken in three ways.

(i) $R \overset{|}{\underset{|}{\div}} X \rightarrow R\cdot + X\cdot$ *Homolytic fission* to give free radicals (atoms or groups with an unpaired electron).

(ii) $R - | X \rightarrow R^- + X^+$ ⎫
(iii) $R | - X \rightarrow R^+ + X^-$ ⎬ *Heterolytic fission.*
                           ⎭

Process (ii) gives carbanions, i.e. species with a negatively charged carbon atom, whilst (iii) produces carbocations (carbonium ions), i.e. species with a positively charged carbon atom. Process (i) usually occurs in the gas phase or in non-polar solvents and is often catalysed by light or by the addition of free radical initiators. Fission by routes (ii) and (iii) occurs in polar solvents. Reversal of any of these processes results in bond formation.

2. Inductive effect

The electrons in covalent bonds between unlike atoms are not equally shared owing to the differing electronegativities of the atoms involved. This process of attraction or repulsion of electrons in single covalent bonds is known as the *inductive* or *I effect*. Most atoms or groups attached to carbon pull electrons away from it and are said to have a negative inductive effect (–I) but alkyl groups (see page 296) have +I effects. The order of +I effects in alkyl groups is:

tertiary, e.g. $(CH_3)_3C-$ > secondary, e.g. $(CH_3)_2CH-$ >
primary, e.g. $CH_3 \cdot CH_2-$

The inductive effect is indicated by an arrow in the bond pointing in the direction in which the electrons have moved, e.g. $\overset{\delta+}{CH_3} \rightarrow \overset{\delta-}{Cl}$. The δ sign indicates a small charge.

It should be noted that these electron movements also occur in π bonds (double bonds). The process here is known as the *mesomeric effect*, and it occurs more readily than in σ bonds (page 287) because of the delocalisation of π electrons (page 288).

3. Electrophilic and nucleophilic attack

The inductive effect in molecules renders some sites electron-rich and others electron-deficient and this will influence the type of reagent which will attack the molecule. For example, the carbon atom in

$$\overset{\delta+}{CH_3} - \overset{\delta-}{Cl}$$

will attract *nucleophiles*, i.e. negatively charged reagents or reagents with an electron-rich atom, e.g. HO^-, CN^-, and $\overset{\cdot\cdot}{N}H_3$ and is said to undergo *nucleophilic attack*. Conversely, the nitrogen atom in methylamine, CH_3NH_2 will attract *electrophiles*, i.e. positively charged or electron-deficient reagents, e.g. H^+, and is said to undergo electrophilic attack.

NOMENCLATURE

In the nineteenth century, very few organic compounds were known and they were named at the whim of the discoverer with no heed to their structure. As more and more organic compounds were discovered, it became apparent that this use of trivial names would lead to chaos and so over the years attempts have been made to name the compounds systematically. However, different chemists and chemical bodies had their own ideas as to how this should be done with the result that some compounds were known by two, three, or even more names. It was not until 1931 that a system of nomenclature was universally agreed upon and this is being constantly updated by the International Union of Pure and Applied Chemistry. According to the I.U.P.A.C. system, the structure of any compound can be derived from the name once a few simple rules have been learnt. Unfortunately, however, many people have been resistant to change and quite a large number of compounds, particularly very common ones, are still invariably known by their trivial names. In this text, therefore, trivial names will be given as well as the I.U.P.A.C. names, the former being the ones in brackets. The rules for I.U.P.A.C. nomenclature are discussed under each class of compound mentioned.

Questions

14.1 Calculate the empirical formula of an organic compound X given that, on combustion 0.206 g of it yielded 0.5666 g of carbon dioxide and 0.4635 g of water.

14.2 Qualitative analysis of a compound Y showed the presence of carbon and hydrogen only. Combustion of 0.196 g of Y yielded 0.4312 g of carbon dioxide and 0.2352 g of water. Determine the empirical formula of Y.

14.3 Calculate the empirical formula of the compound which yielded the following data. Combustion of 0.236 g of it gave 0.352 g of carbon dioxide and 0.180 g of water.

In a determination of nitrogen by the Kjeldahl method, 1.537 g of the compound evolved ammonia which required 26.0 cm^3 of 0.5 M sulphuric(VI) acid for neutralisation.

14.4 0.204 g of an organic compound gave on combustion 0.165 g of carbon dioxide and 0.0844 g of water whilst in a Carius determination for bromine, 0.363 g of the compound gave 0.626 g of silver bromide. Determine the empirical formula of the compound.

14.5 Analysis of two organic compounds both gave the same figures: 0.270 g of each compound yielded 0.240 g of carbon dioxide and 0.098 g of water on combustion.

0.220 g of each compound gave 0.638 g of silver chloride in a Carius determination.

The relative density of each compound was found to be 49. Determine the empirical and molecular formulae of the compounds and suggest a possible structural formula for each.

14.6   A substance X contains   C = 62.07%, H = 10.34%, and   O = 27.59%   (by mass). In an experiment it was found that 0.1704 g of the vapour occupied a volume of 94.8 cm³ at 100 °C and atmospheric pressure.

(a)   Determine the empirical formula of X.

(b)   Determine the molecular formula of X.

(c)   On oxidation, X yields a monobasic acid Y, containing   C = 48.65%, H = 8.11%,   and   O = 43.24%   (by mass). What is the empirical formula of Y?

(d)   Give a possible structural formula of Y.   [J.M.B.]

14.7   The following figures were obtained in the analysis of a compound X. On combustion 0.282 g yielded 0.132 g of carbon dioxide and 0.054 g of water. In a bromine estimation 0.1128 g gave 0.2256 g of silver bromide. In a relative molecular mass determination, 0.2068 g of X occupied 83.9 cm³ at 95 °C and 300 mm mercury pressure. Calculate the molecular formula for X and write two possible structural formulae for it.   [University of London]

14.8   Write an essay on the general methods used for the purification and characterisation of organic compounds, giving examples of the use of the methods you discuss.   [J.M.B.]

15 Molecular Geometry

Tetrahedral distribution of bonds round saturated carbon atoms

Carbon atoms possess six electrons, the detailed electronic structure being $1s^2 2s^2 2p^2$. Since the $2s$ orbital is full, it may have been expected that carbon would be divalent, the two unpaired $2p$ electrons forming covalent bonds by overlapping with the orbitals of two electrons from other atoms. However, carbon is usually tetravalent and so some change obviously occurs. In fact, one of the $2s$ electrons is promoted to the vacant $2p$ orbital and the s and p orbitals then blend or *hybridise* to give four identical sp^3 hybrid orbitals if four single covalent bonds are to be formed (thus giving a saturated carbon atom).

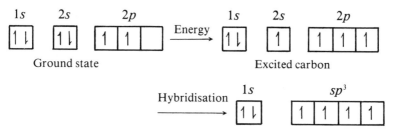

Since electrons repel one another, the hybrid orbitals try to get as far apart as possible and so the tetrahedral configuration is adopted. If the four orbitals overlap with the s orbitals from four hydrogen atoms, then methane is formed (Figure 15.1). Throughout the organic chemistry section the anti-bonding molecular orbitals (page 58) are ignored since they are unoccupied under normal conditions.

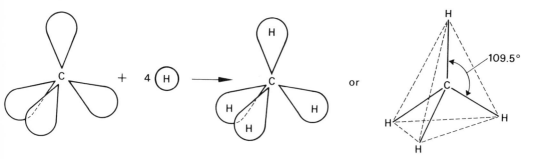

Figure 15.1 Structure of methane

Similarly, the tetrahedral configuration will occur in any compound where carbon forms four single covalent bonds with other atoms. These strong covalent bonds are known as σ (sigma) bonds and the electrons in them are said to be localised since they are concentrated along a line joining the two nuclei.

Planarity of the $\diagup C=C \diagdown$ in alkenes and the $\diagup C=O$ in carbonyl compounds

In the case of alkenes and carbonyl compounds (see pages 299 and 322 respectively) the carbon combines with only three other atoms and so, after excitation of the carbon atom, the $2s$ and only two of the $2p$ orbitals hybridise to give three sp^2 hybrid orbitals.

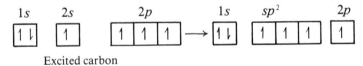

Excited carbon

The three sp^2 orbitals repel one another and so point towards the three corners of an equilateral triangle; i.e. the angle between them is 120°, and the unhybridised $2p$ orbital is in a plane at right angles to them.

The formation of ethene (ethylene) is as shown in Figure 15.2. The diagonally shaded orbitals are the unhybridised $2p$ orbitals and subsequent overlap of these gives a π bond. This lateral overlap of the $2p$ orbitals is less effective than the linear overlap which results in σ bonds. Hence π bonds are weaker than σ bonds. For maximum overlap of the unhybridised $2p$ orbitals, it is necessary for the four hydrogen and two carbon atoms to be exactly coplanar. Any rotation about the σ bond joining the two carbon atoms would result in a reduction of the overlap with consequent weakening of the π bond and so this is resisted. The electrons in the π bond are said to be delocalised because they can move freely throughout the orbital which embraces the two carbon atoms.

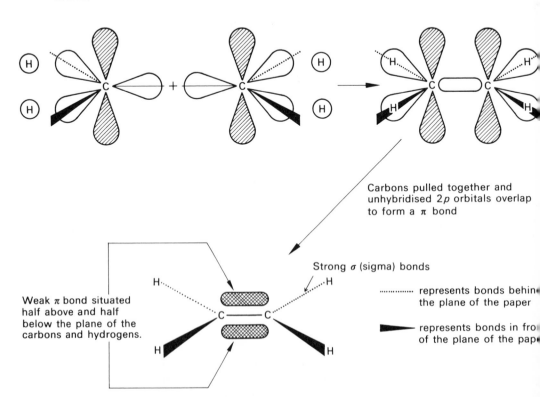

Carbons pulled together and unhybridised $2p$ orbitals overlap to form a π bond

Strong σ (sigma) bonds

Weak π bond situated half above and half below the plane of the carbons and hydrogens.

·········· represents bonds behind the plane of the paper

◀━━ represents bonds in front of the plane of the paper

Figure 15.2   Bond formation in ethene

The C—C bond length in ethane, CH_3—CH_3, is 0.154 nm and in ethene, CH_2=CH_2, is 0.133 nm. Alkenes and carbonyl compounds (ketones, acids, amides, aldehydes, etc) are planar in order that maximum overlap of the un-hybridised $2p$ orbitals can occur.

Note that in carbonyl compounds the unhybridised $2p$ orbital of the carbon atom overlaps with a $2p$ orbital of the oxygen (see page 323).

Linearity of —C≡C— in alkynes

In alkynes, e.g. ethyne (acetylene), CH≡CH, the carbon combines with only two other atoms and so hybridisation of the s and one p orbital occurs to give two sp orbitals. The sp hybrid orbitals are at 180° to one another (linear) whilst the two unhybridised $2p$ orbitals are in planes mutually at right angles.

The formation of ethyne (acetylene) is illustrated in Figure 15.3.

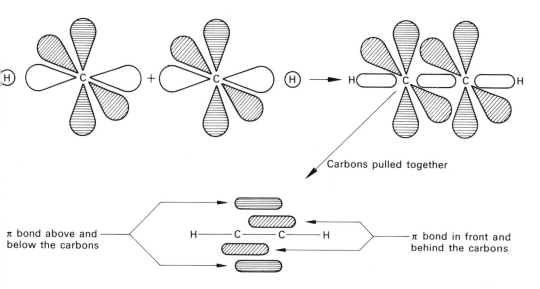

Figure 15.3   Bond formation in ethyne (acetylene)

The C—C bond length in ethyne is 0.120 nm and so the carbon atoms are pulled together even closer than in alkenes.

Shapes of compounds containing five and six saturated carbon atoms in a ring

Cyclohexane has the structural formula

$$CH_2 \quad \overset{CH_2}{\diagup \diagdown} \quad CH_2$$
$$CH_2 \qquad\qquad CH_2$$
$$CH_2 \quad \diagdown_{CH_2}\diagup \quad CH_2$$

If the ring had a planar structure the bond angles between the carbon atoms would be 120° compared with the normal value of 109.5° in saturated compounds. However, this would cause some degree of strain in the molecule because the hydrogen atoms would be forced closer together. In fact, electron diffraction studies, etc. show that cyclohexane exists in several forms; the carbon skeletons of two well-known forms are shown below:

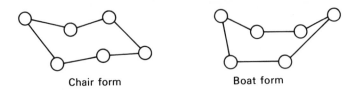

Chair form   Boat form

The various forms are readily interconvertible and cannot be isolated. The chair form is by far the most stable form because it allows the hydrogen atoms to be further apart — this should be demonstrated with models. Cyclopentane

$$\begin{array}{c} CH_2 \\ CH_2 \qquad CH_2 \\ CH_2 - CH_2 \end{array}$$

would have bond angles of 108° between the carbons if it existed as a regular pentagon. However, the ring is slightly puckered because this allows the hydrogen atoms a little more room.

Influence of size and shape of alkanes on their melting and boiling points

The alkanes are a homologous series of saturated hydrocarbons (page 297) and the melting and boiling points of the first six unbranched chain isomers are as follows:

Alkane	m.p./°C	b.p./°C	Alkane	m.p./°C	b.p./°C
CH_4	−182	−161	C_4H_{10}	−138	0
C_2H_6	−183	−88	C_5H_{12}	−130	36
C_3H_8	−187	−42	C_6H_{14}	−95	69

There is a steady increase in melting point and boiling point owing to the increasing relative molecular mass and more particularly to the increased van der Waals attractive forces (page 36) in the molecules with more electrons. Shape also plays a part, however, as illustrated by the two C_5H_{12} isomers below:

$$\begin{array}{ccc} & CH_2 \qquad CH_2 & \\ CH_3 \quad CH_2 \quad CH_3 & \end{array}$$

$$\begin{array}{c} CH_3 \qquad CH_3 \\ C \\ CH_3 \qquad CH_3 \end{array}$$

Pentane, m.p. −130 °C,   2,2-Dimethylpropane, m.p. −16 °C,
b.p. 36 °C.   b.p. 10 °C.

Pentane has a long zig-zag chain with a large surface area and so there is considerable interaction between its molecules. It, therefore, has bigger van der Waals forces and so a higher boiling point than the more spherical 2,2-

dimethylpropane with its smaller surface area. Since 2,2-dimethylpropane is more compact and symmetrical it packs more efficiently in its crystal lattice and so has a higher lattice energy (page 76) and melting point than pentane.

Influence of functional groups on physical properties

Hydrogen bonding (page 37) will occur in any compound which contains hydroxyl groups, for example alcohols:

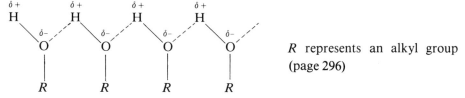

R represents an alkyl group (page 296)

The molecules will exist in pairs or clusters and so energy will be required to break the hydrogen bonds before the molecules can escape as vapour. The boiling point is therefore higher than expected from the relative molecular mass.

Hydrogen bonding affects viscosity, etc. as well as the boiling point. The effect on viscosity of increasing the number of hydroxyl groups in a compound may be readily demonstrated using three lengths of glass tubing about 40 cm long and 1 cm external diameter and fitted with a bung at each end (Figure 15.4). The tubes are filled to within about 2 cm with propan-1-ol, $CH_3 \cdot CH_2 \cdot CH_2OH$, propane-1,2-diol, $CH_3 \cdot CHOH \cdot CH_2OH$, and propane-1,2,3-triol, $CH_2OH \cdot CHOH \cdot CH_2OH$ respectively and the bungs replaced. A fourth tube containing the triethanoate (triacetate) of propane-1,2,3-triol,

$$CH_2 \text{———} CH \text{———} CH_2$$
$$\underset{OOC \cdot CH_3}{\vert} \qquad \underset{OOC \cdot CH_3}{\vert} \qquad \underset{OOC \cdot CH_3}{\vert}$$

may also be used for contrast — the hydroxyl groups have been converted to ester groups and so hydrogen bonding cannot occur, but the relative molecular mass is considerably increased. If the tubes are inverted, the air bubbles in each will rise at different rates depending on the viscosity of the liquids. The observed order of ascent is

propan-1-ol > the triester > propane-1,2-diol > propane-1,2,3-triol

thus showing that the viscosity increases as the opportunity for hydrogen bonding increases and this greatly outweighs the effect of the increase in relative molecular mass.

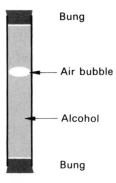

Bung

Air bubble

Alcohol

Bung

Figure 15.4   Comparing viscosities of alcohols

Organic acids also have much higher boiling points than expected from their relative molecular masses, the reason again being hydrogen bonding. For example, relative molecular mass determinations show that ethanoic acid (acetic acid) and benzene carboxylic acid (benzoic acid) exist as dimers (double molecules) in benzene solution:

$$
CH_3-C
\begin{matrix}
O\text{--------}H-O \\
\\
O-H\text{-------}O
\end{matrix}
C-CH_3; \qquad
C_6H_5-C
\begin{matrix}
O\text{--------}H-O \\
\\
O-H\text{-------}O
\end{matrix}
C-C_6H_5
$$

It should be noted that these acids exist as monomers in dilute alcoholic and aqueous solutions because hydrogen bonding occurs with the solvent.

STEREOISOMERISM

Stereoisomerism is exhibited by isomers having the same structure, but different spatial arrangements. The different spatial arrangements are possible because covalent bonds have direction in space. Stereoisomerism may be sub-divided into *geometrical* and *optical isomerism*.

Geometrical isomerism

Two common geometrical isomers are *cis* and *trans*-butenedioic acids (maleic and fumaric acids).

H COOH \ / C ‖ C / \ H COOH *cis*-Butenedioic acid, m.p. 130 °C.	HOOC H \ / C ‖ C / \ H COOH *trans*-Butenedioic acid, m.p. 228 °C.

These compounds are different because the two carbon atoms cannot rotate about the double bond. Atoms joined by single bonds can rotate freely relative to one another, but atoms joined by a double bond cannot, since this would disrupt the π bond

(see structure of $\quad$ $>C=C<$ $\quad$ page 288)

Compounds with identical groups on the same side of the double bond are called *cis*-isomers whilst those with identical groups on opposite sides are called *trans*-isomers.

Geometrical isomers differ in their physical properties and also in many of their chemical properties, particularly in reactions involving ring formation. *cis*-Butenedioic acid (maleic acid) has a much lower melting point than *trans*-butenedioic acid (fumaric acid) because it has two large carboxyl groups on the same side of the double bond and these interfere with one another so decreasing the compound's stability. Also, the *cis*-isomer has a less compact crystal lattice than the *trans*-isomer. This relative instability of *cis*-butenedioic acid (maleic acid) is also reflected in the fact that on heating to 150 °C it loses water to give the acid anhydride. The *trans*-isomer also gives *cis*-butenedioic anhydride (maleic anhydride) if heated to 250 °C, probably because at this temperature the π bond between the two carbon atoms breaks and rotation about the resultant single bond gives *cis*-butenedioic acid (maleic acid) which is then dehydrated.

cis-Butenedioic anhydride (Maleic anhydride)

Optical isomerism

It was stated earlier that if a carbon atom is joined to four atoms or groups then the bonds from it point towards the four corners of a regular tetrahedron. If the four atoms are different, the compound and its mirror image are non-superimposable, i.e. two forms exist analogous to right- and left-handed gloves (models should be constructed to confirm this).

--------, ◀ , and ——— represent bonds behind, in front, and in the plane of the paper respectively and *d*, *e*, *f*, and *g* different atoms or groups.

Such compounds are said to be *optically active* because they rotate the plane of plane polarised light (light vibrating in one plane only), the two forms deflecting the light the same amount but in opposite directions. If the light is rotated to the right as the observer looks towards the beam the substance is said to be *dextrorotatory*, and if it is rotated to the left it is said to be *laevorotatory*, the two forms being known as the + or *d* and the – or *l* forms respectively. This must not be confused with D and L isomers; the D and L here refer to absolute configurations which are beyond the scope of this text.

Optical isomerism occurs whenever a compound has a non-superimposable mirror image and this condition is always fulfilled if a compound has one *asymmetric carbon atom*, i.e. a carbon atom with four different atoms or groups attached to it. Common examples of optically active compounds are:

$$CH_3$$
$$H \blacktriangleright C—COOH$$
$$HO$$
2-Hydroxypropanoic acid.
(Lactic acid)

$$CH_3$$
$$H \blacktriangleright C—COOH$$
$$NH_2$$
2-Aminopropanoic acid.
(Alanine)

$$CH_3 \cdot CH_2$$
$$H \blacktriangleright C—CH_3$$
$$HO$$
Butan-2-ol.

The non-superimposable mirror images are called *enantiomers* or *enantiomorphs*. An equimolar mixture of the two forms is optically inactive, because the rotation of one form is cancelled out by that of the other. Such a mixture is said to be a *racemic* or a ± or a *dl* mixture. Enantiomers have the same chemical and physical properties except the direction in which they rotate plane polarised light.

Separation of a racemic mixture into the pure enantiomers is known as *resolution*. Since the physical properties of enantiomers are identical they cannot be separated by such techniques as fractional crystallisation or distillation. However, enantiomers do behave differently in the presence of another optically active substance and almost all methods of resolution are based upon this fact. Consider, for example, a racemic mixture of an acid. If it is treated with say a dextrorotatory base, two salts will be formed, one from the (+) acid and (+) base, and one from the (−) acid and (+) base. Optical isomers which are not mirror images are said to be *diastereoisomers*; they have different physical properties and so can be separated by physical means. Hydrolysis (decomposition by water) of the separated salts will give the pure (+) or (−) acids:

$$
\begin{array}{l}
(\pm)\,\text{acid} \\
+ \\
(+)\,\text{base}
\end{array}
\left\{
\begin{array}{l}
(+)\,\text{acid} + (+)\,\text{base} \xrightarrow{\text{Hydrolysis}} (+)\,\text{acid} \\
(-)\,\text{acid} + (+)\,\text{base} \xrightarrow{\text{Hydrolysis}} (-)\,\text{acid}
\end{array}
\right.
$$

Alcohols may be resolved by converting them to diastereomeric esters. (Alcohols and acids are discussed in Chapter 16.)

Chromatographic methods have limited utility. The racemic mixture in solution may be passed over an optically active adsorbant in which case one enantiomer may be more strongly adsorbed than the other.

In the mid-nineteenth century, the French scientist Pasteur succeeded in resolving a racemic mixture of ammonium sodium 2,3-dihydroxybutanedioate (ammonium sodium tartrate),

$$
\begin{array}{c}
H \\
| \\
HO—C—COONH_4 \\
| \\
HO—C—COONa \\
| \\
H
\end{array}
$$

simply by separating by hand the two types of crystal — one enantiomer gives crystals which are the mirror image of the other. However, this is really only of historic interest since most enantiomers give crystals identical in appearance.

Finally, note that a synthesis such as

$$CH_3 \cdot CHO + HCN \rightarrow CH_3 - \overset{\overset{\displaystyle H}{|}}{\underset{\underset{\displaystyle CN}{|}}{C}} - OH$$

(see page 324) always gives a racemic mixture because the cyanide group can attack the carbonyl carbon equally well from either side and statistically half will attack from each side:

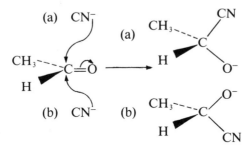

Questions

15.1 Discuss the bond formation and shape of (a) methane (b) ethene (ethylene). Explain the existence of three compounds with the molecular formula $C_2H_2Br_2$.

15.2 Briefly discuss the effect of shape and functional groups on the physical properties of organic compounds.

The following compounds have very similar relative molecular masses; arrange them in order of increasing boiling point, giving reasons for your choice:

$CH_3 \cdot CH_2 \cdot CH_2OH$, $CH_3 \cdot CH_2Cl$, $CH_3 \cdot COOH$, and $CH_3 \cdot CH_2 \cdot CH_2 \cdot NH_2$.

15.3 Define the terms enantiomers (enantiomorphs) and racemic mixtures. Discuss the stereochemistry of the compounds with the formulae:

(a) $CH_3 \cdot CH_2 \cdot CHBr \cdot CH_3$ (b) $CHCl = CHCl$.

16 The Chemistry of the Functional Groups

1   THE C−H BOND − ALKANES

The alkanes form a homologous series with the general formula C_nH_{2n+2} where $n = 1$ or more. The first four members have trivial names, but thereafter the name indicates the number of carbon atoms present in the molecule:

CH_4	methane	C_4H_{10}	butane
C_2H_6	ethane	C_5H_{12}	pentane
C_3H_8	propane	C_6H_{14}	hexane, etc.

Note that all the names end in '-ane'.

The group obtained by removing a hydrogen atom from an alkane is called an *alkyl* group; these are named by removing the ending '-ane' from the alkane name and adding '-yl', e.g.

CH_4	methane	$CH_3−$	methyl
C_2H_6	ethane	$C_2H_5−$	ethyl, etc.

Branched chain alkanes may now be named according to the alkyl substituents present. Thus, the longest unbranched chain is found and named after the corresponding alkane and then the chain is numbered to give the substituents the lowest possible number. Each substituent is specified by name and number as illustrated in the examples below:

$$\overset{1}{C}H_3 - \overset{2}{C}H - \overset{3}{C}H_3 \qquad \text{2-methylpropane}$$
$$\quad\quad\quad | $$
$$\quad\quad\; CH_3$$

$$\overset{4}{C}H_3 - \overset{3}{C}H_2 - \overset{2}{C}H - \overset{1}{C}H_3 \qquad \text{2-methylbutane (not 3-methylbutane)}$$
$$\quad\quad\quad\quad\quad | $$
$$\quad\quad\quad\quad\; CH_3$$

$$\quad\quad\quad CH_3$$
$$\overset{1}{C}H_3 - \overset{2}{\underset{|}{C}} - \overset{3}{C}H_3 \qquad \text{2,2-dimethylpropane}$$
$$\quad\quad\; CH_3$$

$$\overset{1}{C}H_3 - \overset{2}{C}H - \overset{3}{C}H - \overset{4}{C}H_2 - \overset{5}{C}H_2 - \overset{6}{C}H_3 \quad \text{3-ethyl-2-methylhexane}$$
$$\quad\quad\quad | \quad\quad \backslash $$
$$\quad\quad\; CH_3 \quad\; C_2H_5$$

Note that if more than one substituent is present they are written in alphabetical order and so the name 2-methyl-3-ethylhexane for the above example would be incorrect.

Alkanes may be prepared in the laboratory by heating the sodium salts of carboxylic acids (page 330) with soda-lime (NaOH–CaO), e.g. sodium ethanoate (acetate) and soda-lime give methane:

$$CH_3 \cdot COONa(s) + NaOH(s) \rightarrow Na_2CO_3(s) + CH_4(g)$$

The C_1 to C_4 alkanes are colourless gases, C_5 to C_{16} are liquids, and the higher homologues are solids. They are insoluble in water, but soluble in organic solvents, e.g. alcohols and ethers. Alkanes are fairly unreactive since they are saturated (contain no double or triple bonds) and contain only strong C—C and C—H bonds. However, reaction will occur if the conditions are sufficiently forcing.

Reactions of alkanes

(a) Chlorination (substitution of chlorine for hydrogen)

The reaction is confined to methane in view of the number of products formed, and it occurs when methane and chlorine mixtures are irradiated with ultraviolet light.

$$CH_4(g) + Cl_2(g) \xrightarrow{u.v.} HCl(g) + CH_3Cl(g)$$ chloromethane (methyl chloride)

$$CH_3Cl(g) + Cl_2(g) \xrightarrow{u.v.} HCl(g) + CH_2Cl_2(l)$$ dichloromethane

$$CH_2Cl_2(l) + Cl_2(g) \xrightarrow{u.v.} HCl(g) + CHCl_3(l)$$ trichloromethane (chloroform)

$$CHCl_3(l) + Cl_2(g) \xrightarrow{u.v.} HCl(g) + CCl_4(l)$$ tetrachloromethane (carbon tetrachloride)

The function of the u.v. light is to split the chlorine molecules into the much more reactive chlorine atoms, i.e. a homolytic reaction occurs. All atoms joined by covalent bonds vibrate about certain fixed positions. In the case of chlorine molecules, the frequency of the vibration is similar to the frequency of u.v. light and so energy is absorbed, the vibrations become larger and eventually the bonds break. The mechanism of the chlorination is:

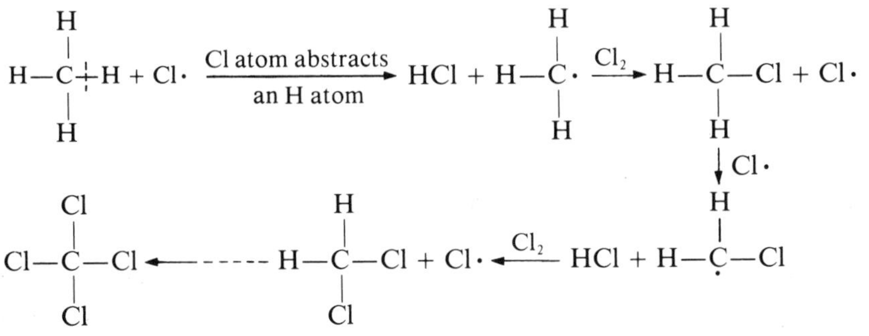

No more C—H bonds and so reaction stops.

Some chlorinated product may also be formed by reactions such as

$$CH_3 \cdot + Cl \cdot \rightarrow CH_3Cl$$

The chlorination is classed as a *free radical* reaction. Free radicals are atoms or groups containing an unpaired electron.

(b) Sulphonation (substitution of a sulphonic acid group, SO₃H, for hydrogen)

This is important with petroleum fractions containing C_{14}–C_{16} alkanes. These fractions are treated with sulphur dichloride dioxide (sulphuryl chloride), or sulphur dioxide and chlorine, in the presence of u.v. light, to give sulphuryl chlorides which are hydrolysed with sodium hydroxide solution to give sodium alkyl sulphonate detergents:

$$R—H(l) + SO_2Cl_2(l) \xrightarrow{\text{u.v.}} R \cdot SO_2Cl(l) + HCl(g)$$
$$R \cdot SO_2Cl(l) + 2NaOH(aq) \longrightarrow R \cdot SO_3Na(aq) + NaCl(aq) + H_2O(l)$$

where R is an alkyl group.

These compounds behave as detergents because the alkyl chain is an oil soluble or *lyophilic* group and the sulphonate group is a water soluble or *hydrophilic* group. The detergent is therefore able to bridge the interface between water and oils or greases.

(c) Nitration (substitution of a hydrogen by a nitro group, NO₂)

Alkanes react with concentrated nitric(V) acid at 400–500 °C to give nitroalkanes. Degradation products are obtained in addition to the simplest substitution product, e.g.

$$CH_3 \cdot CH_2 \cdot CH_3 \xrightarrow[420\,°C]{HNO_3} CH_3 \cdot CH_2 \cdot NO_2 + CH_3 \cdot \overset{\overset{\displaystyle NO_2}{|}}{CH} \cdot CH_3$$

$$+ CH_3 \cdot CH_2 \cdot NO_2 + CH_3 \cdot NO_2$$

The nitroalkanes are used as fuels, solvents, and in synthetic work.

(d) Oxidation

Alkanes readily burn in air giving carbon dioxide and water, e.g.

$$CH_4(g) + 2O_2(g) \rightarrow CO_2(g) + 2H_2O(l)$$

Methane is the major constituent of natural gas, butane is sold as calor gas, and C_6—C_{11} alkanes are used in petrol.

Natural gas and crude petroleum

Natural gas generally occurs in conjunction with crude petroleum; both are the remains of marine and plant life from 100 to 200 million years ago. Natural gas is mainly methane together with progressively smaller amounts of the alkanes up to about C_7, and it is used to supplement or replace coal gas. Crude petroleum contains large numbers of alkanes, mainly in the C_5–C_{40} range, with small amounts of cycloalkanes and aromatic hydrocarbons. It is split into various fractions by fractional distillation. Fractions containing C_5–C_8 alkanes are used as solvents, C_6–C_{11} as motor fuel, C_{11}–C_{18} as fuel oil, C_{16}–C_{20} as lubricants, and C_{18} upwards as greases, waxes, etc. Considerable overlap occurs in the chain length of the alkanes in the various fractions because the numerous structural isomers have different boiling points.

The demand for C_{12}–C_{20} alkanes has increased as diesel and jet engines have been developed. Nevertheless, vast quantities are still converted into gasoline and the very useful short chain (C_1–C_4) alkanes and alkenes by a process known as *cracking*. Cracking involves thermal decomposition of the C_{12}–C_{20} alkanes at 500–600 °C.

The cracking process is an example of homolytic fission and two types of reaction occur.

Dehydrogenation (removal of hydrogen) converts alkanes into alkenes:

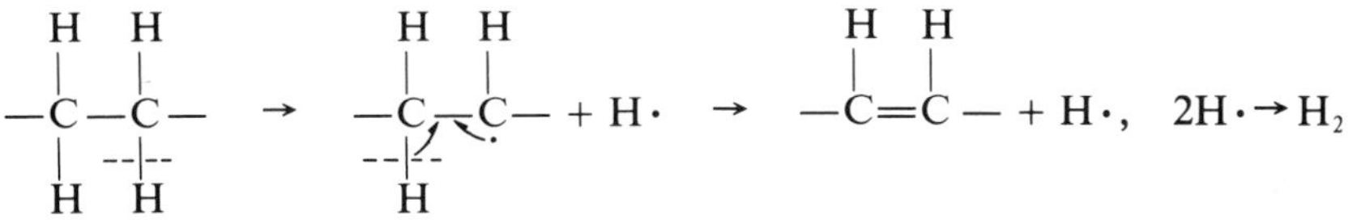

C—C bond fission converts alkanes into shorter chain alkanes and alkenes:

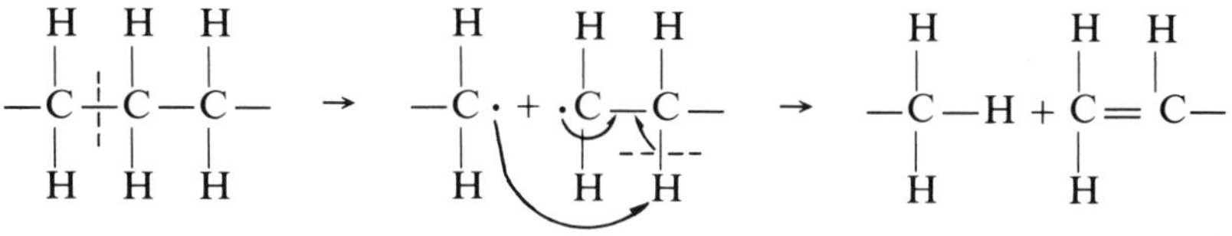

Note the use of curved arrows in reaction mechanisms. The head of the arrow always indicates the destination of the electrons whilst the tail shows their origin. The usual convention is that arrows of the types ⌒ and ⌒ are used to represent the movement of one electron or a pair of electrons respectively.

If alkanes containing six or more carbon atoms are passed over a chromium(III) oxide catalyst at about 500 °C, cyclisation occurs, e.g. hexane gives benzene:

$$C_6H_{14}(l) \rightarrow C_6H_6(l) + 4H_2(g)$$

2. THE C=C BOND — ALKENES

The general formula of this homologous series is C_nH_{2n} where $n = 2$ or more. The method of nomenclature is to pick the longest chain containing the double bond, name it after the corresponding alkane, but change the '-ane' to '-ene'. The chain is numbered so that the double bond has the lowest possible number.

$CH_2=CH_2$	ethene (ethylene)
$CH_3 \cdot CH=CH_2$	propene (propylene)
$CH_3 \cdot CH_2 \cdot CH=CH_2$	but-1-ene
$CH_3 \cdot CH=CH \cdot CH_3$	but-2-ene

No number is required with the first two members, since the double bond is bound to involve one of the end carbon atoms. Note that the butenes are structural isomers.

Alkenes are prepared in the laboratory by dehydration of alcohols (page 317) or by dehydrogenation of halogenoalkanes (page 315). The first three members are colourless gases insoluble in water, but soluble in organic solvents.

The structure of the double bond is

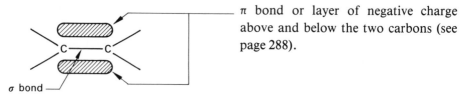

π bond or layer of negative charge above and below the two carbons (see page 288).

The layer of negative charge above and below the two carbon atoms results in the majority of the reactions of alkenes being electrophilic additions, i.e. additions by electron deficient reagents. The alkenes are unsaturated hydrocarbons and, unlike alkanes, are reactive.

Reactions of alkenes

(a) Addition of bromine or chlorine (but not iodine)

The addition occurs at room temperature, e.g.

$$CH_2{=}CH_2(g) + \quad Br_2(l) \quad \rightarrow CH_2Br \cdot CH_2Br(l)$$

   Colourless   Red-brown   Colourless
                                  1, 2-Dibromoethane
                                  (ethylene dibromide)

The reaction is used to distinguish between alkenes and alkanes: alkenes decolourise a solution of bromine in tetrachloromethane (carbon tetrachloride), but there is no reaction with alkanes under these conditions and so the red-brown colour remains. The tetrachloromethane is used to dilute the bromine since if excess of the latter is used no change would be detected in either case.

When the bromine approaches the π bond of the alkene, the outer electrons of the nearest atom are repelled and so the molecule becomes polarised. The resultant slightly positively charged bromine is then attracted by the π bond.

$$\overset{\backslash}{\underset{/}{C}}{=}\overset{/}{\underset{\backslash}{C}} \;+\; Br{-}Br \;\rightarrow\; \overset{\backslash}{\underset{/}{C}}{=}\overset{/}{\underset{\backslash}{C}} \;\rightarrow\; \overset{\backslash}{\underset{/}{C}}{-}\overset{/}{\underset{\backslash}{C}} \;+\; Br^- \;\rightarrow\; \overset{\backslash}{\underset{/}{C}}{-}\overset{/}{\underset{\backslash}{C}}$$

It is immaterial which carbon becomes joined to the bromine first, the other carbon will become positively charged (since it has lost its share of the π electrons) and attract the Br^- ion.

The reaction with chlorine and any alkene is exactly analogous.

If chlorine or bromine water is used then HOX adds on as X^+ and HO^-, e.g.

$$CH_2{=}CH_2(g) + HOCl(aq) \quad \longrightarrow \quad CH_2Cl \cdot CH_2OH(aq)$$

                                              2-Chloroethanol

(b) Addition of hydrogen halide

Hydrogen bromide and hydrogen iodide or the corresponding halogen acids readily add on to alkenes at room temperature, but the more stable hydrogen chloride requires heat and aluminium chloride as catalyst.

$$CH_2{=}CH_2(g) + HBr(g) \rightarrow CH_3 \cdot CH_2Br(l)$$
$$\text{Bromoethane (ethyl bromide)}$$

The mechanism of addition is:

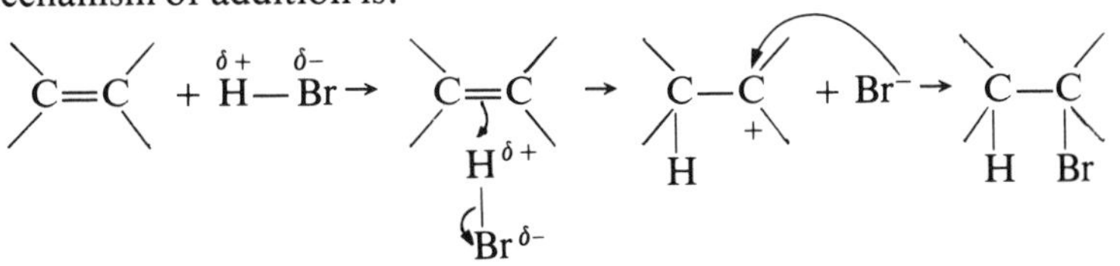

With propene (or other unsymmetrical alkenes) two products could possibly be formed, i.e. $CH_3 \cdot CH_2 \cdot CH_2Br$ or $CH_3 \cdot CHBr \cdot CH_3$. In fact, only the latter compound is formed as predicted by *Markownikoff's rule* which states that: in additions to unsymmetrical alkenes the negative part of the addendum adds on to the carbon with the least number of hydrogen atoms. Markownikoff's rule is purely empirical—it was put forward based on the observed facts in many reactions, without any theoretical justification. It is now known that the deciding factor is the relative stabilities of the possible carbocations (carbonium ions) (page 284). Addition of H^+ to propene could proceed as follows:

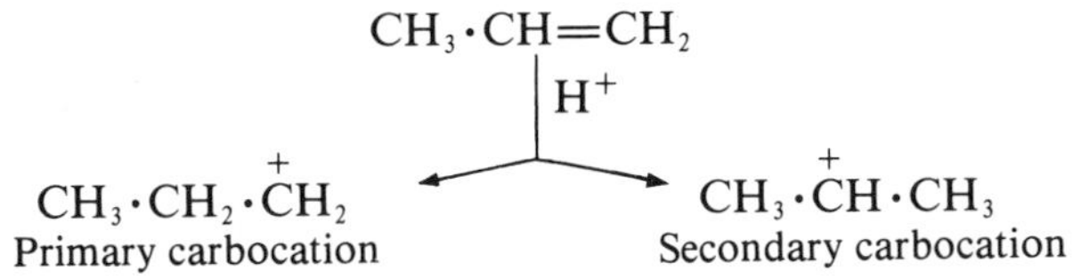

$$CH_3 \cdot CH{=}CH_2$$
$$\downarrow H^+$$

$CH_3 \cdot CH_2 \cdot \overset{+}{C}H_2$	$CH_3 \cdot \overset{+}{C}H \cdot CH_3$
Primary carbocation	Secondary carbocation

The secondary carbocation is the most stable and the one formed because it has two alkyl groups pushing electrons towards the positive carbon and reducing its charge density as against the one alkyl group in the primary carbocation. The final product is therefore 2-bromopropane.

It should be noted that if traces of peroxides,

$$\geqslant C{-}O{-}O{-}H,$$

are present in the alkene then the addition with hydrogen bromide takes place via a free radical mechanism (i.e. $H\cdot$ and $Br\cdot$ add on) and anti-Markownikoff addition occurs, e.g.

$$CH_3 \cdot CH{=}CH_2(g) + HBr(g) \xrightarrow{\text{peroxides}} CH_3 \cdot CH_2 \cdot CH_2Br(l)$$

Peroxides are formed slowly when the alkene is in prolonged contact with air and traces of metal impurities, etc. Thorough purification of the alkene is therefore essential if ionic addition is to be ensured.

(c) Reduction

Alkenes are unaffected by metals and acid but they undergo hydrogenation in the presence of a catalyst such as finely divided platinum, palladium, or Raney nickel at room temperature, or nickel at 200–300 °C. Raney nickel is a very active form of nickel prepared by dissolving the aluminium out of nickel — aluminium alloy with sodium hydroxide solution. The residual nickel is washed and stored under water since the dry powder is pyrophoric (spontaneously flammable in air).

$$\text{C=C} + H_2 \xrightarrow{\text{Raney nickel}} \underset{\underset{H\quad H}{|\quad\;|}}{\text{C}-\text{C}}$$

(d) Oxidation

All alkenes burn in air to give carbon dioxide and water, but controlled oxidation is much more important.

(i) Shaking with cold dilute alkaline potassium manganate(VII) (permanganate) gives a 1, 2-dihydroxy compound (a glycol). The half-reaction for the oxidation is:

$$\text{C=C} + 2HO^- \rightarrow \underset{\underset{OH\quad OH}{|\quad\;\;|}}{\text{C}-\text{C}} + 2e^-$$

or, for simplicity:

$$\text{C=C} + [O] + H_2O \rightarrow \underset{\underset{OH\quad OH}{|\quad\;\;|}}{\text{C}-\text{C}}$$

During the process the purple manganate(VII) (permanganate) colour disappears and a brown precipitate of manganese(IV) oxide is formed.

If stronger oxidising agents are used the dihydroxy compound is oxidised to a variety of products.

Better yields of the dihydroxy compound are obtained if the alkene is treated with hydrogen peroxide and a trace of osmium(VIII) oxide, OsO_4, as catalyst.

(ii) Peroxoacids (peracids) oxidise alkenes to alkene oxides.

$$CF_3 \cdot COOH(l) + H_2O_2(aq) \rightarrow CF_3 \cdot C\underset{O-OH}{\overset{\displaystyle O}{\diagup}} (aq) + H_2O(l)$$

Peroxotrifluoro-
ethanoic acid (pertri-
fluoroacetic acid)

$$\text{C=C} + CF_3 \cdot C\underset{O-OH}{\overset{\displaystyle O}{\diagup}} (l) \rightarrow \underset{\underset{O}{\diagdown\diagup}}{\text{C}-\text{C}} + CF_3 \cdot COOH(l)$$

Alkene oxides are useful in organic syntheses (see industrial reactions below).

(iii) Ozonides are formed when trioxygen (ozone) (formed by passing a silent electric discharge through air or oxygen) is passed through a solution of an alkene dissolved in an inert solvent such as tetrachloromethane (carbon tetrachloride) or ethanoic acid (acetic acid). Since ozonides are explosive they are not isolated, but are hydrolysed with water to give carbonyl compounds (page 322).

$$\underset{/}{\overset{\backslash}{C}}=\underset{\backslash}{\overset{/}{C}} + O_3 \rightarrow \overset{\backslash}{\underset{/}{C}}\overset{O-O}{\underset{O}{\diagdown}}\overset{/}{\underset{\backslash}{C}} \xrightarrow{H_2O} \overset{\backslash}{\underset{/}{C}}=O + O=\overset{/}{\underset{\backslash}{C}} + H_2O_2$$

An ozonide

The hydrogen peroxide formed would tend to oxidise the carbonyl compounds and so the hydrolysis is done in the presence of zinc dust and ethanoic acid (acetic acid) or of hydrogen and platinum so that the hydrogen peroxide is decomposed. The complete process of ozonide formation and the hydrolysis is known as *ozonolysis* and it is used in structure determination of alkenes. Thus, the carbonyl compounds are identified and the alkene has the structure produced when the two groups attached to the oxygens in the carbonyl compounds are joined by a double bond. For example, if the products of ozonolysis are $CH_3 \cdot CH_2 \cdot CO \cdot CH_3$ and $CH_3 \cdot CHO$, the alkene must have had the structure

$$\underset{CH_3}{\overset{CH_3 \cdot CH_2}{\diagdown}}C = CH \cdot CH_3 \quad \text{(3-methylpent-2-ene)}$$

(e) Hydration

Alkenes may be hydrated by adding 85% sulphuric(VI) acid at about $0\,^\circ C$ then diluting the mixture and boiling it, e.g.

$$CH_2 = CH_2(g) \xrightarrow[0\,^\circ C]{85\% \, H_2SO_4} CH_3 \cdot CH_2 \cdot O \cdot SO_3H(aq) \xrightarrow{H_2O/boil}$$

$$CH_3 \cdot CH_2OH(aq) + H_2SO_4(aq)$$

Direct hydration is now very important industrially. Thus the alkene and steam are passed, under pressure, over a hot catalyst of phosphoric(V) acid absorbed in diatomacious earth (an inert solid), e.g.

$$CH_3 \cdot CH = CH_2(g) + H_2O(g) \xrightarrow[300\,^\circ C/pressure]{H_3PO_4} CH_3 \cdot CHOH \cdot CH_3(l)$$
$$\text{propan-2-ol}$$
$$\text{(isopropanol)}$$

(f) Polymerisation

This is discussed on page 359.

Industrial reactions

The petroleum cracking processes (page 299) give large quantities of ethene (ethylene) and propene (propylene) and they are used in the industrial preparation of many products. Some important reactions of ethene (ethylene) are illustrated in Figure 16.1.

Hydration of propene (propylene) gives propan-2-ol (isopropanol) which is used as a solvent, a high octane fuel, and to produce esters and propanone (acetone), etc.

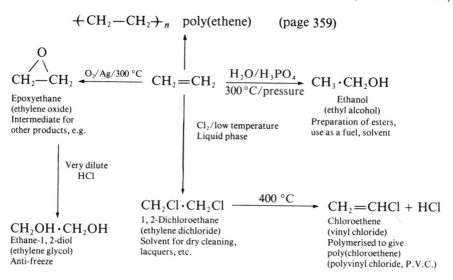

Figure 16.1   Some industrial reactions of ethene

3   THE C≡C BOND — ALKYNES

The alkyne homologous series has the general formula C_nH_{2n-2} where $n = 2$ or more. They are named after the corresponding alkane, but the ending '-ane' is changed to '-yne'. The first and most important member is CH≡CH, ethyne (acetylene).

Ethyne is prepared, industrially and in the laboratory, by adding water to calcium dicarbide (calcium carbide):

$$CaC_2(s) + 2H_2O(l) \rightarrow Ca(OH)_2(aq) + CH{\equiv}CH(g)$$

It is a colourless gas, insoluble in water but soluble in organic solvents.

The formation of the bonds in ethyne is illustrated on page 289. Since the molecule contains two π bonds:

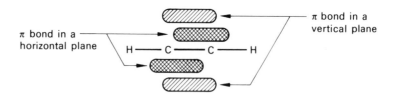

it undergoes electrophilic additions, its reactions being basically similar to those of ethene except that they occur twice, e.g.

$$CH{\equiv}CH(g) + HBr(g) \rightarrow CH_2{=}CHBr \xrightarrow{\text{HBr}} CH_3 \cdot CHBr_2(l)$$

Bromoethene   1, 1-Dibromoethane

(Vinyl bromide)   (Ethylidene bromide)

Ethyne differs from alkenes in that its hydrogen is slightly acidic, e.g. on shaking with an ammoniacal solution of silver nitrate(V) a white precipitate of silver dicarbide (acetylide) is formed:

$$CH{\equiv}CH(g) + 2Ag(NH_3)_2OH(aq) \rightarrow AgC{\equiv}CAg(s) + 4NH_3(aq) + 2H_2O(l)$$

With alkenes there is no reaction.

Ethyne has important industrial applications but these will not be discussed here.

4 AROMATIC HYDROCARBONS

Aromatic hydrocarbons or arenes have the general formula C_nH_{2n-6} where $n = 6$ or more. Only two will be considered: C_6H_6, benzene and $C_6H_5{\cdot}CH_3$, methylbenzene (toluene). Both are colourless liquids.

The structure of benzene has been the subject of much controversy over the years. In 1865, Kekulé suggested the structure

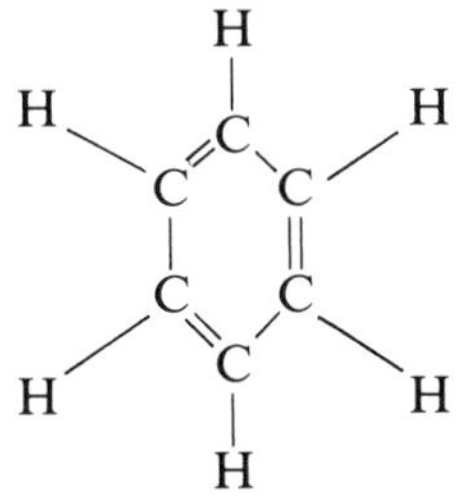

usually written

However, this did not explain the great reluctance of benzene to form addition compounds (e.g. it does not react with hydrogen halides, etc.) or why two isomers cannot be isolated with the structures

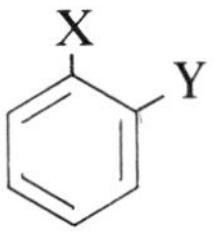

and

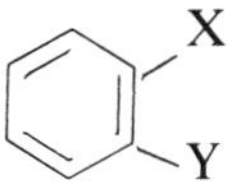

Kekulé's further suggestion that benzene was a mixture of two forms in equilibrium

satisfied the second point but still did not explain why such a highly unsaturated compound failed to undergo many of the addition reactions characteristic of alkenes.

Enthalpies (heats) of hydrogenation indicate that benzene does not contain three alkene type double bonds. Thus, for cyclohexene

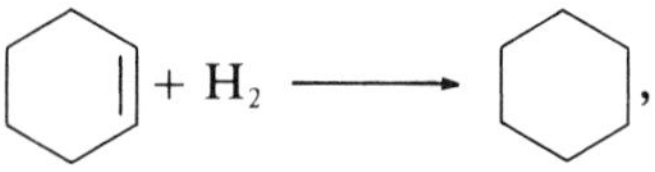

$+ H_2 \longrightarrow$, $\Delta H = -120 \text{ kJ mol}^{-1}$

Now, if benzene contained three double bonds its enthalpy (heat) of hydrogenation would be expected to have a value of three times that for cyclohexene, i.e. $3 \times -120 = -360 \text{ kJ mol}^{-1}$. In fact, for benzene

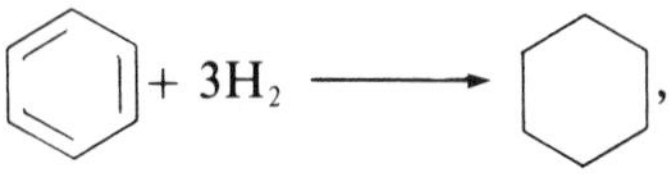

$+ 3H_2 \longrightarrow$, $\Delta H = -208 \text{ kJ mol}^{-1}$

This means that benzene is more stable than a hypothetical cyclic triene by an amount equal to 152 kJ mol^{-1}.

The standard enthalpy (heat) of formation of benzene from its constituent atoms, assuming it to be a cyclic triene, may be calculated by adding the relevant bond enthalpies (energies) and with the sign reversed (bonds are being formed here):

$$3 \times E_{C=C} \quad + 3 \times E_{C-C} \quad + 6 \times E_{C-H}$$

$$= 3 \times -611 \quad + 3 \times -346 \quad + 6 \times -413 \text{ kJ mol}^{-1}$$

$$= -5349 \text{ kJ mol}^{-1}$$

∴ $\Delta H_f^{\ominus}$ from the atoms $= -5349$ kJ mol^{-1}

The experimental value is found, from the energy cycle below, to be −5549 kJ mol^{-1}.

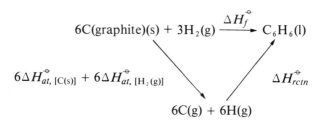

$$6C(\text{graphite})(s) + 3H_2(g) \xrightarrow{\Delta H_f^{\ominus}} C_6H_6(l)$$

$$6\Delta H_{at, [C(s)]}^{\ominus} + 6\Delta H_{at, [H_2(g)]}^{\ominus} \qquad \qquad \Delta H_{rctn}^{\ominus}$$

$$6C(g) + 6H(g)$$

Thus, these figures show that benzene is more stable than the cyclic triene by 200 kJ mol^{-1}. The discrepancy between this figure and the enthalpy of hydrogenation difference is due to the fact that average bond enthalpies (energies) were used.

Electron and X-ray diffraction studies show that the benzene molecule is planar and that all the C—C bond lengths are the same, i.e. 0.139 nm. This indicates that all the bonds are similar and intermediate in character between carbon–carbon double and single bonds which have lengths of 0.133 nm and 0.154 nm respectively.

Benzene may be regarded as being a resonance hybrid of two Kekulé and three Dewar structures, i.e.

Kekulé   Dewar

This does not mean that benzene continually changes from one form into another but rather that all the forms make a contribution to the true structure which is intermediate between these forms. The Kekulé forms make the major contribution.

The structure of benzene is best viewed from the orbital standpoint. Each carbon is joined to three other atoms and so they are *sp*^{2} hybridised (page 288) and the carbon and hydrogen atoms are all co-planar. This leaves an un-hybridised 2*p* orbital on each carbon and they overlap on both sides to give cyclic π orbitals (Figure 16.2). The π electrons are therefore delocalised round the whole system and are not just delocalised between alternate pairs of carbon atoms as in alkenes. Since an orbital can accommodate only two electrons, three cyclic π orbitals will be needed, but the result is rings of negative charge above and below the carbon atoms. The difference between the enthalpy (heat) of hydrogenation of benzene and the value expected for a hypothetical cyclic triene (see above) is known as the *delocalisation energy*, and it is the amount by which the delocalised structure is more stable than a cyclic triene. Benzene is reluctant to form addition compounds because this destroys the cyclic π orbitals.

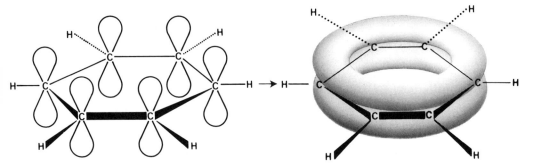

Figure 16.2   Bond formation in benzene

For convenience, benzene is often written as a single Kekulé form, it being understood that this does not imply that it contains three alkene type double bonds. Alternatively, the delocalised nature of the π electrons may be stressed by representing benzene as

and this is the convention used in this text.

Nomenclature of di- and poly-substituted aromatic compounds

When two or more substituents are present in a benzene ring their positions are indicated by numbering the ring, the numbering being such that the substituents have the lowest possible numbers. This is illustrated by the following examples:

1,2-dichlorobenzene

1,3-dichlorobenzene

1,4-dichlorobenzene

1,3,5-trinitrobenzene

The trivial names for disubstituted benzenes are still widely used, for example, the three dichlorobenzene isomers above are referred to as *o*- (short for *ortho*), *m*- (*meta*) and *p*- (*para*) dichlorobenzene respectively.

Reactions of benzene

Benzene does undergo a few addition reactions, e.g. with trioxygen (ozone) it forms a triozonide, $C_6H_6(O_3)_3$, three moles of hydrogen add on in the presence of a catalyst to give cyclohexane, C_6H_{12}, and three moles of chlorine will add on stepwise under the influence of u.v. light to give eventually 1,2,3,4,5,6-hexachlorocyclohexane (benzene hexachloride). However, the cyclic π orbitals and their stability ensure that the majority of its reactions are electrophilic substitution reactions. The general equation for these is:

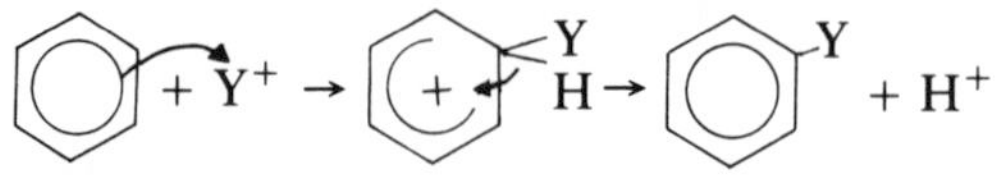

In the intermediate stage the four remaining π electrons (two were used to form the C—Y bond), and the positive charge, are delocalised over five carbon atoms. The substitution is completed by H^+ being eliminated so that the cyclic π orbitals are reformed. Important electrophilic substitutions are discussed below.

(a) Nitration

The introduction of a nitro group (NO_2) into a benzene ring is accomplished by warming the aromatic compound with a mixture of concentrated nitric(V) acid and sulphuric(VI) acid:

$$C_6H_6(l) + HNO_3(conc) \xrightarrow{H_2SO_4/55\ °C} C_6H_5 \cdot NO_2(l) + H_2O(l)$$
$$\text{Nitrobenzene}$$

The function of the sulphuric(VI) acid is to generate the nitryl cation (nitronium ion), NO_2^+.

Mechanism

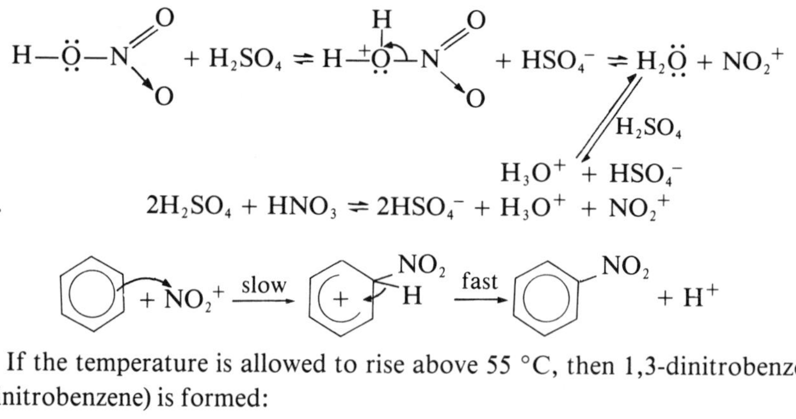

i.e. $\quad 2H_2SO_4 + HNO_3 \rightleftharpoons 2HSO_4^- + H_3O^+ + NO_2^+$

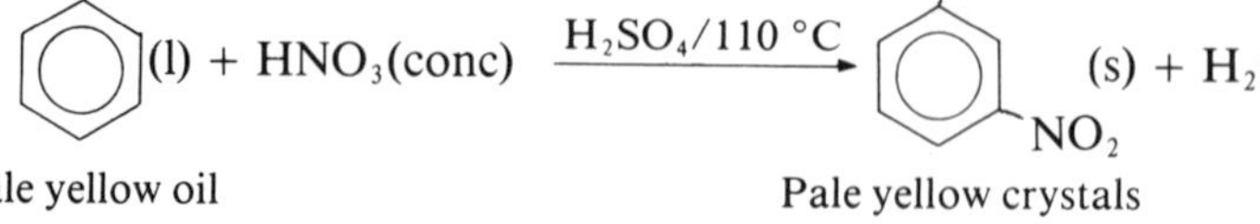

If the temperature is allowed to rise above 55 °C, then 1,3-dinitrobenzene (*m*-dinitrobenzene) is formed:

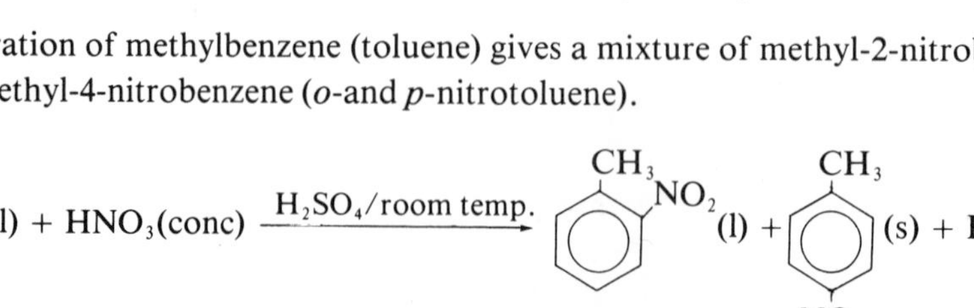

Nitration of methylbenzene (toluene) gives a mixture of methyl-2-nitrobenzene and methyl-4-nitrobenzene (*o*-and *p*-nitrotoluene).

It should be noted that this equation does not contravene the law of conservation of mass. Most organic reactions do not go to completion, and the total yield of the two isomers here is less than 100%. Methylbenzene is nitrated faster, and at a lower temperature, than benzene because the methyl group pushes electrons towards the ring ($+$ I effect) and so facilitates attack by NO_2^+. All aryl (aromatic) compounds with an electron donating group attached to the ring undergo 2- and 4- (*o*- and *p*-) substitution whilst electron withdrawing groups give rise to 3- (*m*-) substitution. The reason for this will be seen by studying the effect that —OH and —COOH groups have on benzene rings (pages 317 and 368 respectively). It should be noted that —OH, —NH$_2$, and halogens attached to a benzene ring give 2- and 4- substitution due to the mesomeric effect. Thus, they are electron donating through overlap of a lone pair of electrons in a *p* orbital with the π orbitals of the benzene ring.

(b) Sulphonation

Sulphonic acids are formed when arenes are warmed with concentrated sulphuric(VI) acid or treated with oleum (sulphuric(VI) acid containing dissolved sulphur(VI) oxide) at room temperature.

$$C_6H_6(l) + H_2SO_4(conc) \xrightarrow{80\,°C} C_6H_5 \cdot SO_3H(aq) + H_2O(l)$$

Benzenesulphonic acid

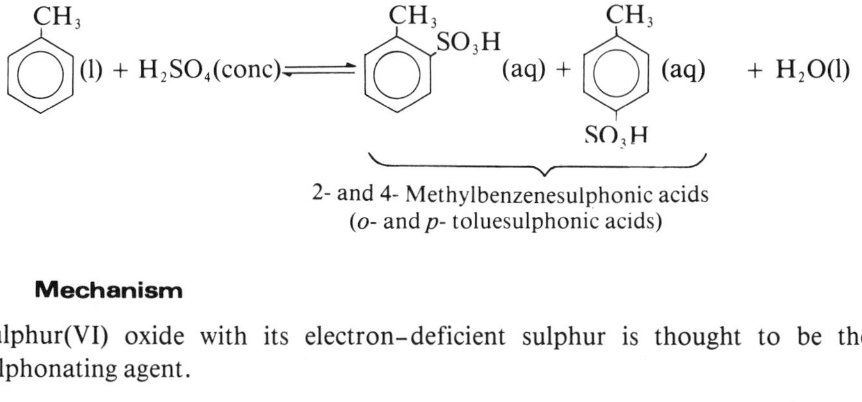

2- and 4- Methylbenzenesulphonic acids
(*o*- and *p*- toluesulphonic acids)

Mechanism

Sulphur(VI) oxide with its electron–deficient sulphur is thought to be the sulphonating agent.

$$2H_2SO_4 \rightleftharpoons SO_3 + H_3O^+ + HSO_4^-$$

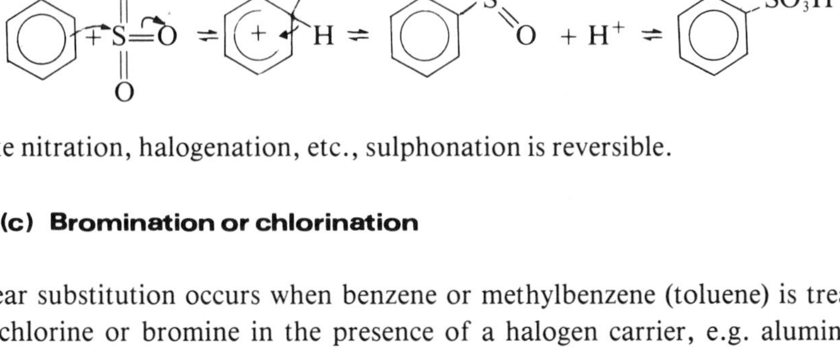

Unlike nitration, halogenation, etc., sulphonation is reversible.

(c) Bromination or chlorination

Nuclear substitution occurs when benzene or methylbenzene (toluene) is treated with chlorine or bromine in the presence of a halogen carrier, e.g. aluminium halides, iron(III) halides, or iodine.

$$C_6H_6(l) + Br_2(l) \xrightarrow[\text{room temperature}]{AlBr_3} C_6H_5Br(l) + HBr(g)$$

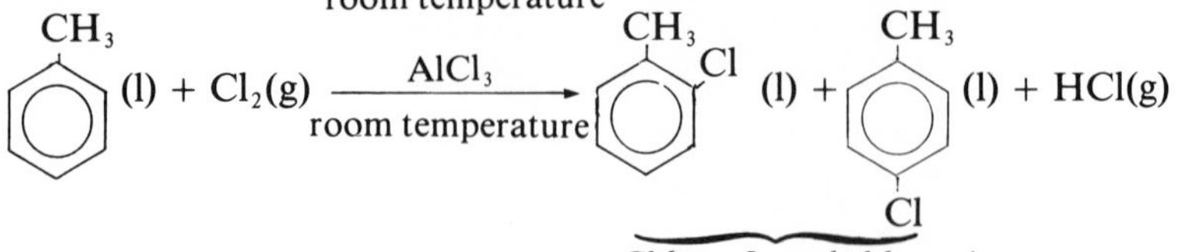

Chloro-2- and chloro-4-
methylbenzene
(*o*- and *p*- chlorotoluene)

Mechanism

The halogen is polarised by the halogen carrier. In the case of iron(III) halides, the iron has vacant *d* orbitals which can accept a pair of electrons, whilst in aluminium halides the aluminium has only six electrons in its outer shell and so is a strong electron acceptor, e.g.

$$:\ddot{Br}-\ddot{Br}: + AlBr_3 \rightarrow :\overset{\delta+}{\ddot{Br}}-\ddot{Br}:\text{-------}\overset{\delta-}{AlBr_3}$$

The aluminium attracts a lone pair of electrons from the nearest bromine atom and so becomes slightly negatively charged, whilst the bromine makes up its slight deficiency by attracting more strongly the electrons in the Br—Br bond. As a result the end bromine acquires a small positive charge and is attracted by the π electrons of the benzene ring.

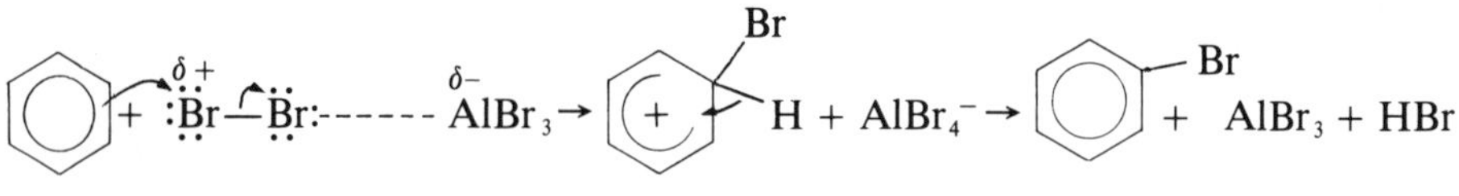

Iodine can act as a halogen carrier because the outer electrons of its atoms are a long way from the nucleus and so readily displaced, i.e. the molecule is easily polarised and this in turn polarises the chlorine or bromine:

$$:\ddot{I}-\ddot{I}: + :\ddot{Cl}-\ddot{Cl}: \rightarrow :\overset{\delta-}{\ddot{I}}-\overset{\delta+}{\ddot{I}}: \quad :\ddot{Cl}-:\ddot{Cl}: \rightarrow :\overset{\delta-}{\ddot{I}}-\ddot{I}:\text{----}:\ddot{Cl}-\overset{\delta+}{\ddot{Cl}}:$$

Side chain substitution occurs if methylbenzene (toluene) under reflux is treated with chlorine or bromine under the influence of ultraviolet light:

$$C_6H_5\cdot CH_3 + Cl_2 \xrightarrow[\text{reflux}]{\text{u.v.}} C_6H_5\cdot CH_2Cl + HCl \xrightarrow[\text{reflux}]{\text{u.v./}Cl_2} C_6H_5\cdot CHCl_2 + HCl$$

(Chloromethyl) benzene (benzyl chloride)

(Dichloromethyl) benzene (benzal or benzylidene chloride)

$$C_6H_5\cdot CHCl_2 \xrightarrow{Cl_2\text{/u.v./reflux}} C_6H_5\cdot CCl_3 + HCl$$

(Trichloromethyl) benzene
(benzotrichloride or benzylidyne chloride)

The mechanism is similar to that for chlorination or bromination of methane (page 297). It has already been mentioned that, under the same conditions, benzene undergoes addition of halogen.

(d) Acylations and alkylations (Friedel - Crafts reactions)

Arenes react with acyl chlorides (page 334), in the presence of aluminium chloride as catalyst, to give aromatic ketones. The reaction stops after one acyl group has entered the ring:

$$C_6H_6(l) + CH_3\cdot COCl(l) \xrightarrow[50\ °C]{AlCl_3} C_6H_5\cdot CO\cdot CH_3(l) + HCl(g)$$

Ethanoyl chloride Phenylethanone
(acetyl chloride)   (acetophenone)

Mechanism

The function of the catalyst is the same as in halogenation.

where R = an alkyl or aryl group.

Similar reactions occur between benzene and halogenoalkanes (alkyl halides) in the presence of aluminium chloride as catalyst, e.g.

$$C_6H_6(l) + CH_3Cl(g) \xrightarrow{AlCl_3} C_6H_5\cdot CH_3(l) + HCl(g)$$

It is, however, difficult to prevent more than one alkyl group entering the benzene ring.

Alkenes may be used instead of halogenoalkanes, but it is then necessary to have an acid (e.g. HCl, H_3PO_4) present as well as the Lewis acid. The function of the acid is to generate a carbonium ion:

$$CH_2\!=\!CH_2 + H^+ \rightarrow CH_3\cdot \overset{+}{C}H_2 \xrightarrow[BF_3\quad or\quad AlCl_3]{C_6H_6} C_6H_5\cdot CH_2\cdot CH_3 + H^+$$

This reaction is extremely important industrially since the product, ethylbenzene, is dehydrogenated to phenylethene (styrene) which is used to make polymers (page 359)

$$C_6H_5\cdot CH_2\cdot CH_3(l) \xrightarrow{ZnO/600\ °C} C_6H_5\cdot CH\!=\!CH_2(l) + H_2(g)$$

(e) Oxidation

Benzene is resistant to oxidation by the usual laboratory reagents, but it is oxidised with air, on an industrial scale, over hot vanadium(V) oxide as catalyst. The product is butenedioic anhydride (maleic anhydride) and it is used for making varnishes and lacquers.

The side chain in arenes is readily oxidised by refluxing with acid or alkaline potassium manganate(VII) (permanganate), acidified potassium dichromate(VI), dilute nitric(V) acid, etc. The side chain, regardless of its length, is oxidised to an acid group attached to the benzene ring.

$$C_6H_5 \cdot CH_3(l) + 3[O] \rightarrow C_6H_5 \cdot COOH(s) + H_2O(l)$$

Benzenecarboxylic acid
(benzoic acid)

The half-reaction for the oxidation is:

$$C_6H_5 \cdot CH_3 + 2H_2O \rightarrow C_6H_5 \cdot COOH + 6H^+ + 6e^-$$

The industrial oxidation of methylbenzene (toluene) in the vapour phase over a heated catalyst gives benzenecarbaldehyde (benzaldehyde):

$$C_6H_5 \cdot CH_3(l) + O_2(g) \xrightarrow[\text{(air)}]{\underset{500\,°C/N_2}{\text{Mn, Mo, and Zn oxides}}} C_6H_5 \cdot CHO(l) + H_2O(l)$$

The air is diluted with nitrogen to prevent the benzenecarbaldehyde (benzaldehyde) being oxidised to benzenecarboxylic acid (benzoic acid). The liquid phase oxidation of methylbenzene with manganese(IV) oxide and 65% sulphuric(VI) acid at 40 °C is also an important industrial method of preparing benzenecarbaldehyde (benzaldehyde).

5 HALOGENO-COMPOUNDS

The halogenoalkanes (alkyl halides) form a homologous series with the general formula $C_nH_{2n+1}X$ where $n = 1$ or more, and $X = Cl$, Br, or I. They are named systematically as halogen substituted alkanes, e.g.

CH_3Br bromomethane; $CH_3 \cdot CHI \cdot CH_3$ 2-iodopropane

The trivial names, in which they are known as the halide of the corrresponding alkyl group, are still fairly popular. The trivial names for the above two compounds are methyl bromide and isopropyl iodide respectively.

Halogenobenzenes are named as halogen substituted benzenes, e.g.

C_6H_5Br bromobenzene.

Halogenoalkanes are generally prepared by heating alcohols with hydrogen halides (page 318) or treating alcohols with phosphorus halides, e.g.

$$3CH_3OH(l) + PI_3(s) \rightarrow 3CH_3I(l) + H_3PO_3(s)$$
$$CH_3 \cdot CH_2OH(l) + PCl_5(s) \rightarrow CH_3 \cdot CH_2Cl(g) + HCl(g) + POCl_3(l)$$

Halogenobenzenes (aryl halides) are prepared by the action of the halogen and a halogen carrier on benzene (page 309).

Chloromethane, bromomethane, and chloroethane are gases whilst the other simple halogenoalkanes and the halogenobenzenes are colourless liquids.

Halogenoalkanes are polarised due to the inductive effect, e.g.

$\overset{\delta+}{CH_3} \rightarrow \overset{\delta-}{Cl.}$

This may be demonstrated by the attraction of a charged rod for a thin stream of the liquid halides (page 64). As a result of the slightly positively charged carbon, halogenoalkanes are susceptible to nucleophilic attack, i.e. attack by electron rich reagents. In all these reactions the halogen is replaced by another group, the general equation being

$$Y^- + R-X \longrightarrow R-Y + X^-$$

and so they are known as *nucleophilic substitution reactions.* Halogenobenzenes are much less reactive than halogenoalkanes, the reason being that the halogens have a lone pair of electrons in a *p* orbital that overlaps with the cyclic π orbitals of the ring. Hence, there is some double bond character between the carbon and the halogen and so the halogen is more firmly held. This is confirmed by the bonds lengths, e.g.

$$C-Cl \text{ in chloroalkanes } = 0.177 \text{ nm}$$
$$C-Cl \text{ in chlorobenzene } = 0.169 \text{ nm}$$

Halogenobenzenes will undergo nucleophilic substitutions but vigorous conditions are required.

The reactivity of the halides may be demonstrated by adding similar amounts of 1-chloro-, 1-bromo-, and 1-iodobutane to a little ethanol (solvent) in each of three tubes. Dilute silver nitrate(V) solution is added and then each mixture is shaken and placed in a constant temperature bath. It is seen that the rate of formation of silver halide by the reaction

$$C_4H_9X(l) + H_2O(l) \rightarrow C_4H_9OH(aq) + HX(aq) \xrightarrow{AgNO_3} AgX(s)$$

is

$$C_4H_9I > C_4H_9Br > C_4H_9Cl$$

This order is in agreement with the bond lengths and strengths, i.e. the C—I bond is the longest and weakest. No reaction occurs with halogenobenzenes under these conditions, but the halogen can be replaced by OH using very vigorous conditions, e.g.

$$C_6H_5Cl(l) \xrightarrow{NaOH \text{ solution}/300\ °C/pressure} C_6H_5OH(s)$$

It is found that the ease of substitution in halogenoalkanes depends upon the environment of the halogen as well as on the particular halogen involved. If, for example, the rates of hydrolysis of the isomeric bromides

$$CH_3 \cdot CH_2 \cdot CH_2 \cdot CH_2Br, \quad \begin{matrix} CH_3 \cdot CH_2 \\ CH_3 \end{matrix}\!\!\!CH-Br, \quad \text{and} \quad \begin{matrix} CH_3 \\ CH_3-C-Br \\ CH_3 \end{matrix}$$

(primary   secondary   and tertiary halogenoalkanes)

are studied in experiments similar to those above, it is found that hydrolysis occurs and silver bromide is precipitated in the order

$$(CH_3)_3CBr > CH_3 \cdot CH_2 \cdot CHBr \cdot CH_3 > CH_3 \cdot CH_2 \cdot CH_2 \cdot CH_2Br$$

Thus, tertiary halogenoalkenes are hydrolysed more readily than secondary which in turn are hydrolysed more rapidly than primary halogenoalkanes. This order reflects the increased polarisation and ionic character when more alkyl groups are attached to the carbon joined to the halogen. Kinetic studies indicate (page 161) that hydrolysis of primary halogenalkanes is a second order reaction, whilst hydrolysis of tertiary halogenoalkanes is first order. The mechanisms are:

$$
HO^- \overset{\delta+}{+} \overset{\delta-}{CH_2} - Br
\quad \xrightarrow{\text{slow}} \quad
\left[HO - - - CH_2 - - - Br \right]^-
\quad \xrightarrow{\text{fast}} \quad
HO - CH_2 + Br^-
$$

with CH_2, CH_2, CH_3 chains attached.

and

$$
(CH_3)_3CBr \xrightarrow{\text{slow}} (CH_3)_3C^+ + Br^- \xrightarrow[\text{fast}]{HO^-} (CH_3)_3COH
$$

Secondary halogenoalkanes tend to be hydrolysed by both mechanisms concurrently.

Nucleophilic substitution reactions of halogenoalkanes have wide synthetic utility since many nucleophiles can be used. The reactions are illustrated with bromoethane (ethyl bromide), but any halogenoalkane may be used.

(a) Refluxing with sodium hydroxide solution gives an alcohol:

$$C_2H_5Br(l) + NaOH(aq) \rightarrow NaBr(aq) + C_2H_5OH(aq)$$
Ethanol
(ethyl alcohol)

This reaction may be used to distinguish between aliphatic and aromatic halides. Thus, the unknown halide is refluxed for a few minutes with dilute sodium hydroxide solution. The aliphatic halide will be hydrolysed, but the aromatic halide will not. Hence acidification of the reaction mixture with dilute nitric(V) acid, followed by addition of silver nitrate(V) solution, will give a precipitate of silver halide in the case of the aliphatic compound but not with the aromatic halide. Note that halogen in the side chain of an aromatic compound gives the reactions of halogenoalkanes (alkyl halides).

(b) Refluxing with an alcoholic solution of an alkoxide gives an ether:

$$C_2H_5Br + CH_3ONa \rightarrow NaBr + C_2H_5 \cdot O \cdot CH_3$$
Sodium   Methoxyethane
methoxide   (ethyl methyl ether)

(c) Refluxing with the sodium salt of an acid gives an ester:

$$C_2H_5Br(l) + CH_3 \cdot COONa(aq) \rightarrow NaBr(aq) + CH_3 \cdot COOC_2H_5(l)$$
Ethyl ethanoate (ethyl acetate)

(d) Refluxing with an aqueous alcoholic solution of sodium cyanide gives a nitrile (cyanide):

$$C_2H_5Br + NaCN \rightarrow NaBr + C_2H_5CN$$
Propanenitrile
(ethyl cyanide)

(e) Heating with an aqueous alcoholic solution of ammonia under pressure gives an amine salt:

$$C_2H_5Br + NH_3 \xrightarrow{180\ ^\circ C/pressure} C_2H_5\overset{+}{N}H_3Br^-$$

Ethylammonium bromide
(ethylamine hydrobromide)

The reaction between many halogenoalkanes and sodium or potassium hydroxide is not as simple as is shown. Thus, hydroxide ion behaves as a strong base (i.e. it has a strong attraction for H$^+$) as well as a nucleophile (attraction for positively charged carbon) and so elimination of halogen acid competes with nucleophilic substitution.

$$CH_3 \cdot CHBr \cdot CH_3 \xrightarrow{HO^-} \begin{cases} \text{substitution} \longrightarrow CH_3 \cdot CHOH \cdot CH_3 + Br^- \\ \text{elimination} \longrightarrow CH_3 \cdot CH{=}CH_2 + H_2O + Br^- \end{cases}$$

The mechanism of the elimination is:

$$CH_3{-}\underset{\curvearrowleft Br}{CH}{-}\underset{\underset{OH}{\overset{|}{H}}}{\overset{|}{C}}{-}H \rightarrow CH_3 \cdot CH{=}CH_2 + H_2O + Br^-$$

the hydroxide ion attacking from the opposite side to the bromine so that they are as far apart as possible. The competition occurs whenever sodium or potassium hydroxide solution is refluxed with any halogenoalkane other than a halogenomethane or ethane. Elimination is the major reaction if alcohol is added to the mixture.

Alkenes other than ethene (ethylene) are prepared in the laboratory by refluxing halogenoalkanes with alcoholic potassium hydroxide, i.e. by dehydro-halogenation (removal of hydrogen halide) of halogenoalkanes. If a dihalogeno-alkane is used, the product is an alkyne, e.g.

$$CH_3 \cdot CHCl_2 + 2KOH \xrightarrow{ethanol/reflux} CH{\equiv}CH(g) + 2KCl + 2H_2O$$

$$CH_2Br \cdot CH_2Br + 2KOH \xrightarrow{ethanol/reflux} CH{\equiv}CH(g) + 2KBr + 2H_2O$$

6 HYDROXY COMPOUNDS

A compound containing one or more hydroxy groups attached to an alkyl group or to the side chain in a benzene ring is classed as an alcohol. However, a compound containing a hydroxyl group attached directly to a benzene ring is called a phenol.

Monohydric alcohols (alcohols containing one hydroxyl group) have the general formula $C_nH_{2n+1}OH$ where $n = 1$ or more. They are named as substituted alkanes, the alkane ending '-e' being removed and '-ol' added, e.g.

CH_3OH methanol (methyl alcohol)

$CH_3 \cdot CHOH \cdot CH_3$ propan-2-ol (isopropyl alcohol)

$$CH_3{-}\underset{\underset{CH_3}{\overset{|}{}}}{\overset{\overset{CH_3}{\overset{|}{}}}{C}}{-}OH \quad \text{2-methylpropan-2-ol (tertiary butyl alcohol)}$$

Alcohols with the functional groups

$$-CH_2OH, \qquad \diagdown CHOH, \qquad \text{and} \quad \diagup COH$$

are termed primary, secondary, and tertiary alcohols respectively.

The only aromatic hydroxy compound which will be considered is phenol, C_6H_5OH.

Alcohols are generally made in the laboratory by hydrolysis of halogeno-alkanes (page 314). Phenol can be made from benzenediazonium chloride (page 344).

The lower alcohols are colourless liquids, but 2-methylpropan-2-ol and phenol are low-melting-point solids. The comparatively high boiling points of the alcohols compared with their relative molecular masses is a consequence of hydrogen bonding (page 37) giving them a higher effective relative molecular mass.

Reactions of hydroxy compounds

(a) Salt formation

Alcohols are very weak acids because the alkyl group pushes electrons towards the hydroxyl group so that the oxygen does not strongly attract the electrons in the O—H bond.

$$R - O^{\delta-} \diagdown H^{\delta+}$$

Further, once a RO^- ion is formed it cannot be stabilised by delocalisation of the charge. Hence alcohols react only to a very small extent with alkalis, but will react with the very electropositive metals, under anhydrous conditions, giving alkoxides, e.g.

$$2CH_3OH(l) + 2Na(s) \rightarrow 2CH_3O^-Na^+(s) + H_2(g)$$
Sodium methoxide
(white solid)

Adding water regenerates the alcohol:

$$CH_3ONa(s) + H_2O(l) \rightarrow CH_3OH(aq) + NaOH(aq)$$

Phenol, on the other hand, is a stronger acid because the O—H bond breaks more readily and the resultant anion is stabilised by delocalisation of its charge. The easier breaking of the O—H bond is a result of one of the lone pairs of electrons of the oxygen overlapping with the π orbitals of the ring, thus making the oxygen slightly electron deficient. This is probably best illustrated by Kekulé type structures as below; the partial double bond character between the carbon and oxygen is confirmed by the bond length being shorter than for normal C—O bonds.

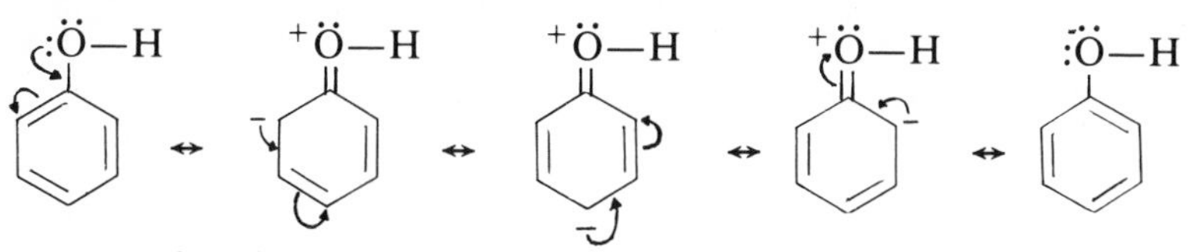

The oxygen, therefore, attracts the electrons in the O—H bond more strongly. The delocalisation of charge in the anion is as follows:

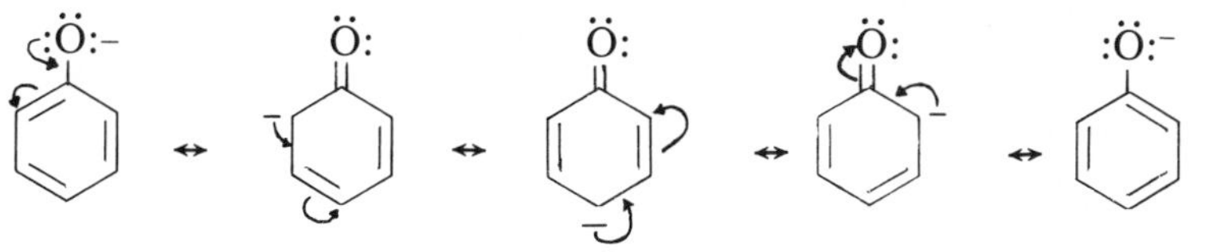

Phenol readily forms the phenoxide ion on treatment with solutions of alkali, e.g.

$$C_6H_5OH(s) + NaOH(aq) \rightarrow C_6H_5ONa(aq) + H_2O(l)$$

but it is not a strong enough acid to liberate carbon dioxide from sodium carbonate, and in this way it may be distinguished from acids. Adding acid to the salt regenerates the phenol:

$$C_6H_5ONa(aq) + HCl(aq) \rightarrow C_6H_5OH(aq) + NaCl(aq)$$

(b) Dehydration

Alcohols are readily dehydrated to alkenes by heating them with concentrated sulphuric(VI) acid or phosphoric(V) acid at 180 °C, or by passing their vapour over aluminium oxide at 350 °C, e.g.

$$CH_3 \cdot CH_2OH(l) \xrightarrow{H_2SO_4/180\ °C} CH_2{=}CH_2(g) + H_2O(l)$$

Mechanism

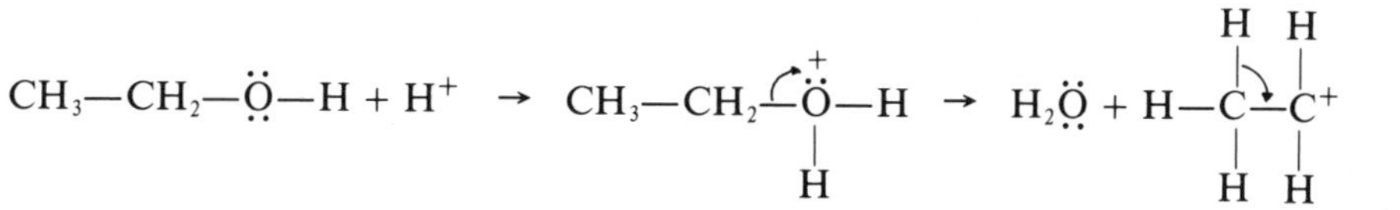

$$\longrightarrow CH_2{=}CH_2 + H^+$$

The ease of dehydration of alcohols is tertiary > secondary > primary because tertiary carbocations (carbonium ions) are more stable than secondary which in turn are more stable than primary carbocations. The stability of a carbocation, and so the ease with which it is formed, depends upon the extent to which its charge may be delocalised. In tertiary carbocations, e.g. $(CH_3)_3C^+$ there are three alkyl groups pushing electrons towards the positively charged carbon atom, but in secondary, e.g.

$$CH_3 \cdot \overset{+}{C}H \cdot CH_3 \text{ and primary carbocations, e.g. } CH_3 \cdot \overset{+}{C}H_2$$

there are only two and one such groups respectively.

If excess alcohol and slightly lower temperatures are used, ether formation occurs rather than dehydration, e.g.

$$CH_3 \cdot CH_2OH(l) + H_2SO_4(conc) \rightarrow CH_3 \cdot CH_2 \cdot O \cdot SO_3H + H_2O$$

$$\downarrow CH_3 \cdot CH_2OH/140\ ^\circ C$$

$$CH_3 \cdot CH_2 \cdot O \cdot CH_2 \cdot CH_3(l) + H_2SO_4(aq)$$
Ethoxyethane (diethyl ether)

Phenol does not undergo dehydration reactions.

(c) Halide formation

Alcohols react with phosphorus halides (page 312), and with hydrogen halides on heating, to give halogenoalkanes.

$$ROH + HBr \rightleftharpoons RBr + H_2O$$

In practice, a dehydrating agent is often also present to drive the reaction over to the right. For example, the hydrogen bromide is made *in situ* by adding excess concentrated sulphuric(VI) acid to sodium bromide.

The reaction is probably due to the basic character of the alcohol, i.e. a lone pair of electrons on the oxygen. Protonating the oxygen makes it more electron attracting and increases the positive charge on the carbon which then attracts the Br^- ion.

$$R-\overset{\delta-}{\underset{..}{O}}-\overset{\delta+}{H} + \overset{\delta+}{H}-\overset{\delta-}{Br} \rightarrow Br^- + R-\overset{+}{\underset{|}{\underset{H}{O}}}-H \rightarrow RBr + H_2\overset{..}{\underset{..}{O}}$$

Phenol does not undergo reaction with hydrogen halides.

(d) Ether formation

The formation of ethers is discussed above (see under dehydration of alcohols) and on page 317.

(e) Esterification

Esters are formed when alcohols, but not phenols, are refluxed with carboxylic acids in the presence of an acid catalyst (page 332), e.g.

$$CH_3 \cdot COOH(l) + C_2H_5OH(l) \underset{}{\overset{H^+}{\rightleftharpoons}} CH_3 \cdot COOC_2H_5(l) + H_2O(l)$$

Ethanoic acid   Ethyl ethanoate
(acetic acid)   (ethyl acetate)

Both alcohols and phenols form esters on reaction with acid chlorides and anhydrides (pages 335 and 339).

(f) Oxidation

Classification of the oxidation products is one method of distinguishing between primary, secondary, and tertiary alcohols. The usual oxidising agents used in the laboratory are acid or alkaline potassium manganate(VII) (permanganate), acidified potassium dichromate(VI), or dilute nitric(V) acid.

Primary alcohols are readily oxidised to aldehydes and then to acids. The alcohols, aldehyde, and acid all contain the same number of carbon atoms, e.g.

$$CH_3 \cdot CH_2OH \xrightarrow{[O]} CH_3 \cdot CHO \xrightarrow{[O]} CH_3 \cdot COOH$$

Ethanol   Ethanal   Ethanoic acid
                (acetaldehyde)   (acetic acid)

The half-reactions for these oxidations are:

$$CH_3 \cdot CH_2OH \rightarrow CH_3 \cdot CHO \ + 2H^+ + 2e^-$$
$$CH_3 \cdot CHO + H_2O \rightarrow CH_3 \cdot COOH + 2H^+ + 2e^-$$

The aldehyde may be detected by its reactions with 2,4-dinitrophenylhydrazine, and Tollen's reagent or Schiff's reagent (pages 326, 327 and 329).

Secondary alcohols are oxidised to ketones and then, on prolonged treatment, to two acids. The ketone has the same number of carbon atoms as the alcohol, but both acids have less, e.g.

$$CH_3 \cdot CHOH \cdot CH_2 \cdot CH_2 \cdot CH_3 \xrightarrow{[O]} CH_3 \cdot CO \cdot CH_2 \cdot CH_2 \cdot CH_3 \xrightarrow{[O]}$$

Pentan-2-ol
                        Pentan-2-one
                        (methyl propyl ketone)

$$CH_3 \cdot COOH + CH_3 \cdot CH_2 \cdot COOH$$
Ethanoic and propanoic acids
(acetic and propionic acids)

The ketone may be detected by its reaction with 2,4-dinitrophenylhydrazine and lack of reaction with Tollen's or Schiff's reagents.

Tertiary alcohols are oxidised, in acidic solutions only, to give a mixture of ketone and acid both with fewer carbon atoms than the alcohol, e.g.

2-Methylbutan-2-ol
                        Propanone   Ethanoic acid
                        (acetone)   (acetic acid)

If the alcohol under test is refluxed with dilute alkaline potassium manganate(VII) (permanganate), an aldehyde will be obtained from a primary alcohol, a ketone from a secondary alcohol, and there will be no reaction with the tertiary alcohol.

Primary and secondary alcohols are dehydrogenated industrially by passing their vapour over copper at 300 °C, e.g.

$$CH_3 \cdot CH_2OH(l) \xrightarrow{Cu/300\ °C} CH_3 \cdot CHO(l) + H_2(g)$$

$$CH_3 \cdot CHOH \cdot CH_3 \xrightarrow{Cu/300\ °C} CH_3 \cdot CO \cdot CH_3(l) + H_2(g)$$

Tertiary alcohols are dehydrated to alkenes under these conditions, e.g.

2-Methylpropan-2-ol
(t-butyl alcohol)
                        2-Methylpropene
                        (isobutene)

Oxidation of phenols generally results in breakdown of the benzene ring with the formation of unimportant products.

(g) The haloform reaction

This reaction is characteristic of alcohols containing the $CH_3 \cdot CHOH$ group.

Alcohols with this grouping give a yellow precipitate of triiodomethane (iodo-form) when warmed with iodine and dilute sodium hydroxide solution, e.g.

$$CH_3 \cdot CH_2OH(l) + 4I_2(aq) + 6NaOH(aq) \rightarrow$$
$$CHI_3(s) + HCOONa(aq) + 5NaI(aq) + 5H_2O(l)$$

Similarly, sodium chlorate(V) and bromate(V) convert these alcohols to trichloromethane (chloroform) and tribromomethane (bromoform) respectively.

The reaction may be used to distinguish between methanol and ethanol.

Dihydric alcohols

Di- and polyhydric alcohols have increasingly high boiling points and viscosities (page 291) as a result of the greater opportunities for hydrogen bonding.

Ethane-1,2-diol (ethylene glycol) is a very important dihydric alcohol since it is the major constituent of many anti-freeze mixtures and it is used in the manufacture of terylene (page 361). Its preparation has been discussed previously (pages 302 and 304) and its general reactions are similar to those of monohydric alcohols.

Uses of alcohols

Alcohols are used as solvents for various varnishes and paints etc. and in the manufacture of aldehydes, ketones, and esters. Methanol and ethanol, as well as ethane-1,2-diol, are used in anti-freeze mixtures. Large quantities of ethanol are consumed in the form of alcoholic drinks.

Electrophilic substitution reactions of phenol

It was seen above (page 317) that interaction of a lone pair of electrons on the oxygen in phenol with the π electrons of the ring causes the 2-, 4-, and 6- positions to become electron rich. Phenol is therefore considerably more susceptible to electrophilic attack than benzene. For example, benzene is brominated by bromine in the presence of a halogen carrier but addition of bromine water to phenol results in an immediate white precipitate of 2,4,6-tribromophenol:

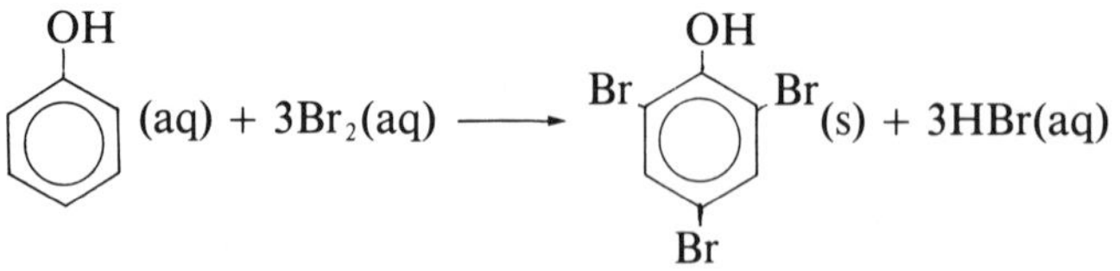

This reaction may be used to distinguish phenol from alcohols.

Similarly, benzene is nitrated by a mixture of concentrated nitric(V) and sulphuric(VI) acids at 55 °C, but phenol reacts with dilute nitric(V) acid at about 0 °C to give a mixture of 2- and 4-nitrophenol.

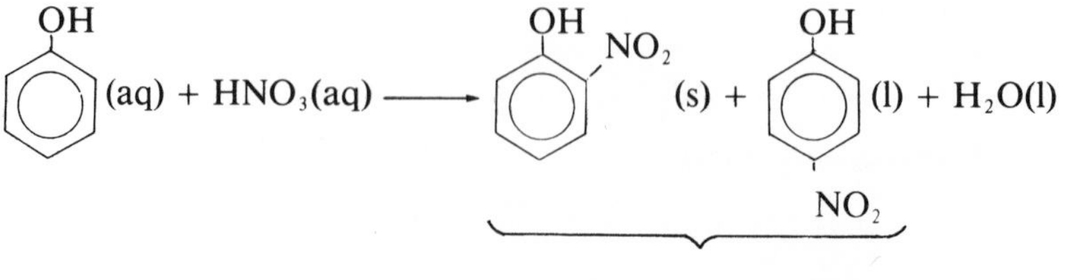

Uses of phenol

Phenol is used in the manufacture of Bakelite. It is also the starting material for making cyclohexanol which in turn is used in the manufacture of nylon (page 362).

7 ETHERS

Ethers have the structure R—O—R′ where R and R′ are alkyl or aryl groups which may be the same or different. The functional group is C—O—C. They are named as alkoxyhydrocarbons, e.g.

$CH_3 \cdot CH_2 \cdot O \cdot CH_2 \cdot CH_3$ ethoxyethane (diethyl ether)
$CH_3 \cdot O \cdot C_6H_5$ methoxybenzene (anisole)

Structural isomerism occurs as with other classes of compound, e.g. there are three ethers with molecular formula $C_4H_{10}O$:

$$CH_3 \cdot CH_2 \cdot O \cdot CH_2 \cdot CH_3, \qquad \underset{CH_3}{\overset{CH_3}{\diagdown}} CH \cdot O \cdot CH_3, \quad \text{and} \quad CH_3 \cdot O \cdot CH_2 \cdot CH_2 \cdot CH_3$$

This type of structural isomerism, where the alkyl groups in the molecule differ but the functional group remains the same, is known as *metamerism* and the isomers are called *metamers*.

Ethers may be prepared from alcohols and sulphuric(VI) acid (page 317), or by the reaction between alkoxides and halogenoalkanes. For example, refluxing an alcoholic solution of sodium methoxide with iodomethane (methyl iodide) gives methoxymethane (dimethyl ether):

$$CH_3ONa(s) + CH_3I(l) \xrightarrow{\text{Alcohol}} CH_3 \cdot O \cdot CH_3(g) + NaI(s)$$

A phenoxide may be used instead of an alkoxide, e.g.

$$\underset{\substack{\text{Sodium phenoxide} \\ \text{(phenate)}}}{C_6H_5ONa(s)} + CH_3 \cdot CH_2I(l) \xrightarrow{\text{Alcohol}} \underset{\substack{\text{Ethoxybenzene} \\ \text{(phenetole)}}}{C_6H_5 \cdot O \cdot CH_2 \cdot CH_3(l)} + NaI(s)$$

Ethers have much lower boiling points than alcohols because their molecules are not associated. Methoxymethane (dimethyl ether) is a gas and it, and the other lower members, are highly flammable. They are chemically unreactive since, unlike in alcohols and phenols, the oxygen is joined to two alkyl or aryl groups and not an acidic hydrogen that can be substituted by other groups.

Ethers are basic due to the lone pairs of electrons on the oxygen atom and so they will dissolve in strong acids, e.g.

$$CH_3 \cdot CH_2 \cdot \overset{..}{\underset{..}{O}} \cdot CH_3(l) + H_2SO_4(\text{conc}) \rightarrow [CH_3 \cdot CH_2 \cdot \overset{..}{O} \cdot CH_3]^+ \ HSO_4^-$$
$$\underset{H}{\overset{|}{}}$$

The salts decompose on addition of water because water is a stronger base than ethers.

Ethers decompose on refluxing with acids, e.g.

$$R—\ddot{O}—R + HBr \rightarrow Br^- + R—\overset{+}{\underset{\underset{H}{|}}{\ddot{O}}}—R \rightarrow RBr + R\ddot{O}H$$

Uses of ethers

In view of their unreactivity and their ability to dissolve many oganic substances, ethers are widely used as solvents. Ethoxyethane is an anaesthetic.

8 CARBONYL COMPOUNDS

The carbonyl group, $\overset{\diagdown}{\diagup}C{=}O$, is present in aldehydes, ketones, and carboxylic acids and their derivatives, but acids will be discussed later.

Aldehydes have the structure

$$R—C\overset{\diagup O}{\diagdown H} \qquad \text{where} \quad R = H \text{ or an alkyl or aryl group.}$$

Ketones have the structure

$$\overset{R\diagdown}{\underset{R\diagup}{}}C{=}O \qquad \text{where} \quad R = \text{alkyl or aryl groups.}$$

Aliphatic aldehydes are named after the alkane with the same number of carbon atoms, the end '-e' being replaced by '-al', e.g.

HCHO	CH$_3$·CHO	CH$_3$·CH$_2$·CHO
Methanal (formaldehyde)	Ethanal (acetaldehyde)	Propanal (propionaldehyde)

Aromatic aldehydes are named as the hydrocarbon carbaldehyde or the trivial name is still used, e.g. C$_6$H$_5$·CHO is benzenecarbaldehyde (benzaldehyde).

Aliphatic ketones are named by changing the corresponding alkane ending from '-e' to '-one', e.g.

CH$_3$·CO·CH$_3$	CH$_3$·CH$_2$·CH$_2$·CO·CH$_3$
Propanone (acetone)	Pentan-2-one (methyl propyl ketone)

The two common aromatic ketones, C$_6$H$_5$·CO·CH$_3$ and C$_6$H$_5$·CO·C$_6$H$_5$, are known as phenylethanone (acetophenone) and diphenylmethanone (benzophenone) respectively.

Aldehydes are generally prepared by oxidation of primary alcohols (page 318) but reduction of acid chlorides with hydrogen in the presence of palladium on a barium sulphate(VI) support (the Rosenmund reduction) is also used, e.g.

$$CH_3·COCl(l) + H_2(g) \xrightarrow{\text{Pd/BaSO}_4} CH_3·CHO(l) + HCl(g)$$

Ethanoyl chloride	Ethanal
(acetyl chloride)	(acetaldehyde)

Oxidation of methyl side chains in aromatic compounds also gives aldehydes (page 312).

Ketones are prepared by oxidation of secondary alcohols (page 319). Aromatic ketones may also be prepared by the Friedel–Crafts reaction (page 311).

Methanal (formaldehyde) is a gas which is very soluble in water, but the majority of the simple aldehydes and ketones are liquids. A 40% solution of methanal in water is commonly known as formalin.

Structure of the carbonyl groups

Since the carbonyl carbon atom bonds to three other atoms it is sp^2 hybridised and three sigma bonds are formed. The unhybridised $2p$ orbital of the carbon atom then overlaps with a $2p$ orbital of the oxygen atom to give a π bond:

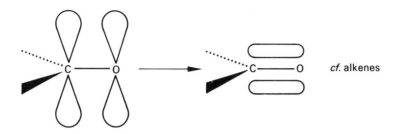

cf. alkenes

The π electrons are drawn towards the oxygen since it is more electronegative than carbon, i.e:

$$\overset{\delta+}{\underset{}{>}}\overset{\delta-}{C=O}$$

Hence, carbonyl compounds can undergo reaction via nucleophilic attack on the carbon or electrophilic attack on the oxygen. In fact, the former type constitutes the majority of its reactions, the latter type usually being important only when the electrophile is H^+, and this corresponds to acid catalysed addition of a nucleophile.

Alkenes do not undergo the same addition reactions as carbonyl compounds, because the C=C bond is not polarised and so electrophilic attack is much more common than nucleophilic attack.

At first sight, the addition reactions of carbonyl compounds may have been expected to be strongly acid catalysed, protonated oxygen making the carbon atom more positive and susceptible to nucleophilic attack. However, the position is complicated by the fact that the H^+ may interfere with the rest of the reaction. For instance, the dissociation of the weak acid, HCN:

$$HCN(aq) \rightleftharpoons H^+(aq) + CN^-(aq)$$

is repressed by adding H^+, the concentration of CN^- is lowered and so the addition to the carbonyl group is retarded. Similarly, H^+ converts the nucleophile $R\ddot{N}H_2$ into unreactive $R\overset{+}{N}H_3$. The concentration of H^+ must therefore be carefully controlled and so buffered solutions are often used.

The rate of nucleophilic attack on the carbonyl group depends on the size of the positive charge on the carbon atom. Electron-repelling groups will reduce this charge and so the following order of reactivity is observed:

$$\begin{matrix} H \\ \diagdown \\ \diagup \\ H \end{matrix} C{=}O > \begin{matrix} R \\ \diagdown \\ \diagup \\ H \end{matrix} C{=}O > \begin{matrix} R \\ \diagdown \\ \diagup \\ R \end{matrix} C{=}O$$

Addition reactions of carbonyl compounds

(a) Addition of hydrogen cyanide

Hydrogen cyanide adds on to aldehydes and ketones to give hydroxynitriles (cyanohydrins):

$$\begin{matrix} \diagdown \\ \diagup \end{matrix} C{=}O + HCN \rightarrow \begin{matrix} \diagdown \\ \diagup \end{matrix} C \begin{matrix} \diagup OH \\ \diagdown CN \end{matrix}$$

The reaction is base catalysed, the function of the base being to increase the cyanide ion concentration, e.g.

$$HCN(aq) + \overset{+}{Na}\overset{-}{O}H(aq) \rightarrow \overset{+}{Na}\overset{-}{C}N(aq) + H_2O(l)$$

Buffered sodium hydroxide or ammonia solution has to be used because strong bases promote condensation reactions. The mechanism is:

$$CN^{\frown} + \begin{matrix} \diagdown \\ \diagup \end{matrix} \overset{\delta+}{C} {\underset{\frown}{=}} \overset{\delta-}{O} \rightarrow \begin{matrix} \diagdown \\ C{-}O^- \\ | \\ CN \end{matrix} \xrightarrow{\text{HCN or H}_2O} \begin{matrix} \diagdown \\ C{-}OH \\ | \\ CN \end{matrix}$$

The reaction is important as a means of preparing hydroxycarboxylic acids, the cyanide group being readily hydrolysed to —COOH by refluxing with 70% sulphuric(VI) acid or dilute sodium hydroxide solution, e.g.

$$CH_3{-}C{\underset{H}{\overset{O}{\diagup}}}(l) + HCN(aq) \rightarrow CH_3{-}C{\underset{H}{\overset{OH}{\diagup}}}CN(aq) \xrightarrow[H_2SO_4]{70\%} CH_3 \cdot CHOH \cdot COOH(aq)$$

2-Hydroxypropane-
nitrile (acetaldehyde
cyanohydrin)   2-Hydroxypropanoic acid
(lactic acid)

(b) Addition of sodium hydrogensulphate(IV) (sodium hydrogen-sulphite)

White crystalline addition compounds are slowly precipitated when aldehydes are shaken with saturated solutions of sodium hydrogensulphate(IV) (sodium hydrogensulphite), e.g.

$$C_6H_5 \cdot CHO(l) + NaHSO_3(aq) \rightarrow C_6H_5 \cdot \overset{\displaystyle H}{\underset{\displaystyle SO_3Na}{\overset{|}{\underset{|}{C}}} {-}OH(s)}$$

Benzenecarbaldehyde sodium hydrogensulphate(IV)
(Benzaldehyde sodium hydrogensulphite)

The reaction with ketones is slow or non-existent unless the ketone is cyclic or a methyl ketone, i.e. $CH_3 \cdot CO \cdot R$. With larger groups *steric hindrance* occurs, i.e. the large groups get in the way of the bulky SO_3^{2-} ion and prevent it getting at the positive carbon.

The addition compounds (hydroxysulphonic acids) may be decomposed by warming with dilute acid or alkali, the carbonyl compound being regenerated. Hence the reaction is useful for separating carbonyl compounds from other organic compounds.

(c) Addition of sodium tetrahydridoborate(III) (sodium borohydride) or lithium tetrahydridoaluminate(III) (lithium aluminium hydride)

Sodium tetrahydridoborate(III), $NaBH_4$, and lithium tetrahydridoaluminate(III), $LiAlH_4$, act as hydride ion (H^-) carriers and reduce carbonyl compounds to alcohols. The reduction is completed by adding water to decompose the organometallic compound formed:

Lithium tetrahydridoaluminate(III) (lithium aluminium hydride) is used in an ether as solvent, but the less reactive sodium tetrahydridoborate(III) (sodium borohydride) may be used in water or alcohol. Aldehydes and ketones are reduced to primary and secondary alcohols respectively.

Addition – elimination reactions of carbonyl compounds

These reactions involve the addition of a compound to aldehydes or ketones followed by elimination of water; they are also known as *condensation reactions*. Two main types of compound are involved: (a) substituted ammonia compounds, and (b) alcohols.

(a) Substituted ammonia compounds react with aldehydes and ketones via a mechanism which involves nucleophilic attack on the carbonyl carbon atom by the lone pair of electrons on a nitrogen atom, followed by elimination of water:

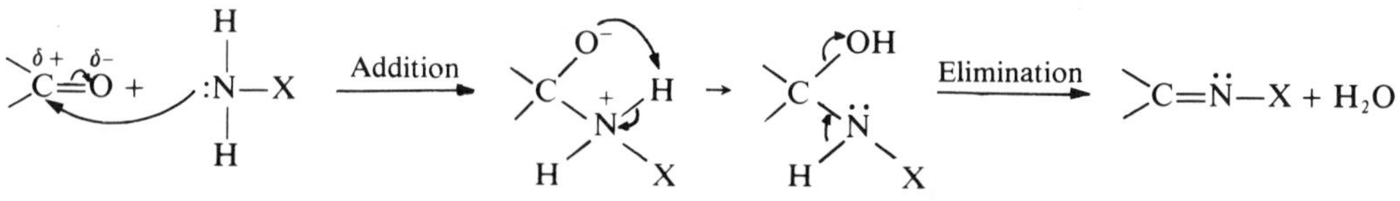

where   X = NH_2, $C_6H_5 \cdot NH$, [structure], OH, $NH \cdot CO \cdot NH_2$ or alkyl or aryl groups.

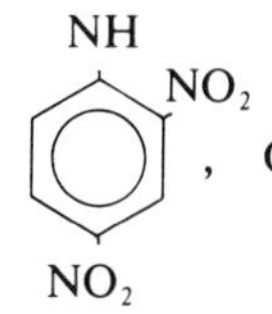

These derivatives are important in characterising carbonyl compounds and they may be hydrolysed back to the carbonyl compound by refluxing them with dilute mineral acid. The individual reactions are given below. Note that the substituted ammonia compounds are generally obtained as their hydrochlorides and it is necessary to obtain the free base without making the solution strongly alkaline since this could lead to polymerisation of the carbonyl compound. Sodium ethanoate (acetate) is therefore added to free the base and buffer the solution, e.g.

$$NH_2OH \cdot HCl(aq) + CH_3 \cdot COONa(aq) \rightarrow$$
$$CH_3 \cdot COOH(aq) + NaCl(aq) + NH_2OH(aq)$$

(i) Hydrazine, $NH_2 \cdot NH_2$, can react with aldehydes and ketones to give monoderivatives called hydrazones, i.e.

$$\begin{array}{c}\diagup\\\diagdown\end{array}C{=}N \cdot NH_2 \quad \text{and diderivatives called azines, i.e.} \quad \begin{array}{c}\diagup\\\diagdown\end{array}C{=}N \cdot N{=}C\begin{array}{c}\diagdown\\\diagup\end{array}$$

However, if substituted hydrazines are used, azine formation does not arise and the higher relative molecular mass products are more likely to be crystalline solids. For this reason, phenylhydrazine, $C_6H_5 \cdot NH \cdot NH_2$, or more particularly 2,4-dinitrophenylhydrazine is used, e.g.

$$CH_3 \cdot CHO(l) + \underset{NO_2}{\underset{|}{\overset{NH \cdot NH_2}{\overset{|}{\bigcirc}}}} NO_2 \longrightarrow \underset{NO_2}{\underset{|}{\overset{NH \cdot N=CH \cdot CH_3}{\overset{|}{\bigcirc}}}} NO_2 \;(s) + H_2O(l)$$

Ethanal 2,4-dinitrophenylhydrazone
(acetaldehyde 2,4-dinitrophenylhydrazone)

2,4-Dinitrophenylhydrazones are formed as yellow–orange precipitates.

(ii) Hydroxylamine, NH_2OH, reacts with carbonyl compounds to give oximes, e.g.

$$\underset{CH_3}{\overset{CH_3}{>}}C{=}O(l) + NH_2OH(aq) \rightarrow \underset{CH_3}{\overset{CH_3}{>}}C{=}NOH(s) + H_2O(l)$$

Propanone oxime (acetoxime)

(iii) Aldehydes and ketones form semicarbazones with semicarbazide and these are more likely to be crystalline solids than are oximes.

$$\bigcirc{=}O(l) + NH_2 \cdot NH \cdot CO \cdot NH_2(aq) \rightarrow \bigcirc{=}N \cdot NH \cdot CO \cdot NH_2(s) + H_2O(l)$$

Cyclohexanone   Cyclohexanone semicarbazone

(iv) Carbonyl compounds will react with primary aliphatic and aromatic amines, but the reactions with the former class of compound are unimportant. Primary aromatic amines however, react with aldehydes and ketones to give stable crystalline solids known as anils or Schiff's bases, e.g.

$$C_6H_5 \cdot CHO(l) + C_6H_5 \cdot NH_2(l) \rightarrow C_6H_5{-}\underset{\underset{OH}{|}}{CH}{-}\underset{\underset{H}{|}}{N}{-}C_6H_5$$

$$\rightarrow C_6H_5 \cdot CH{=}N \cdot C_6H_5(s) + H_2O(l)$$
N-Benzylidene phenylammine (Benzylidene aniline)

(b) Aldehydes react with alcohols in two stages, in the presence of dry hydrogen chloride or calcium chloride as catalyst, to give hemiacetals and then acetals e.g.

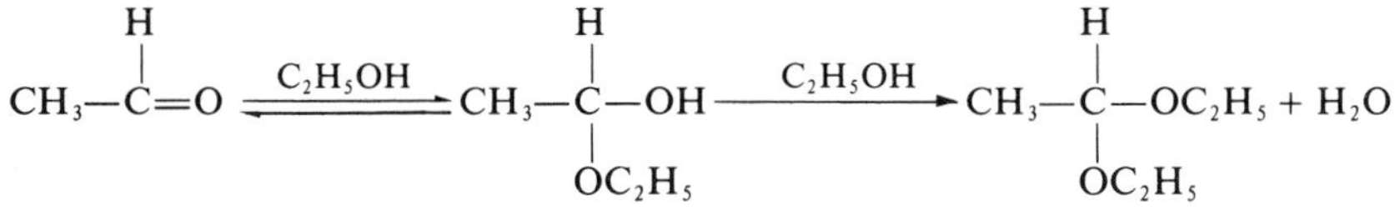

$$CH_3-\overset{\overset{\displaystyle H}{|}}{C}=O \xrightleftharpoons{C_2H_5OH} CH_3-\overset{\overset{\displaystyle H}{|}}{\underset{\underset{\displaystyle OC_2H_5}{|}}{C}}-OH \xrightarrow{C_2H_5OH} CH_3-\overset{\overset{\displaystyle H}{|}}{\underset{\underset{\displaystyle OC_2H_5}{|}}{C}}-OC_2H_5 + H_2O$$

A hemiacetal   An acetal
1,1-Diethoxyethane
(diethyl acetal)

The mechanism is:

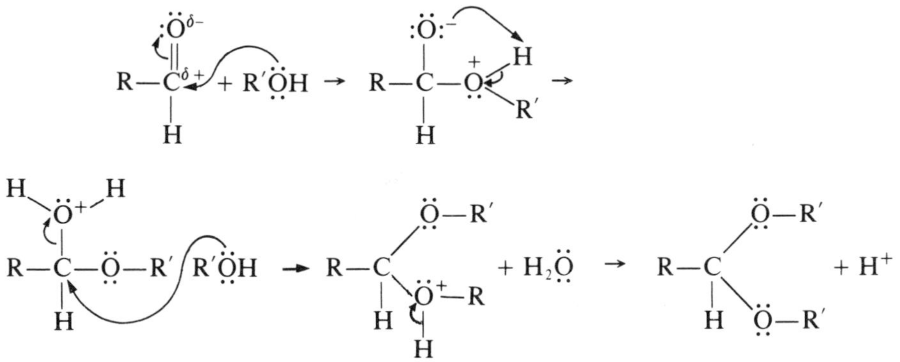

Acetals are hydrolysed back to aldehydes on refluxing with dilute acid, but are unaffected by alkalis and oxidising agents and so they are formed to protect aldehyde groups during reaction on other parts of the molecule.

Other reactions of carbonyl compounds

(a) Oxidation

Aldehydes are readily oxidised to carboxylic acids, aromatic aldehydes being oxidised even on standing in air. The ease of oxidation is due to the fact that the C—H bond can be broken without breaking any C—C bonds, e.g.

$$CH_3-\overset{\overset{\displaystyle O}{||}}{C}-H \xrightarrow{[O]} CH_3-\overset{\overset{\displaystyle O}{||}}{C}-OH$$

Ketones are considerably more difficult to oxidise because this entails breaking C—C and C—H bonds; a mixture of acids is formed, e.g.

$$CH_3 \cdot CO \cdot CH_3 \xrightarrow{[O]} CH_3 \cdot COOH + CO_2 + H_2O$$

$$CH_3 \cdot CH_2 \cdot CH_2 \cdot CO \cdot CH_3 \xrightarrow{[O]} CH_3 \cdot CH_2 \cdot COOH + CH_3 \cdot COOH$$

Both aldehydes and ketones are oxidised by strong oxidising agents such as acidified potassium manganate(VII) (potassium permanganate), acidified potassium dichromate(VI), nitric(V) acid, etc. However, aldehydes are also oxidised by very mild reagents such as Tollen's reagent and Fehling's solution, and this affords a method of distinguishing between aldehydes and ketones.

Tollen's reagent is made by adding ammonia solution to silver nitrate(V) solution until the initial precipitate of silver oxide is nearly, or just dissolved. A silver mirror is deposited on the walls of the container when Tollen's reagent is warmed with an aldehyde:

$$AgNO_3(aq) \xrightarrow{\text{NaOH(aq)}} [AgOH] \rightarrow Ag_2O(s) \xrightarrow{\text{NH}_3\text{(aq)}} [Ag(NH_3)_2]^+OH^-(aq)$$
Diamminesilver hydroxide

$$R \cdot CHO(l) + 2[Ag(NH_3)_2]OH(aq) \rightarrow RCOONH_4(aq) + 2Ag(s) + 3NH_3(aq) + H_2O(l)$$

For simplicity, Fehling's solution may be regarded as being copper(II) oxide in solution as a complex. When it is boiled with an aliphatic aldehyde a red precipitate of copper(I) oxide is formed, e.g.

$$CH_3 \cdot CHO(l) + 2CuO(aq) \rightarrow CH_3 \cdot COOH(aq) + Cu_2O(s)$$

The reaction with aromatic aldehydes is negligible, possibly due to their low solubility in water.

(b) Reduction

Aldehydes and ketones may be reduced to primary and secondary alcohols respectively, e.g.

$$CH_3 \cdot CHO \rightarrow CH_3 \cdot CH_2OH$$
$$CH_3 \cdot CO \cdot CH_3 \rightarrow CH_3 \cdot CHOH \cdot CH_3$$

The reduction may be effected with sodium amalgam and water, sodium and ethanol, zinc and ethanoic acid (acetic acid), or catalytically with hydrogen and nickel or dichromium(III) copper(II) oxide (copper chromite) at 200–300 °C. When the reduction is accomplished with dissolving metals, the half-reaction is:

$$\text{>C=O} + 2e^- + 2H^+ \rightarrow \overset{\displaystyle H}{\underset{\displaystyle |}{\text{>C}}}-OH$$

The reduction with lithium tetrahydridoaluminate(III) (lithium aluminium hydride), and sodium tetrahydridoborate(III) (sodium borohydride) was discussed above under addition reactions.

Ketones may be reduced to hydrocarbons by refluxing them with zinc amalgam and concentrated hydrochloric acid, e.g.

$$C_6H_5 \cdot CO \cdot CH_3(l) \xrightarrow{\text{Zn–Hg/HCl}} C_6H_5 \cdot CH_2 \cdot CH_3(l)$$
Ethylbenzene

The reaction does not work very well with aldehydes.

(c) Polymerisation

Aliphatic aldehydes readily polymerise forming brown resinous products of unknown composition when warmed with concentrated sodium hydroxide solution.

Ethanal (acetaldehyde) forms a trimer, $(CH_3 \cdot CHO)_3$, (used as a hypnotic) on addition of a little concentrated sulphuric(VI) acid at room temperature. If the reaction is done at 0 °C, a tetramer, $(CH_3 \cdot CHO)_4$, known as metaldehyde, (used in slug bait) is formed.

Aliphatic aldehydes, with the exception of methanal (formaldehyde), undergo the *aldol condensation* on treatment with dilute sodium hydroxide solution, e.g.

$$CH_3 \cdot CHO + CH_3 \cdot CHO \xrightarrow{NaOH} CH_3 \cdot \overset{\overset{\displaystyle OH}{|}}{CH} \cdot CH_2 \cdot CHO$$

3-Hydroxybutanal (aldol)

Ketones have little tendency to polymerise although they do undergo aldol type condensations.

(d) Schiff's reagent

Aldehydes restore the violet colour to Schiff's reagent (a dye which has been decolourised by passing sulphur dioxide through it). The reagent is unaffected by ketones, except propanone (acetone) which restores the colour very slowly, and so the reaction is used as a method of distinguishing between aldehydes and ketones.

(e) The triiodomethane (iodoform) reaction

Ethanal (acetaldehyde) and ketones containing the $CH_3 \cdot CO$ group (or alcohols which are readily oxidised to carbonyl compounds containing this group, e.g. ethanol and propan-2-ol) undergo the triiodomethane (iodoform) reaction. Thus, warming these compounds with iodine and sodium hydroxide solution leads to a yellow precipitate of triiodomethane, e.g.

$$CH_3 \cdot CO \cdot CH_3(l) + 3I_2(aq) + 4NaOH(aq) \rightarrow CHI_3(s) + CH_3 \cdot COONa(aq) + 3NaI(aq) + 3H_2O(l)$$

The reactions of aldehydes and ketones are summarised in Table 16.1.

Table 16.1.   Summary of the reactions of aldehydes and ketones

Reagent	Aldehydes	Ketones
1. HCN	Addition occurs	Addition occurs
2. NaHSO$_3$	Addition occurs	Reaction slow if at all
3. LiAlH$_4$ (i.e. reduction)	Reduced to primary alcohols	Reduced to secondary alcohols
4. Substituted ammonias	Undergo condensation reactions	Undergo condensation reactions
5. Alcohols	Form acetals	Little reaction
6. Acidified KMnO$_4$	Readily give acids with the same number of C atoms	More difficult to oxidise; give acids with fewer C atoms
7. Tollen's reagent	Silver mirror formed	No reaction
8. Fehling's solution	Red precipitate with aliphatic aldehydes	No reaction
9. Polymerisation	Occurs readily with aliphatic aldehydes	Little tendency to do this
10. Schiff's reagent	Restore the violet colour	Only $CH_3 \cdot CO \cdot CH_3$ gives this reaction
11. Triiodomethane (iodoform) reaction	Occurs only with $CH_3 \cdot CHO$	Occurs with ketones containing the $CH_3 \cdot CO$ group.

Uses of aldehydes and ketones

Aldehydes are used to make acids, etc. Methanal is used to preserve biological specimens and in the manufacture of Bakelite and various resins.

Ketones are used as solvents for plastics and varnishes. Propanone is used in the manufacture of Perspex.

9   CARBOXYLIC ACIDS AND THEIR DERIVATIVES

Carboxylic acids have the functional group $-C\underset{O-H}{\overset{O}{\diagup}}$

The aliphatic acids are named after the corresponding alkane, the ending '-e' being replaced by '-oic acid', e.g.

HCOOH   methanoic acid (formic acid)
$CH_3 \cdot COOH$ ethanoic acid (acetic acid)

Aromatic acids are named as the hydrocarbon carboxylic acid, e.g.

$C_6H_5 \cdot COOH$ benzenecarboxylic acid (benzoic acid)

Carboxylic acids may be prepared by oxidation of alcohols (page 318) and aldehydes (page 327) and by hydrolysis of nitriles (page 340). Aromatic acids can also be prepared by oxidation of alkyl side chains (page 312).

The lower aliphatic acids are colourless liquids, their boiling points being higher than expected from their relative molecular masses owing to association or hydrogen bonding (page 37):

$$R-C \begin{array}{c} \overset{\delta^+}{} \\ \end{array} \begin{array}{c} \overset{\delta^-}{O}-\overset{\delta^+}{H}-----\overset{\delta^-}{O} \\ O-----H-O \\ \overset{\delta^-}{} \quad \overset{\delta^+}{} \; \overset{\delta^-}{} \end{array} \begin{array}{c} \overset{\delta^+}{} \\ \end{array} C-R$$

The solubility in water decreases as the relative molecular mass increases, the solubility being due to hydrogen bonding between the water and the carboxyl group. Benzenecarboxylic acid (benzoic acid) is a white crystalline solid, sparingly soluble in cold water but quite soluble in hot water.

The reason for the acidity of the carboxyl group, and the effect of substituents on the acidity of acids, is discussed in Chapter 20.

The carbonyl group in acids is relatively unreactive. This may be attributed to the carboxyl group being a resonance hybrid:

$$-C \overset{\ddot{O}:}{\underset{\ddot{O}-H}{\diagup}} \longleftrightarrow -C \overset{\ddot{O}:^-}{\underset{\overset{+}{O}-H}{\diagup}}$$

i.e. a lone pair of electrons on the hydroxyl oxygen interacts with the π orbitals of the carbonyl group. The carbonyl carbon atom is therefore far less electron deficient than in aldehydes and ketones and, in contrast to them, shows far less vulnerability to nucleophilic attack.

Methanoic acid (formic acid) differs from the other acids in a number of ways.

(i)   It does not form an acid chloride or anhydride.

(ii) It is dehydrated by concentrated sulphuric(VI) acid (see page 198).

(iii) It reduces Fehling's solution and Tollen's reagent (cf. aldehydes) because it

contains the readily oxidisable $-C\begin{smallmatrix}\nearrow O \\ \searrow H\end{smallmatrix}$ group. The half-reaction is:

$$HCOOH \rightarrow CO_2 + 2H^+ + 2e^-$$

Reactions of carboxylic acids

(a) Acidity

The carboxylic acids behave as weak acids:

$$R \cdot COOH + H_2\ddot{O} \rightleftharpoons R \cdot COO^- + H_3O^+$$

and react with alkalis to give salts. They liberate carbon dioxide from sodium carbonate solution and this is used as a test for them:

$$2CH_3 \cdot COOH(l) + Na_2CO_3(aq) \rightarrow 2CH_3 \cdot COONa(aq) + H_2O(l) + CO_2(g)$$
$$\text{Sodium ethanoate}$$
$$\text{(sodium acetate)}$$

The acids are displaced from their salts by stronger acids, e.g.

$$R \cdot COONa + HCl \rightarrow R \cdot COOH + NaCl$$

(b) Decarboxylation

Decarboxylation is the removal of carbon dioxide from a carboxylic acid group. The usual method is to heat the sodium salt of the acid with soda-lime (NaOH–CaO); the product is a hydrocarbon, e.g.

$$CH_3 \cdot COONa(s) + NaOH(s) \rightarrow Na_2CO_3(s) + CH_4(g)$$
$$C_6H_5 \cdot COONa(s) + NaOH(s) \rightarrow Na_2CO_3(s) + C_6H_6(l)$$

The mechanism may be

$$R-\overset{O^-}{\underset{O}{\overset{|}{C}}}\underset{\|}{+ OH^-} \rightarrow R^- + \overset{O^-}{\underset{O}{\overset{|}{C}}}\underset{\|}{-OH} \rightarrow RH + \overset{O^-}{\underset{O}{\overset{|}{C}}}\underset{\|}{-O^-} \; (CO_3{}^{2-})$$

Heating calcium methanoate (calcium formate) gives methanal (formaldehyde) but heating the calcium or barium salts of the other acids gives ketones, e.g.

$$(HCOO)_2Ca(s) \rightarrow CaCO_3(s) + HCHO(g)$$

$$(CH_3 \cdot COO)_2Ca(s) \rightarrow CaCO_3(s) + CH_3 \cdot CO \cdot CH_3(l)$$

(c) Reduction

Acids are resistant to most of the usual reducing agents but lithium tetrahydridoaluminate(III) (lithium aluminium hydride) reduces them to primary alcohols:

$$R \cdot COOH \xrightarrow[\text{(ii) } H_2O]{\text{(i) } LiAlH_4} R \cdot CH_2OH$$

The reagent is highly selective and, for example, the $C{=}C$ bond in unsaturated acids is unaffected by it.

(d) Esterification

Esters have the structure
$$\overset{\displaystyle O}{\overset{\displaystyle \|}{R-C-O-R'}}$$
where $R = H$, alkyl, or aryl and $R' = $ alkyl or aryl. They are named as alkyl or aryl salts of the acid, e.g.

$CH_3 \cdot COOC_2H_5$	ethyl ethanoate	(ethyl acetate)
$C_6H_5 \cdot COOC_6H_5$	phenyl benzenecarboxylate	(phenyl benzoate)

Esterification is the process of converting acids into esters by refluxing them with an alcohol. The reaction is an equilibrium process and it is acid catalysed — a few drops of concentrated sulphuric(VI) acid are added or alternatively hydrogen chloride is passed through the mixture:

$$R \cdot COOH + R'OH \underset{}{\overset{H^+/\text{reflux}}{\rightleftharpoons}} R \cdot COOR' + H_2O$$

The acid not only catalyses the reaction but also increases the yield by tying up the water to some extent and so hindering the reverse reaction. The reverse reaction is also retarded by using a large excess of alcohol (or acid) so that the concentration of the water is effectively reduced.

The mechanism is:

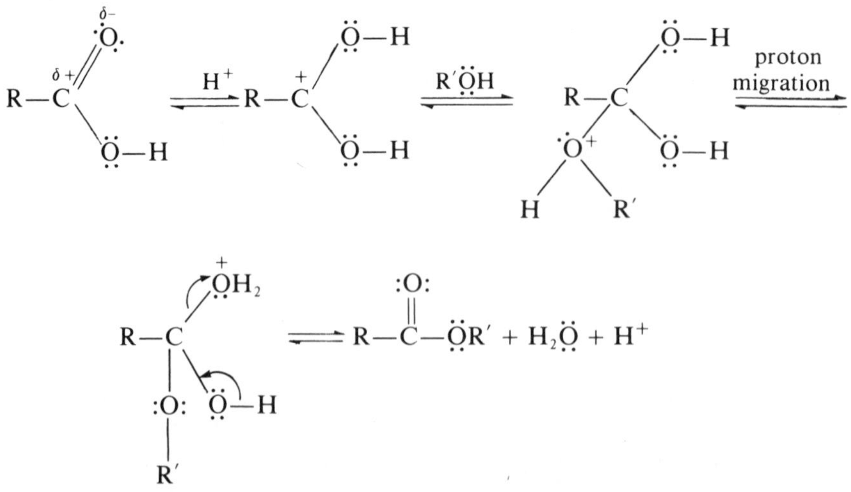

If the esterification is done using an alcohol labelled with the ^{18}O isotope, it is found that the ester contains the labelled oxygen and not the water:

$$R \cdot COOH + R'^{18}OH \underset{}{\overset{H^+}{\rightleftharpoons}} R - \overset{\overset{\displaystyle O}{\|}}{C} - {}^{18}OR' + H_2O$$

By this means it has been shown that the oxygen in the water comes from the acid, i.e. acyl-oxygen not alkyl-oxygen fission occurs.

Esters have pleasant fruity smells. They are readily hydrolysed, the hydrolysis being catalysed by acids and alkalis. Acid catalysed hydrolysis is the reverse of esterification:

$$R - \overset{\overset{\displaystyle O}{\|}}{C} - OR' + H_2O \underset{}{\overset{H^+/reflux}{\rightleftharpoons}} R - \overset{\overset{\displaystyle O}{\|}}{C} - OH + R'OH$$

A large excess of water is used but, as the reaction is reversible, the hydrolysis never goes to completion and so it is not very convenient. Alkaline hydrolysis is much faster than acid catalysed hydrolysis and it goes to completion. The process is known as *saponification* since soaps are made by this reaction.

$$R - \overset{\overset{\displaystyle O}{\|}}{C} - OR' + NaOH \overset{reflux}{\longrightarrow} R - \overset{\overset{\displaystyle O}{\|}}{C} - ONa + R'OH$$

The mechanism is:

$$R - \overset{\overset{\displaystyle O^{\delta-}}{\|}}{\underset{\underset{\displaystyle OR'}{\,}}{C^{\delta+}}} + HO^- \rightarrow R - \overset{\overset{\displaystyle O^-}{\|}}{\underset{\underset{\displaystyle OR'}{\,}}{C}} - OH \rightarrow R - \overset{\overset{\displaystyle O}{\|}}{C} - O - H + R'O^-$$

$$\rightarrow R - \overset{\overset{\displaystyle O}{\|}}{C} - O^- + R'OH$$

It should be pointed out that, when the carbonyl atom gets a share of the electrons from the HO$^-$, it releases its hold on two other electrons and it is the π electrons which can move the easiest. The electron-attracting R'O group is then pulling electrons away from a carbon which is attached to an electron rich atom, O$^-$. The π electrons are therefore pulled back to reform the π bond and R'O$^-$ is eliminated. The final stage is proton transfer since the RO$^-$ ion is less stable than the RCOO$^-$ ion which exists as a resonance hybrid:

$$R - \overset{\overset{\displaystyle O}{\|}}{C} - O^- \longleftrightarrow R - \overset{\overset{\displaystyle O^-}{|}}{C} = O$$

i.e. the charge is delocalised between the two oxygen atoms whereas in R'O$^-$ it remains firmly on the single oxygen atom.

(e) Reaction with phosphorus chlorides and sulphur dichloride oxide (thionyl chloride)

Acids react vigorously with phosphorus pentachloride at room temperature, and with the trichloride on warming, to give acid chlorides, e.g.

$$PCl_5(s) + CH_3 \cdot COOH(l) \rightarrow CH_3 \cdot COCl(l) + POCl_3(l) + HCl(g)$$

Ethanoyl chloride
(acetyl chloride)

$$PCl_3(l) + 3CH_3 \cdot COOH(l) \rightarrow 3CH_3 \cdot COCl(l) + H_3PO_3(s)$$

Alternatively, the acid may be refluxed with sulphur dichloride oxide (thionyl chloride), e.g.

$$C_6H_5 \cdot COOH(s) + SOCl_2(l) \rightarrow C_6H_5 \cdot COCl(l) + HCl(g) + SO_2(g)$$
Benzenecarbonyl
chloride (benzoyl
chloride)

Methanoic acid (formic acid) is an exception in that it does not form an acid chloride.

(f) Halogenation

Boiling acids react with chlorine or bromine in the presence of red phosphorus as catalyst, the hydrogen atoms on the carbon adjacent to the acid group being successively replaced, e.g.

$$CH_3 \cdot COOH(l) + Cl_2(g) \xrightarrow{P} CH_2Cl \cdot COOH(l) + HCl(g)$$
Chloroethanoic acid
(chloroacetic acid)

$$CH_2Cl \cdot COOH(l) + Cl_2(g) \xrightarrow{P} CHCl_2 \cdot COOH(l) + HCl(g)$$
Dichloroethanoic acid
(dichloroacetic acid)

$$CHCl_2 \cdot COOH(l) + Cl_2(g) \xrightarrow{P} CCl_3 \cdot COOH(s) + HCl(g)$$
Trichloroethanoic
acid (trichloro-
acetic acid)

Uses of acids

Acids are used in the manufacture of esters for use as synthetic perfumes and as flavourings in foods. Terylene (page 361) and Rayon (cellulose ethanoate) are very important esters. Some long chain acids are converted into soaps and detergents. Vinegar is essentially a dilute solution of ethanoic acid.

ACID OR ACYL CHLORIDES

The acid or acyl chlorides are prepared from acids as described above. They are named by removing the '-ic acid' from the acid name and adding '-yl chloride'. They have the functional group

$$-C \underset{Cl}{\overset{O}{\big\backslash}}$$

and are generally fuming, colourless liquids.

Reactions of acyl chlorides

(a) Hydrolysis

The lower aliphatic acyl chlorides are hydrolysed rapidly by cold water, e.g.

$$CH_3 \cdot COCl(l) + H_2O(l) \rightarrow CH_3 \cdot COOH(aq) + HCl(aq)$$

Aromatic acyl chlorides hydrolyse much more slowly due to their lower solubility and their carbonyl carbon atom being less susceptible to nucleophilic attack since it can attract electrons from the benzene ring (its unhybridised $2p$ orbital overlaps with the π orbitals of the ring).

The mechanism of hydrolysis is:

The hydrolysis is faster with alkali because HO^- is a stronger nucleophile than H_2O.

(b) Alcoholysis

The reaction with alcohols is basically similar to hydrolysis and is known as alcoholysis. The product is an ester, e.g.

$$CH_3 \cdot COCl(l) + C_2H_5OH(l) \rightarrow CH_3 \cdot COOC_2H_5(l) + HCl(g)$$

Phenol behaves similarly but a base catalyst is required for it to react with aromatic acyl chlorides. The base used is sodium hydroxide solution or pyridine, C_5H_5N, and its function is to generate the more nucleophilic ion, $C_6H_5O^-$.

$$C_6H_5 \cdot COCl(l) + C_6H_5OH(s) + NaOH(aq) \rightarrow C_6H_5 \cdot COOC_6H_5(s) + NaCl(aq) + H_2O(l)$$
$$\text{Phenyl}$$
$$\text{benzenecarboxylate}$$
$$\text{(phenyl benzoate)}$$

The process of replacing a hydrogen atom in a hydroxy or amino group by an ethanoyl (acetyl) group, $CH_3 \cdot CO—$, is known as *ethanoylation (acetylation)* and replacement by a benzenecarbonyl (benzoyl), group, $C_6H_5 \cdot CO—$, is called *benzenecarbonylation (benzoylation)*. The general reaction is referred to as *acylation*.

(c) Reaction with ammonia and amines

Ammonia reacts with acyl chlorides to give amides, the reaction being known as ammonolysis. Low temperatures and/or a large excess of ammonia solution are used to moderate the reaction.

$$CH_3 \cdot COCl(l) + 2NH_3(aq) \rightarrow CH_3 \cdot CO \cdot NH_2(aq) + NH_4Cl(aq)$$
$$\text{Ethanamide}$$
$$\text{(acetamide)}$$

$$C_6H_5 \cdot COCl(l) + 2NH_3(aq) \rightarrow C_6H_5 \cdot CO \cdot NH_2(s) + NH_4Cl(aq)$$
Benzenecarboxamide
(benzamide)

Reaction with primary and secondary amines gives *N*- and *N,N*-disubstituted amides respectively, e.g.

$$C_6H_5 \cdot NH_2(l) + C_6H_5 \cdot COCl(l) + NaOH(aq) \rightarrow$$
Phenylamine
(aniline)

$$C_6H_5 \cdot NH \cdot CO \cdot C_6H_5(s) + NaCl(aq) + H_2O(l)$$
N-phenylbenzenecarboxamide
(benzanilide)

$$(CH_3)_2NH(l) + CH_3 \cdot COCl(l) \rightarrow CH_3 \cdot CO \cdot N(CH_3)_2(s) + HCl(g)$$
Dimethylamine
N,N-dimethylethanamide
(*N,N*-dimethylacetamide)

The mechanism for these reactions is similar to that for hydrolysis and alcoholysis.

(d) Other reactions

Acyl chlorides react with the sodium salts of acids to give acid anhydrides (see page 338) and with benzene to give ketones (see Friedel–Crafts reactions, page 311).

AMIDES

Amides have the functional group $-C\overset{\displaystyle O}{\underset{\displaystyle NH_2}{\Big\langle}}$

They are named after the corresponding acid, the ending '-oic acid' being removed and 'amide' being added. They may be prepared from acyl chlorides as above or by heating ammonium salts of carboxylic acids, e.g.

$$CH_3 \cdot COONH_4(s) \rightarrow CH_3 \cdot CO \cdot NH_2(s) + H_2O(l)$$
Ethanamide
(acetamide)

$$C_6H_5 \cdot COONH_4(s) \rightarrow C_6H_5 \cdot CO \cdot NH_2(s) + H_2O(l)$$
Benzenecarboxamide
(benzamide)

Methanamide (formamide) is a high boiling liquid whilst the remainder are white solids, the high boiling points being due to association. The lower members are water-soluble.

Properties of amides

(a) Hydrolysis

The amides are readily hydrolysed on refluxing with dilute acid or alkali:

$$R \cdot CO \cdot NH_2 + H_2O \xrightarrow{\quad HO^- \text{ or } H^+ \quad} R \cdot COOH + NH_3$$

The mechanisms are:

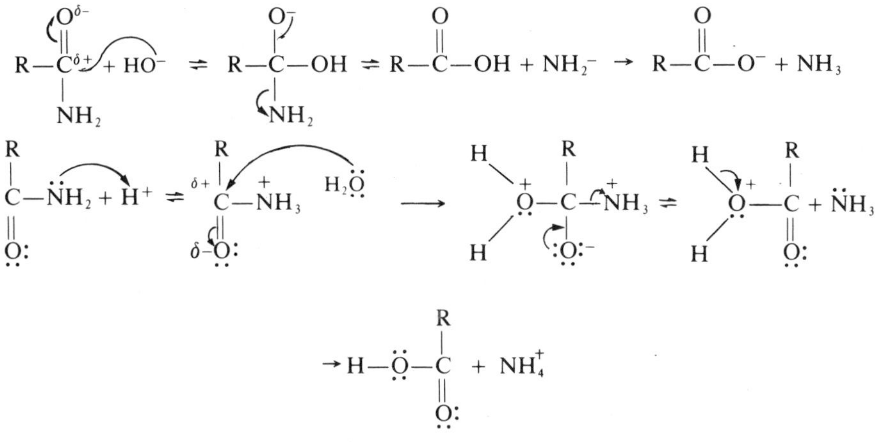

The explanatory notes on the alkaline hydrolysis of esters are also applicable to these mechanisms.

(b) Dehydration

Heating amides with a dehydrating agent such as phosphorus(V) oxide leads to nitriles, e.g.

$$C_6H_5 \cdot CO \cdot NH_2(s) \xrightarrow{P_2O_5} C_6H_5 \cdot CN(l) + H_2O(l)$$

Benzene-
carbonitrile
(benzonitrile)

(c) Hofmann degradation of amides

Amides react with potassium hydroxide solution and bromine in a rearrangement reaction to give amines containing one less carbon atom, e.g.

$$CH_3 \cdot CO \cdot NH_2(s) + Br_2(l) + 4KOH(aq) \rightarrow CH_3 \cdot NH_2(g) + K_2CO_3(aq) + 2KBr(aq)$$

The reaction is known as the Hofmann degradation of amides and is useful for preparing pure primary amines. It can also be used for descending a homologous series, for example, ethanoic acid (acetic acid) may be converted into methanoic acid (formic acid) by the following series of reactions but the yield is very poor.

$$CH_3 \cdot COOH \xrightarrow{NH_3} CH_3 \cdot COONH_4 \xrightarrow{heat} CH_3 \cdot CO \cdot NH_2$$
$$\xrightarrow{KOH/Br_2} CH_3 \cdot NH_2 \xrightarrow{HNO_2} CH_3 \cdot OH \xrightarrow{[O]} HCOOH$$

(d) Reduction

Amides are reduced to primary amines by lithium tetrahydridoaluminate(III) (lithium aluminium hydride), sodium and ethanol, etc.

$$R \cdot CO \cdot NH_2 \xrightarrow{4[H]} R \cdot CH_2 \cdot NH_2 + H_2O$$

The basicity of amines is discussed in Chapter 20.

ACID ANHYDRIDES

Acid anhydrides have the structure

$$R-C \overset{O}{\underset{\displaystyle R-C}{\diagdown}} O$$

where R = alkyl or aryl. They are named by replacing the word 'acid' in the name of the acid by 'anhydride', e.g.

$$CH_3 \cdot COOH \qquad \text{ethanoid acid, (acetic acid)}$$
$$(CH_3 \cdot CO)_2 O \qquad \text{ethanoic anhydride (acetic anhydride)}$$

Methanoic acid (formic acid) does not form an anhydride.

Acid anhydrides are prepared by refluxing an acyl chloride with the sodium salt of an acid, e.g.

$$CH_3 \cdot COCl(l) + CH_3 \cdot COONa(s) \rightarrow (CH_3 \cdot CO)_2O(l) + NaCl(s)$$

The mechanism is:

$$CH_3 - C \overset{\delta+}{\underset{Cl}{\overset{\delta-}{\diagup}}} O \quad + CH_3 \cdot COO^- \quad \rightarrow \quad CH_3 - \overset{O^-}{\underset{Cl}{\overset{|}{C}}} - O - \overset{O}{\overset{||}{C}} - CH_3$$

$$\rightarrow \quad CH_3 - C \overset{O}{\underset{\displaystyle CH_3 - C}{\diagdown}} O + Cl^-$$

Ethanoic anhydride (acetic anhydride) is a high boiling liquid which is not very soluble in water.

Acid anhydrides are less reactive than acyl chlorides because they exist as resonance hybrids and so their carbonyl carbons have a smaller positive charge and are less susceptible to nucleophilic attack.

$$R-C \overset{O}{\diagdown} \quad \longleftrightarrow \quad R-C \overset{O^-}{\diagdown} \quad \longleftrightarrow \quad R-C \overset{O}{\diagdown}$$

Their reactions are similar to those of acyl chlorides. Only half of the molecule is used in acylations, the other half being converted into the acid.

Reactions of acid anhydrides

(a) Hydrolysis

Acid anhydrides are slowly hydrolysed by water to the acid, e.g.

$$(CH_3 \cdot CO)_2O(l) + H_2O(l) \rightarrow 2CH_3 \cdot COOH(aq)$$

The mechanism is:

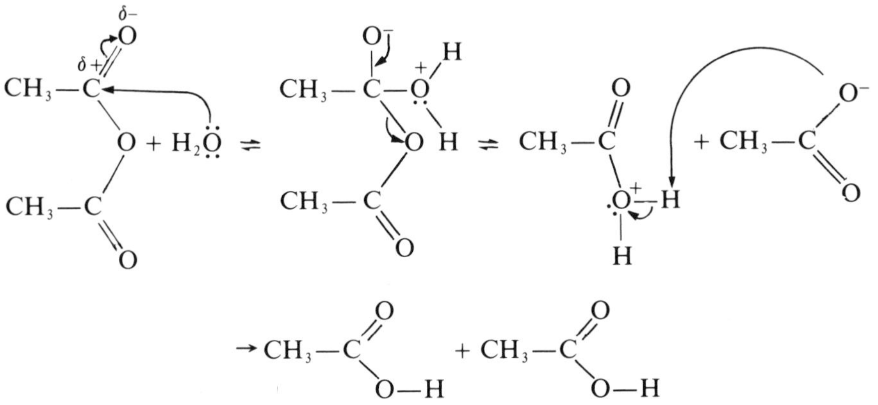

(b) Reaction with alcohols and phenols

Acid anhydrides react with alcohols and phenols to give esters, heating generally being required, e.g.

$$(CH_3 \cdot CO)_2O(l) + C_2H_5OH(l) \rightarrow CH_3 \cdot COOC_2H_5(l) + CH_3 \cdot COOH(l)$$

Ethyl ethanoate
(ethyl acetate)

The mechanism is similar to that for hydrolysis.

(c) Reaction with ammonia and amines

Acid anhydrides react with ammonia solution and amines to give amides, e.g.

$$(CH_3 \cdot CO)_2O(l) + 2NH_3(aq) \rightarrow CH_3 \cdot CO \cdot NH_2(aq) + CH_3 \cdot COONH_4(aq)$$

Ethanamide
(acetamide)

$$(CH_3 \cdot CO)_2O(l) + CH_3 \cdot NH_2(aq) \rightarrow CH_3 \cdot NH \cdot CO \cdot CH_3(s) + CH_3 \cdot COOH(l)$$

N-methylethanamide
(N-methylacetamide)

The mechanism is similar to that for hydrolysis.

DICARBOXYLIC ACIDS

It is possible to have two or more acid groups in the same molecule. A common dicarboxylic acid is ethanedioic acid (oxalic acid); it may be synthesised from ethene (ethylene) by the following series of reactions:

$$CH_2{=}CH_2(g) \xrightarrow{Br_2} CH_2Br \cdot CH_2Br(l) \xrightarrow[\text{reflux}]{NaOH\ (aq)} CH_2OH \cdot CH_2OH(aq)$$

$$\xrightarrow{[O]} COOH \cdot COOH\ (aq)$$

Both acid groups undergo the normal acid reactions.

Butenedioic acid exists as geometrical isomers (page 292).

10 NITRILES

Nitriles have the functional group—C≡N. They are named by adding 'nitrile' to the name of the parent hydrocarbon, e.g. $CH_3 \cdot CN$ is called ethanenitrile.

Aliphatic nitriles can be prepared by refluxing halogenoalkanes (alkyl halides) with aqueous alcoholic sodium or potassium cyanide solution (page 314) whilst both aliphatic and aromatic nitriles may be obtained by dehydration of amides (page 337). The lower aliphatic nitriles and benzenecarbonitrile (benzonitrile) are colourless liquids.

Reactions of nitriles

(a) Hydrolysis

Refluxing nitriles with dilute acid or alkali hydrolyses them via the amide to carboxylic acids, e.g. benzenecarbonitrile (benzonitrile) is hydrolysed to benzene-carboxylic acid (benzoic acid):

$$C_6H_5 \cdot CN(l) + 2H_2O(l) \xrightarrow[\text{(hydrolysis)}]{H^+ \text{ or } HO^-} C_6H_5 \cdot COOH + NH_3$$

(b) Reduction

Nitriles are reduced to primary amines by sodium and ethanol, lithium tetra-hydridoaluminate(III) (lithium aluminium hydride), and by catalytic hydrogenation, e.g.

$$CH_3 \cdot C \equiv N(l) + 4[H] \rightarrow CH_3 \cdot CH_2 \cdot NH_2(g)$$

The introduction of a cyanide group into a compound may be used to ascend a homologous series. Thus, ethanoic acid (acetic acid) may be converted into propanoic acid (propionic acid) by the following series of reactions:

$$CH_3 \cdot COOH \xrightarrow{LiAlH_4} CH_3 \cdot CH_2OH \xrightarrow{HI} CH_3 \cdot CH_2I \xrightarrow{KCN} CH_3 \cdot CH_2 \cdot CN$$
$$\xrightarrow{H^+} CH_3 \cdot CH_2 \cdot COOH$$

11 AMINES

Amines are organic compounds derived from ammonia, and they can be divided into three classes:

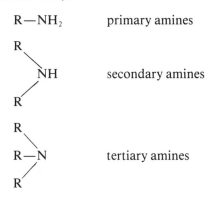

$$R-NH_2 \qquad \text{primary amines}$$

$$\begin{array}{c} R \\ \diagdown \\ NH \qquad \text{secondary amines} \\ \diagup \\ R \end{array}$$

$$\begin{array}{c} R \\ \diagdown \\ R-N \qquad \text{tertiary amines} \\ \diagup \\ R \end{array}$$

where the R groups can be alkyl or aryl or any possible combination of these. The organic analogue of ammonium salts is quaternary ammonium salts, the general formula being $R_4N^+X^-$ where X^- is an anion such as Cl^-, HO^-, etc.

The amines are named according to the alkyl or aryl groups present, e.g.

$CH_3 \cdot NH_2$	methylamine
$C_6H_5 \cdot NH_2$	phenylamine (aniline)
$C_6H_5 \cdot CH_2 \cdot NH_2$	(phenylmethyl)amine (benzylamine)
$(CH_3)_2NH$	dimethylamine
$C_6H_5 \cdot NH \cdot CH_3$	N-methylphenylamine (N-methylaniline)
$(C_2H_5)_3N$	triethylamine

Amines in which only alkyl groups or hydrogen atoms are attached to the nitrogen are known as alkylamines whilst those with the nitrogen attached directly to a benzene ring are known as arylamines. Thus, (phenylmethyl)amine (benzylamine), $C_6H_5 \cdot CH_2 \cdot NH_2$, is classed as, and behaves as, an alkylamine.

Primary amines may be prepared by reduction of the corresponding nitro compound. This method is invariably used to prepare arylamines, e.g.

$$C_6H_5 \cdot NO_2 + 6[H] \rightarrow C_6H_5 \cdot NH_2 + 2H_2O$$

i.e. $\qquad C_6H_5 \cdot NO_2 + 6H^+ + 6e^- \rightarrow C_6H_5 \cdot NH_2 + 2H_2O$

The reduction is done with tin, zinc, or iron and hydrochloric acid, or catalytically with nickel and hydrogen at 300 °C. If more than one nitro group is present in the molecule, they may be reduced one at a time by using aqueous alcoholic ammonium hydrogensulphide, aqueous sodium sulphide, or tin(II) chloride in hydrochloric acid as the reducing agent, e.g.

$$\underset{\substack{\\ NO_2}}{\overset{\substack{NO_2 \\ }}{\bigcirc}} + 3SnCl_2 + 6HCl \rightarrow \underset{\substack{\\ NO_2}}{\overset{\substack{NH_2 \\ }}{\bigcirc}} + 3SnCl_4 + 2H_2O$$

Other methods of preparing primary amines are reduction of nitriles (page 340) and the Hofmann degradation of amides (page 337).

Primary, secondary, and tertiary amines and quaternary ammonium salts can be made by heating halogenoalkanes (alkyl halides) with ammonia under pressure, e.g.

$$CH_3Cl + NH_3 \rightarrow CH_3 \cdot NH_2 + HCl$$

$$CH_3Cl + CH_3 \cdot NH_2 \rightarrow (CH_3)_2NH + HCl$$

$$CH_3Cl + (CH_3)_2NH \rightarrow (CH_3)_3N + HCl$$

$$CH_3Cl + (CH_3)_3N \rightarrow (CH_3)_4N^+Cl^-$$
tetramethylammonium chloride

For simplicity, the free amine has been given as the product in the first three equations but, in fact, it is the amine salt which is formed, e.g. $[CH_3NH_3]^+ Cl^-$. These alkylation reactions occur via nucleophilic attack by the lone pair of electrons of the nitrogen atom on the slightly positively charged carbon atom of the halogenoalkane, e.g.

$$(CH_3)_2\ddot{N}H + \overset{\delta+}{C}H_3 \overset{\delta-}{-}Cl \rightarrow (CH_3)_3\overset{+}{N}HCl^-$$

Methylamine, ethylamine, and dimethylamine are gases at room temperature while most of the other common amines are liquids. Association occurs with primary and secondary amines owing to hydrogen bonding,

$$\begin{array}{ccccc}
 & R & & R & & R \\
 & | & & | & & | \\
\overset{\delta+}{H}-\overset{\delta-}{N}--------\overset{\delta+}{H}-\overset{\delta-}{N}--------\overset{\delta+}{H}-\overset{\delta-}{N}--------- \\
 & | & & | & & | \\
 & H & & H & & H
\end{array}$$

and the polarisation of the N—H bonds also results in the lower-relative molecular mass amines being water-soluble.

Reactions of amines

(a) As bases

Primary, secondary, and tertiary amines are all basic since the lone pair of electrons on the nitrogen atom can attract a proton from an acid, e.g.

$$C_6H_5 \cdot \ddot{N}H_2(l) + HCl(aq) \rightarrow C_6H_5 \cdot \overset{+}{N}H_3Cl^-(aq)$$
Phenylammonium chloride
(aniline hydrochloride)

Stronger bases displace weaker ones from their salts and so the free amines are regenerated by adding alkali, e.g.

$$C_6H_5 \cdot \overset{+}{N}H_3Cl^-(aq) + NaOH(aq) \rightarrow C_6H_5 \cdot NH_2(l) + NaCl(aq) + H_2O$$

The relative basic strengths of the amines are discussed in Chapter 20.

(b) Acylations

Primary and secondary amines are acylated by acyl chlorides (page 335) and acid anhydrides (page 339). The preparation of acylated derivatives is an important method of characterising amines; with phenylamine (aniline) the equations are:

$$C_6H_5 \cdot NH_2(l) + C_6H_5 \cdot COCl(l) + NaOH(aq) \rightarrow$$
$$C_6H_5 \cdot NH \cdot CO \cdot C_6H_5(s) + NaCl(aq) + H_2O(l)$$

N-phenylbenzenecarbox
-amide (benzanilide)

$$C_6H_5 \cdot NH_2(l) + CH_3 \cdot COCl(l) \rightarrow C_6H_5 \cdot NH \cdot CO \cdot CH_3(s) + HCl(g)$$

N-phenylethanamide
(acetanilide)

(c) With halogenoalkanes (alkyl halides)

Secondary and tertiary amines and quaternary ammonium salts are prepared by alkylation of primary amines as described above.

(d) With nitric(III) acid (nitrous acid)

The reaction between primary alkylamines and nitric(III) acid (nitrous acid) results in the evolution of nitrogen, e.g.

$$CH_3 \cdot CH_2 \cdot NH_2(aq) + HNO_2(aq) \rightarrow CH_3 \cdot CH_2OH(l) + H_2O(l) + N_2(g)$$

The equation is, however, a simplification of the true state of affairs since a variety of other products are formed besides the alcohol. Primary arylamines react with nitric(III) acid (nitrous acid) to give diazonium salts. The amine is dissolved or suspended in excess dilute hydrochloric acid or sulphuric(VI) acid and cooled in ice-water to about 5 °C. Cold dilute sodium nitrate(III) (sodium nitrite) solution is slowly added until a slight excess is present as shown by the reaction mixture turning starch-iodide paper blue.

$$C_6H_5 \cdot NH_2(l) + NaNO_2(aq) + 2HCl(aq) \rightarrow C_6H_5 \cdot \overset{+}{N}{=}NCl^-(aq) + NaCl(aq) + 2H_2O(l)$$

Benzenediazonium
chloride

The temperature must be carefully controlled because at or below 0 °C the reaction is very slow and above 10 °C the diazonium salt decomposes to phenol.

This important difference in reaction of primary alkyl- and arylamines is utilised to differentiate between them (see page 344).

The diazonium salts exist as resonance hybrids, e.g.

$$C_6H_5{-}\overset{\cdot\cdot}{N}{=}\overset{+}{N}: \;\; \leftrightarrow \;\; C_6H_5{-}\overset{+}{N}{\equiv}N: \quad \text{etc.}$$

The first structure will be used to illustrate the reactions since it is the terminal nitrogen which joins to other atoms in the coupling processes (see below).

Reactions of diazonium salts

Diazonium salts are important synthetic reagents. Their reactions may be divided into two types according to whether the nitrogen is retained or replaced.

Reactions in which the nitrogen is replaced

(a) By hydroxyl

When a diazonium salt solution is boiled or steam distilled, a phenol is formed:

$$C_6H_5 \cdot N{\overset{+}{=}}N \, HSO_4^-(aq) + H_2O(l) \rightarrow C_6H_5OH(aq) + H_2SO_4(aq) + N_2(g)$$

(b) By chlorine or bromine

The diazonium salt solution may either be warmed with the halogen acid and the corresponding copper(I) halide:

$$C_6H_5 \cdot N{\overset{+}{=}}NCl^-(aq) \xrightarrow{\text{CuCl/HCl}} C_6H_5Cl(l) + N_2(g)$$

or the diazonium halide solution is warmed with copper powder:

$$C_6H_5 \cdot N{\overset{+}{=}}N \, Br^-(aq) \xrightarrow{\text{Cu}} C_6H_5Br(l) + N_2(g)$$

(c) By iodine

Here no catalyst is necessary and the diazonium salt solution is simply warmed with potassium iodide solution:

$$C_6H_5 \cdot N{\overset{+}{=}}N \, Cl^-(aq) + KI(aq) \rightarrow C_6H_5I(l) + KCl(aq) + N_2(g)$$

Reactions in which the nitrogen is retained

Diazonium salts couple with phenols and aromatic amines giving coloured azo compounds. Reactions of this type form the basis of the dye industry.

(a) Coupling with phenols

The diazonium salt solution is added to an alkaline solution of the phenol. With phenol, an orange precipitate of (4 -hydroxyphenyl)azobenzene (*p*-hydroxyazobenzene) is formed:

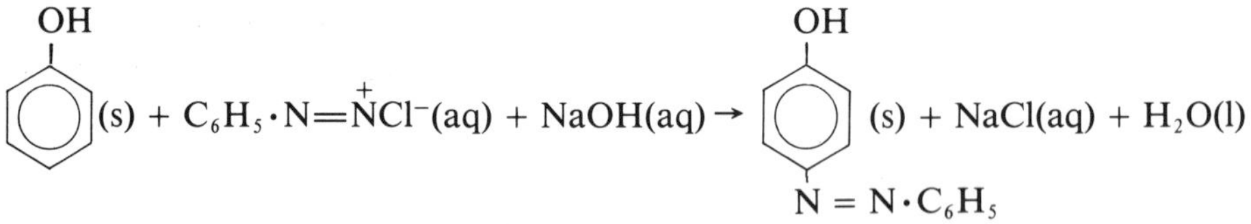

If napthalen-2-ol (β-napthol is used, a red precipitate of 1-(phenylazo) napthalen-2-ol (1-phenylazo-2-napthol) is formed:

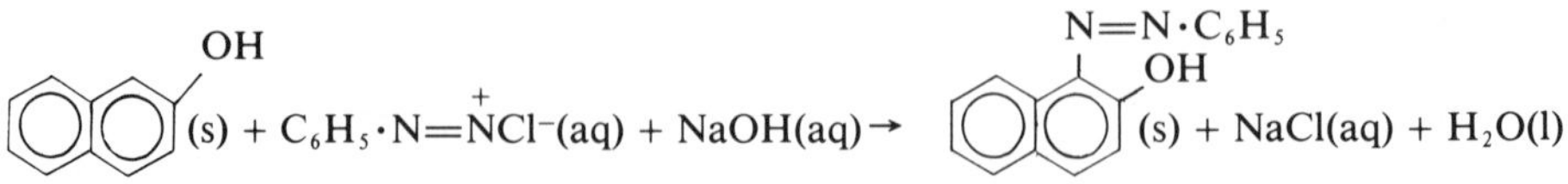

This reaction is used as a test for primary arylamines. Thus, the compound is treated with dilute hydrochloric acid and sodium nitrate(III) (sodium nitrite) at 5 °C and then added to an alkaline solution of naphthalen-2-ol (β-naphthol). A red precipitate indicates that the initial compound was a primary arylamine.

(b) Coupling with primary amines

Attack occurs on the amino group if the reaction is performed at 5 °C but rearrangement occurs to give the ring substituted product on warming to 40 °C. With phenylamine (aniline) at 5 °C, a yellow precipitate of N-(phenylazo) phenylamine (diazoaminobenzene) is obtained and, when heated, this rearranges to the orange coloured 4-(phenylazo)phenylamine (p-aminoazo-benzene):

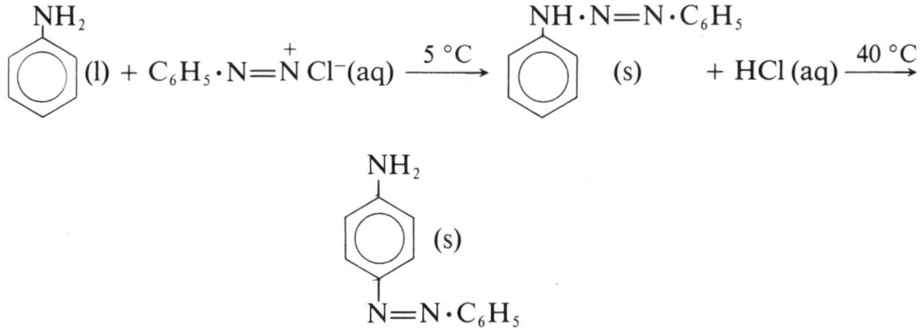

(c) Coupling with secondary amines

In weakly acidic solution, attack occurs on both the nitrogen and carbon, but in more acidic solutions, substitution takes place only at the carbon. Under the latter conditions, N-methylphenylamine (N-methylaniline) gives 4-(phenylazo)-N-methylphenylamine (p-N-methylaminoazobenzene):

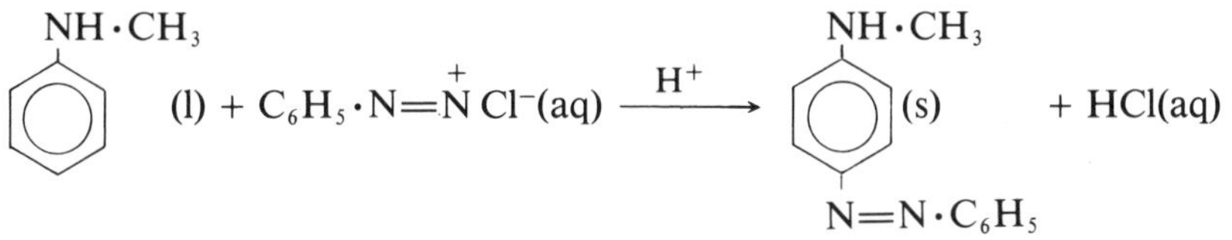

(d) Coupling with tertiary amines

In this case, substitution can occur only in the benzene ring. With N,N-dimethyl-phenylamine (N,N-dimethylaniline), a green precipitate of 4-(phenylazo)-N,N-dimethylphenylamine (N,N-dimethylaminoazobenzene) is formed:

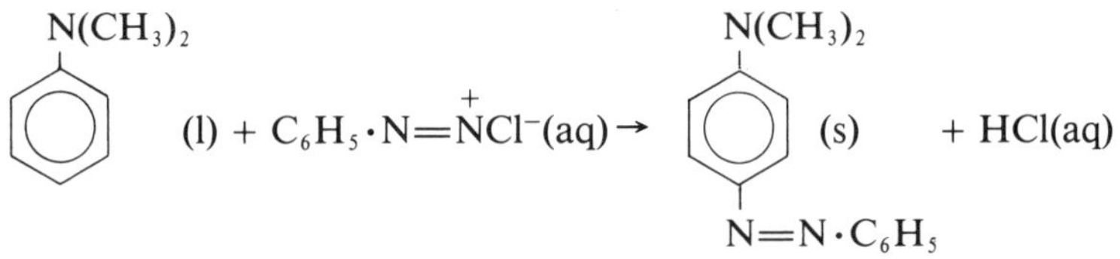

Questions

16.1 (a) A mixture of ethane and ethene (ethylene) was passed through a solution of 4 g of bromine in 150 cm³ of ethanoic (acetic) acid, using conditions such that the bromine added quantitatively to the ethene but did not combine with the ethane. The bromine was completely used up after 1255 cm³ of the gas mixture had passed at a temperature of 27 °C and a pressure of 745 mm Hg. What was the percentage of composition by volume of the mixture of gases?

(b) Name the product and give the structure of the organic compound formed in the following reactions:
(i) ethene with bromine;
(ii) ethene with bromine water;
(iii) ethene with hydrogen bromide;
(iv) propene (propylene) with hydrogen bromide.
Discuss briefly the mechanism in each case.   [J.M.B.]

16.2 Discuss the detailed mechanism of the reaction of hydrogen bromide with (a) ethene (ethylene) and (b) propene (propylene). Show how the use of addition reactions involving ethene leads to the industrial production of each of the following compounds:
ethanol, epoxyethane (ethylene oxide) $(CH_2)_2O$, ethane-1,2-diol (ethylene glycol) $CH_2OH \cdot CH_2OH$, and chloroethene (vinyl chloride) $CH_2 = CHCl$.
Describe briefly the importance of these compounds.   [J.M.B.]

16.3 (a) What deductions can you make from the following information?
(i) Benzene is a hydrocarbon containing 92.3% carbon and has a relative molecular mass of 78.
(ii) Under the influence of ultraviolet light 1 mole of benzene reacts with 3 moles of chlorine without the production of hydrogen chloride.
(iii) Methylbenzene (toluene) reacts with alkaline potassium manganate(VII) (permanganate) but benzene does not.
(iv) Benzene reacts with bromine under suitable conditions to substitute one hydrogen atom per mole for one bromine atom. When this is done, there is only one organic product.
(v) $CH_3 \cdot CH = CH_2(g) + H_2(g) \rightarrow CH_3 \cdot CH_2 \cdot CH_3(g)$ $\Delta H = -126$ kJ

$$\bigcirc\!\!| \;(g) + H_2(g) \; \rightarrow \; \bigcirc \;(l) \qquad \Delta H = -122 \text{ kJ}$$

$$\bigcirc\!\!| \;(g) + 3H_2(g) \rightarrow \; \bigcirc \;(l) \qquad \Delta H = -210 \text{ kJ}$$

(b) Show with detailed mechanistic equations how benzene may be converted into (i) phenylethene (styrene), $(C_6H_5 \cdot CH = CH_2)$, (ii) phenylethanone (acetophenone), $(C_6H_5 \cdot CO \cdot CH_3)$.   [J.M.B.]

16.4 (a) Explain what is meant by the terms (i) nucleophilic and (ii) electrophilic. Explain with the aid of examples, why reagents which will give addition products with propene (propylene) will not, in general, react in the same way with ethanal (acetaldehyde) and vice versa.

(b) Outline the steps by which hydroxylamine reacts with ethanal, naming the type of reaction occurring at each step and also the final product of the reaction.

(c) Benzene and concentrated nitric(V) acid react in the presence of concentrated sulphuric(VI) acid forming nitrobenzene, $C_6H_5 \cdot NO_2$. What is the nature of the species which reacts with benzene in this reaction and how is it formed?   [J.M.B.]

16.5 (a) State the Markownikoff rule for predicting the direction of electrophilic addition of hydrogen bromide to alkenes. Explain in detail the rule in terms of the relative stabilities of primary, secondary, and tertiary carbonium ions.

(b) Propene (propylene) reacts with hydrogen bromide to give a substance A, C_3H_7Br. Substance A, when heated with aqueous potassium hydroxide, gives an alcohol B.
(i) Derive structures for A and B.
(ii) Explain and illustrate the meaning of the terms base, nucleophile, and inductive effect by referring to the reactions of substance A with potassium hydroxide under various conditions.   [J.M.B.]

16.6 Discuss the general utility of sulphuric(VI) acid as a reagent in organic chemistry by considering its reactions with (a) a primary alcohol, (b) an alkene,

(c) an aromatic hydrocarbon. Clearly indicate the mechanism of one reaction in either (a) or (b) above. [J.M.B.]

16.7 Show how the products of oxidation of (a) methylbenzene (toluene) and (b) ethene (ethylene) vary in accordance with the reagents and conditions used. (Mechanistic explanations are not required.)

Explain what happens when (a) ethanol and (b) propan-2-ol (isopropyl alcohol) are warmed gently with acidified potassium dichromate(VI) solution. (Mechanistic explanations are not required.) Describe the mechanism of the reaction of hydrogen cyanide with either one of the organic products formed from these alcohols. [J.M.B.]

16.8 For each of the following reactions, give the products and show, with the aid of suitably annotated equations, the mechanism of the reaction.

(a) The reaction of propanone (acetone) with semicarbazide hydrochloride and sodium ethanoate (acetate) in water.

(b) The reaction of benzene with ethanoyl chloride (acetyl chloride) in the presence of aluminium chloride.

(c) The reaction of propene (propylene) with hydrogen bromide in the absence of air or peroxides.

(d) The reaction of ethanoyl chloride (acetyl chloride) with ethanol.

(e) The reaction of benzene with a mixture of concentrated sulphuric(VI) and nitric(V) acids. [J.M.B.]

16.9 Describe the reactions between each of the following pairs of compounds.

(a) Ethanal (acetaldehyde) and sodium hydrogensulphate(IV) (sodium hydrogen-sulphite).

(b) 2-bromopropane and potassium cyanide. In each case give
(i)   the type of reaction taking place,
(ii) a balanced equation,
(iii) the name of the organic product of the reaction,
(iv) the mechanism of the change.

(c) Compare and contrast the reactions between
(i) hydrogen cyanide and propanone (acetone),
(ii) hydrogen cyanide and ethene (ethylene). [J.M.B.]

16.10 Compound A is an organic liquid containing 52.2% carbon, 13.0% hydrogen, and 34.8% oxygen. Mild oxidation of A yields compound B which, in the presence of dried calcium chloride, reacts with an excess of compound A to give compound C of molecular formula $C_6H_{14}O_2$.

Further oxidation of B leads to the formation of a compound D which reacts with A in an equimolar ratio giving compound E of molecular formula $C_4H_8O_2$.

Explain the above processes and identify the compounds A to E inclusive.

Suggest how compound B may be obtained from a simple hydrocarbon.
[University of London]

16.11 Discuss the use of the following inorganic reagents in preparative organic chemistry: (i) bromine, (ii) ammonia, (iii) anhydrous aluminium chloride, (iv) sodium tetrahydridoborate(III) (sodium borohydride). In each case write equations to illustrate as many types of reaction as possible, clearly indicating the principal mechanistic features of each reaction. [J.M.B.]

16.12 State what is meant by five of the following terms as used in organic chemistry:

(a) sulphonation,   (b) polymerisation,   (c) esterification,

(d) ozonolysis,   (e) decarboxylation,   (f) hydrogenation.

Illustrate each answer by an example, specifying the reagents and approximate conditions. [University of London]

16.13 The following substance is known as serine (2-amino-3-hydroxypropanoic acid): $HO-CH_2-CH(NH_2)-COOH$

From your knowledge of organic chemistry predict what reactions you would expect to take place when this compound is treated with:

(a) nitric(III) acid (nitrous acid);
(b) ethanol;
(c) lithium tetrahydridoaluminate(III) (lithium aluminium hydride);
(d) potassium dichromate(VI);
(e) hydrochloric acid;
(f) phosphorus pentachloride.

In each case, state the formula of the reaction products and clearly indicate the experimental conditions required for the reactions.

[University of London]

16.14   The characteristic properties of a functional group in an organic compound are influenced by the nature of the remainder of the molecule.
    Discuss this statement using as examples

(a) the hydroxyl group,
(b) the amino group.   [J.M.B.]

16.15   Discuss in detail the differences between the reaction of ethanal (acetaldehyde) with (a) sodium cyanide followed by dilute hydrochloric acid and with (b) hydroxylamine in the presence of sodium ethanoate (acetate) and ethanoic (acetic) acid.
    Compare these reactions with the reactions of
(i)   ethanoic (acetic) anhydride with ethanol, and
(ii) ethanoyl (acetyl) chloride with phenol.   [J.M.B.]

16.16   A compound A of relative molecular mass 94.5 reacts with phosphorus pentachloride to give a compound B of relative molecular mass 113. When B was treated with potassium carbonate solution, a compound C of relative molecular mass 132.5 was formed. When C was heated with aqueous potassium cyanide, a compound D of relative molecular mass 123 was obtained and D on boiling with methanol and sulphuric(VI) acid gave E as $C_5H_8O_4$.
    Deduce the identities of A, B, C, D, E and elucidate the changes above. How would A be prepared from ethanoic acid (acetic acid)?

[University of London]

16.17   Describe typical reactions of benzene with electrophilic reagents. Write a mechanism for the nitration of benzene and explain why benzene preferentially undergoes substitution rather than addition.
    Suggest a method of preparing the alcohol $C_6H_5CH(OH)CH_3$, using ethanoyl (acetyl) chloride and any other common reagents.   [J.M.B.]

16.18   Describe briefly how each of the following can be prepared in the laboratory, starting from methylbenzene (toluene) and any other organic or inorganic reagents. Give equations and outline the mechanisms of the reactions where possible.

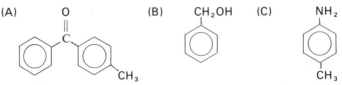

State briefly how you would carry out experiments to demonstrate each of the following:
(i)   that the hydroxyl group of substance (B) is in the side chain and not directly attached to the ring;
(ii) that the substance (C) is a primary aromatic amine.   [J.M.B.]

16.19   Give a systematic account of the reactions by which a chlorine atom may be introduced into an organic molecule. Comment briefly on the mechanisms of the reactions you discuss.   [University of London]

16.20 An organic substance X of molecular formula C_5H_8O is found to react readily with the following: bromine, hydroxylamine (NH_2OH), and sodium tetra-hydridoborate(III) ($NaBH_4$).

(a) For each of the reagents listed name one functional group within X which would give a positive reaction with it.

(b) Bearing in mind the above reactions, explain which functional group must be present in X if it gives a carboxylic acid when treated with chromic(VI) acid.

(c) What can you deduce from the observation that propanone is formed when X is suitably treated with trioxygen (ozone)?

(d) Suggest a structure for X. [J.M.B.]

16.21 A compound A, whose composition is 80.0% C, 6.67% H and 13.33% O, has a relative molecular mass of about 120. When treated with hydrogen cyanide, it is converted into B (C_9H_9ON). B, on treatment with boiling dilute sulphuric(VI) acid, gives C ($C_9H_{10}O_3$) and C, on oxidation, gives an acid D. When A reacts with iodine and sodium hydroxide, it gives triiodomethane and the sodium salt of D.
 Identify A, B, C, and D and give equations for the reactions mentioned. What intermediate compound would be formed in the conversion of B into C? [University of London]

16.22 Aldehydes ($R \cdot CO \cdot H$), ketones ($R \cdot CO \cdot R'$), acids ($R \cdot CO \cdot OH$) and esters ($R \cdot CO \cdot OR'$) all have a $>CO$ group in their formulae. Select one example from each class and give a reaction by which it could conveniently be prepared in the laboratory. Describe one reaction for each of your examples which is not undergone by any of the others. In one case discuss the likely mechanism of the reaction. [University of London]

16.23 By means of equations and essential experimental conditions, outline how you would prepare a sample of the substance $CH_3CH(OH)COOH$ starting with ethanal, CH_3CHO.
 Describe experiments which you could undertake to show that the substance you have prepared:

(a) contains the elements carbon and hydrogen,

(b) behaves as a secondary alcohol,

(c) is acidic in nature.

 What would be the effect of a portion of your sample on the plane of polarisation of light? Explain your answer. [J.M.B.]

16.24 A haloalkane, A, of formula C_3H_7Cl, on heating with an alcoholic solution of potassium cyanide yields B, of formula C_4H_7N. When B is reduced with lithium tetrahydridoaluminate(III) a substance C, of formula $C_4H_{11}N$ is formed. Treatment of C with an aqueous solution of sodium nitrate(III) (nitrite) followed by dilute hydrochloric acid gives D, of formula $C_4H_{10}O$. Mild oxidation of D produces a carbonyl compound with reducing properties.
 From the above reactions deduce possible structures for A, B, C, and D.
 When A is treated with dilute sodium hydroxide and the product is gently oxidised, the resulting compound is a non-reducing carbonyl compound. Using this additional information choose between the possible structures of A. [J.M.B.]

16.25 The following is a route to the formation of 2,2-dimethylpropanoic acid, $C_5H_{10}O_2$.

$$C_4H_8 \xrightarrow{\text{Concentrated HBr}} ? \xrightarrow{\text{KCN/dilute } H_2SO_4} C_5H_9N \rightarrow C_5H_{10}O_2$$
$$A \qquad\qquad\qquad B \qquad\qquad\qquad C \qquad D$$

(a) B has the composition — carbon 35.04%, hydrogen 6.57%, bromine 58.39%. Calculate the empirical formula of B.

(b) Give the structures of A, B, C, and D.

(c) Give the reagents and conditions of reaction for the conversion of C into D.

(d) A student who carried out the above sequence of reactions started with 5.6 litres of A at s.t.p. and obtained 8.5 g of D. Calculate the overall percentage yield of D. [University of London]

16.26 Describe and explain how you would distinguish by means of chemical reactions between

(a) 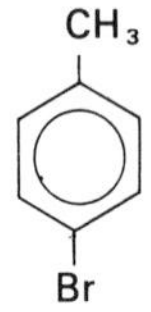 and 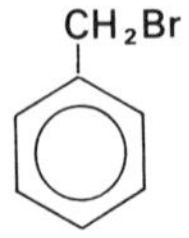

(b)  and $CH_3(CH_2)_3CH{=}CH_2$

(c)  and 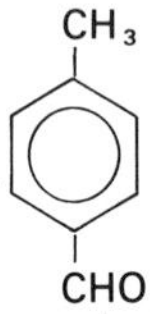

(d) $HCO_2C_2H_5$ and $CH_3CO_2CH_3$

(e) 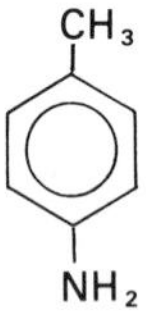 and 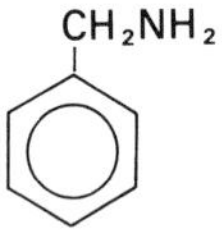

Give the results of your experiments on each member of the pair.
[University of London]

16.27 Many methods are available for the reduction of organic compounds. Select any four such methods and for each one describe one example of its use. Choose a different type of reagent in each case and give reaction conditions and the structures of the major products. You may like to select from the following reagents but equal credit will be given for other examples: (a) hydrogen and a catalyst, (b) zinc (or zinc amalgam) and hydrochloric acid, (c) sodium (or sodium amalgam) and ethanol, (d) lithium tetrahydridoaluminate(III) (lithium aluminium hydride). [University of London]

16.28 Starting from benzene or methylbenzene (toluene), synthesise, in four steps or less, each of the following:

(a) C_6H_5OH (b) $C_6H_5-\underset{\underset{NOH}{\|}}{C}-CH_3$ (c) $C_6H_5\cdot CH_2OH$

(d) $C_6H_5\cdot COOCH_3$ (e) $C_6H_5\cdot N(CH_3)_2$ (f) $C_6H_5\cdot CN$

16.29 Give equations and conditions for the following conversions, each of which can be done in four or less steps:

(a) $CH_3\cdot CH{=}CH_2 \rightarrow CH_3\cdot CO\cdot CH_3$ (b) $CH_4 \rightarrow CH_3\cdot COOC_2H_5$
(c) $C_6H_6 \rightarrow C_6H_5\cdot CH{=}CH_2$ (d) $CH_3\cdot COOH \rightarrow CH_3\cdot CH_2\cdot NH_2$
(e) $CH_3CN \rightarrow (CH_3\cdot CH_2)_2NH$

16.30 Suggest how each of the following conversions may be performed. A maximum of four steps is required in each case.

(a) $C_6H_6 \rightarrow C_6H_5I$ (b) $CH_3I \rightarrow CH_3\cdot CO\cdot NH_2$
(c) $C_6H_5\cdot CH_3 \rightarrow (C_6H_5\cdot CO)_2O$ (d) $CH_3\cdot CH_2\cdot NH_2 \rightarrow CH_3\cdot CH_2\cdot O\cdot CH_2\cdot CH_3$
(e) $CH_3\cdot CH_2OH \rightarrow CH_3\cdot CHOH\cdot COOH$

17 Carbohydrates

Carbohydrates have the general formula $C_x(H_2O)_y$. They are polyhydroxy aldehydes and ketones, the former being classed as *aldoses* whilst the latter are referred to as *ketoses*. They can be divided into two classes, *sugars* and *polysaccharides*. Sugars are sweet, crystalline, and soluble in water, whilst polysaccharides are usually more complex, non-crystalline, and not sweet.

The sugars may be sub-divided into mono-, di-, tri-, and tetrasaccharides. Monosaccharides cannot be hydrolysed to smaller molecules, but hydrolysis of di-, tri-, and tetrasaccharides gives two, three, and four moles of monosaccharide respectively.

Glucose and fructose are important monosaccharides, whilst sucrose and maltose are disaccharides.

Starch and cellulose are polysaccharides and on hydrolysis they give large numbers of monosaccharide molecules, i.e. they are polymers of monosaccharides.

Glucose

The structure of glucose is:

$$
\begin{array}{l}
\text{CHO} \\
|\\
\text{(CHOH)}_4 \\
|\\
\text{CH}_2\text{OH}
\end{array}
$$

It will be noted that there are four asymmetric carbon atoms in the molecule and so various optical isomers (page 293) exist depending upon the relative positions of the H and OH groups. In fact, + (*d*-) and – (*l*-) glucose are two of the sixteen optical isomers with the formula $C_6H_{12}O_6$. Note that the *d*- and *l*- here indicate the direction of rotation of plane polarised light and not the absolute configuration (written as D- and L-) which is beyond the scope of this text.

Glucose is manufactured by heating starch with dilute hydrochloric acid under pressure:

$$(C_6H_{10}O_5)_n(s) + nH_2O(l) \xrightarrow{\ H^+\ } nC_6H_{12}O_6(aq)$$

There is considerable evidence to show that glucose exists as an equilibrium mixture of the open chain form as above and a ring form. The ring form results from internal hemiacetal formation (see page 327).

$$
\begin{array}{l}
\text{CHO} \\
|\\
\text{CHOH} \\
|\\
\text{CHOH} \\
|\\
\text{CHOH} \\
|\\
\text{CHOH} \\
|\\
\text{CH}_2\text{OH}
\end{array}
\rightleftharpoons
\begin{array}{l}
\text{CHOH} \\
|\\
\text{CHOH} \\
|\\
\text{CHOH} \\
|\\
\text{CHOH} \\
|\\
\text{CH} \\
|\\
\text{CH}_2\text{OH}
\end{array}
\quad \text{O}
$$

The equilibrium is almost completely on the right hand side and this modifies the properties somewhat. Nevertheless, glucose gives some of the typical reactions of aldehydes and alcohols.

1. It reacts with ethanoic anhydride (acetic anhydride) giving a pentaethanoate (pentaacetate):

$$CHO \cdot (CHOH)_4 \cdot CH_2OH \xrightarrow{(CH_3 \cdot CO)_2O} CHO \cdot (CHOOC \cdot CH_3)_4 \cdot CH_2OOC \cdot CH_3$$

2. The aldehyde group is reduced to a primary alcohol group by sodium and ethanol etc.:

$$CHO \cdot (CHOH)_4 \cdot CH_2OH \xrightarrow{Na/C_2H_5OH} CH_2OH \cdot (CHOH)_4 \cdot CH_2OH$$

3. Fehling's solution and Tollen's reagent oxidise the aldehyde group to acid:

$$CHO \cdot (CHOH)_4 \cdot CH_2OH \xrightarrow{[O]} COOH \cdot (CHOH)_4 \cdot CH_2OH$$

4. Hydrogen cyanide adds on to give a hydroxynitrile (cyanohydrin):

$$CHO \cdot (CHOH)_4 \cdot CH_2OH + HCN \rightarrow CH(OH)CN \cdot (CHOH)_4 \cdot CH_2OH$$

5. Reaction with hydroxylamine gives an oxime:

$$CHO \cdot (CHOH)_4 \cdot CH_2OH + NH_2OH \rightarrow$$
$$CH{=}NOH \cdot (CHOH)_4 \cdot CH_2OH + H_2O$$

6. Glucose like all other sugars is dehydrated by concentrated sulphuric(VI) acid:

$$C_6H_{12}O_6 \xrightarrow{H_2SO_4} 6C + 6H_2O$$

Glucose does not form an addition compound with ammonia or with sodium hydrogensulphate(IV) (sodium hydrogensulphite).

Fructose

Fructose, like glucose, has the molecular formula $C_6H_{12}O_6$ and its structure is

$$
\begin{array}{c}
CH_2OH \\
| \\
C{=}O \\
| \\
(CHOH)_3 \\
| \\
CH_2OH
\end{array}
$$

Since there are three asymmetric carbon atoms + (*d*-) and − (*l*-) fructose are two of eight optical isomers with this structure.

Fructose occurs in fruit and honey and is also obtained from the hydrolysis of sucrose (see below). It undergoes internal hemiacetal formation and exists mainly in the ring form.

$$
\begin{array}{ccc}
\mathrm{CH_2OH} & & \mathrm{CH_2OH} \\
| & & | \quad \diagup \mathrm{OH} \\
\mathrm{C{=}O} & & \mathrm{C} \diagdown \\
| & & | \\
\mathrm{CHOH} & \rightleftharpoons & \mathrm{CHOH} \\
| & & | \\
\mathrm{CHOH} & & \mathrm{CHOH} \\
| & & | \qquad \mathrm{O} \\
\mathrm{CHOH} & & \mathrm{CHOH} \\
| & & | \\
\mathrm{CH_2OH} & & \mathrm{CH_2{-}}
\end{array}
$$

Its reactions are generally similar to those of glucose including the reducing reactions with Tollen's reagent and Fehling's solution. Although ketones do not normally reduce these reagents, α-hydroxyketones (ketones with a hydroxyl group on a carbon atom adjacent to the carbonyl carbon), are exceptions.

Sucrose

Sucrose has the molecular formula $C_{12}H_{22}O_{11}$ and it is obtained from cane sugar and sugar beet.

Sucrose is readily hydrolysed by dilute acid giving equimolar amounts of glucose and fructose. However, it has no effect on Fehling's solution or on Tollen's reagent and so it is said to be a *non-reducing* sugar. Since it is non-reducing, the aldehyde group of the glucose must be linked to the keto group of the fructose. Sucrose can be regarded as being made from glucose and a five-membered ring form of fructose as follows:

$$
\begin{array}{cc}
\mathrm{CHOH} & \mathrm{CH_2OH} \\
| & | \\
\mathrm{CHOH} \quad \mathrm{HO{-}C} & \\
| & | \\
\mathrm{CHOH} \quad \mathrm{O} \quad \mathrm{CHOH} & \\
| & | \quad \mathrm{O} \\
\mathrm{CHOH} \qquad \mathrm{CHOH} & \\
| & | \\
\mathrm{CH} \qquad\qquad \mathrm{CH} & \\
| & | \\
\mathrm{CH_2OH} \qquad \mathrm{CH_2OH} &
\end{array}
\;\; \xrightarrow{-H_2O} \;\;
\begin{array}{cc}
\mathrm{CH} & \mathrm{CH_2OH} \\
| \quad \mathrm{O} & | \\
\mathrm{CHOH} \qquad \mathrm{C} & \\
| & | \\
\mathrm{CHOH} \; \mathrm{O} \quad \mathrm{CHOH} & \\
| & | \quad \mathrm{O} \\
\mathrm{CHOH} \qquad \mathrm{CHOH} & \\
| & | \\
\mathrm{CH} \qquad\qquad \mathrm{CH} & \\
| & | \\
\mathrm{CH_2OH} \qquad \mathrm{CH_2OH} &
\end{array}
$$

Maltose

Hydrolysis of starch with the enzyme diastase (an enzyme is an organic catalyst formed by living cells) gives maltose which, like sucrose, has a molecular formula of $C_{12}H_{22}O_{11}$.

Maltose reduces Fehling's solution and Tollen's reagent and is hydrolysed by dilute acid to glucose. It is, therefore, made from two glucose units and, as it is a *reducing sugar*, at least one of the carbonyl carbon atoms is not involved in linking the molecules. The structure is derived from two molecules of glucose as follows:

Reducing half   Non-reducing half

Cellulose

Cellulose occurs in the cell walls of plants. It has the molecular formula $(C_6H_{10}O_5)_n$ and, on hydrolysis with dilute acid, it gives glucose, i.e. it is a polymer of glucose. It does not reduce Fehling's solution or Tollen's reagent because the glucose units are linked via their aldehyde carbon atoms:

Cellulose is used in the manufacture of paper, rayon, cellophane, etc.

Starch

Starch is present in all green plants, wheat, barley, rice, potatoes, etc. Its molecular formula is $(C_6H_{10}O_5)_n$ and on hydrolysis it gives glucose. Starch is non-reducing and its structure is similar to that of cellulose, the difference being the relative positions in which the oxygens, linking the glucose units, and the hydrogen atoms are joined to the top carbon atoms.

Starch gives a blue-black colour with iodine and this is used as a test for iodine (or starch). It is used in the manufacture of alcoholic drinks and to make glucose.

Question

17.1   How do polysaccharides differ from sugars? Give the formulae of the organic compounds obtained when glucose reacts with (a) hydrogen cyanide, (b) lithium tetrahydridoaluminate(III) (lithium aluminium hydride), (c) ethanoyl (acetyl) chloride, (d) an ammoniacal solution of silver nitrate(V). With which of these reagents would you expect fructose to react?

Explain why cellulose does not react with Fehling's solution or Tollen's reagent but its hydrolysis products do.

State what would be observed when starch is treated with (i) cold concentrated sulphuric(VI) acid (ii) iodine.

18 Aminoacids, Polypeptides, and Proteins

Aminoacids

Aminoacids contain an acid and an amino group in the same molecule and so they are said to be difunctional compounds. Aminoacids of various types may be synthesised, but the naturally occurring ones are 2-aminoacids (α-aminoacids) and this discussion is confined to these.

Two common aminoacids are

$$
\begin{array}{cc}
\overset{\displaystyle H}{\underset{\displaystyle NH_2}{H-C-COOH}} & \text{and} \quad \overset{\displaystyle H}{\underset{\displaystyle NH_2}{CH_3-C-COOH}} \\
\text{Aminoethanoic acid} & \text{2-Aminopropanoic acid} \\
\text{(glycine)} & \text{(alanine)}
\end{array}
$$

It is apparent that the 2-aminoacids other than aminoethanoic acid (glycine) will contain an asymmetric carbon atom and so will exhibit optical isomerism (see page 293).

The 2-aminoacids may be prepared by the action of concentrated ammonia solution on 2-chloroacids, e.g.

$$CH_3 \cdot COOH(l) + Cl_2(g) \xrightarrow{\text{red phosphorus}} CH_2Cl \cdot COOH(s) + HCl(g)$$
$$CH_2Cl \cdot COOH(s) + 2NH_3(aq) \rightarrow NH_2 \cdot CH_2 \cdot COOH(aq) + NH_4Cl(aq)$$

Alternatively, they can be made via hydroxynitriles (cyanohydrins), e.g.

$$CH_3-C\overset{\displaystyle O}{\underset{\displaystyle H}{\diagup}} + HCN \rightarrow CH_3-C\overset{\displaystyle OH}{\underset{\displaystyle H}{\diagdown}}CN \xrightarrow{PCl_5} CH_3-C\overset{\displaystyle Cl}{\underset{\displaystyle H}{\diagdown}}CN \xrightarrow[H^+]{\text{hydrolysis}}$$

$$CH_3-C\overset{\displaystyle Cl}{\underset{\displaystyle H}{\diagdown}}COOH \xrightarrow{NH_3} CH_3-C\overset{\displaystyle NH_2}{\underset{\displaystyle H}{\diagdown}}COOH$$

The aminoacids have high melting points, and low solubility in non-polar solvents. For this and other reasons (e.g. nuclear magnetic resonance spectroscopy) it is thought that, both in the solid state and in solution, they exist as inner salts or *zwitterions*, e.g.

$$\overset{+}{N}H_3 \cdot CH_2 \cdot COO^-$$

The aminoacids give reactions characteristic of acids and bases.

1. Refluxing with alcohols in the presence of an acid catalyst gives esters, e.g.

$$CH_3 \cdot CH(NH_2) \cdot COOH(s) + C_2H_5OH(l) \xrightarrow{HCl}$$

$$[CH_3 \cdot CH(\overset{+}{N}H_3) \cdot COOC_2H_5]Cl^-(aq) + H_2O(l)$$

The free aminoester is obtained by adding dilute alkali.

2. Reaction with acyl chlorides or acid anhydrides gives acylated products, e.g.

$$NH_2 \cdot CH_2 \cdot COOH(s) + CH_3 \cdot COCl(l) \rightarrow$$

$$CH_3 \cdot CO \cdot NH \cdot CH_2 \cdot COOH(s) + HCl(g)$$

3. Reaction with nitric(III) acid (nitrous acid) leads to the replacement of the amino group by a hydroxyl group, e.g.

$$NH_2 \cdot CH_2 \cdot COOH(s) + HNO_2(aq) \rightarrow HO \cdot CH_2 \cdot COOH(aq) + H_2O(l) + N_2(g)$$
Hydroxyethanoic acid
(glycollic acid)

4. They form salts with acids, e.g.

$$NH_2 \cdot CH_2 \cdot COOH(s) + HCl(aq) \rightarrow [\overset{+}{N}H_3 \cdot CH_2 \cdot COOH]Cl^-(aq)$$

Proteins and polypeptides

Proteins and polypeptides are constituents of all living cells. They are made up of aminoacids which are joined through amide linkages involving the carboxyl group of one aminoacid and the amino group of the neighbouring one, e.g.

$$NH_2 \cdot CH_2 \cdot CO\overline{OH} + NH_2 \cdot CH_2 \cdot COOH \rightarrow$$

$$NH_2 \cdot CH_2 \cdot CO \cdot NH \cdot CH_2 \cdot COOH + H_2O$$

Each end of the chain can now react with more aminoacid and so a polymer (a polyamide) will result. The amide linkages, —CO—NH—, are known as peptide linkages.

Aminoacid polymers of the above type with a relative molecular mass of 10 000 or less are known as polypeptides whereas those with higher relative molecular masses are called proteins.

A particular protein may contain many different aminoacids and these can be identified by hydrolysing the protein and analysing the resultant aminoacids by paper chromatography (page 275). Proteins (or any compound containing the —CO—NH— linkage) may be detected by the violet colour they produce on treatment with dilute sodium hydroxide solution and dilute copper(II) sulphate(VI) solution.

Synthetic polyamides (nylons) have greater commercial utility; their formation is discussed in Chapter 19.

Question

18.1   Outline a synthesis of aminoethanoic acid (glycine) from methane. Write equations for the reaction between aminoethanoic acid and

(a)   benzenecarbonyl chloride (benzoyl chloride),

(b)   methanol in the presence of a little concentrated sulphuric(VI) acid,

(c)   lithium tetrahydridoaluminate(III) (lithium aluminium hydride).

Explain why aminoacids dissolve in water but not in ethoxyethane (diethyl ether).

Explain the relationship between aminoacids, polypeptides, and proteins.

19 Synthetic Macromolecules

Polymerisation is the process of forming large molecules by joining together many small molecules. The small molecules are known as the *monomers* and the macromolecules are called *polymers*.

There are two types of polymerisation process: *addition polymerisation* and *condensation polymerisation*. In addition polymerisation, all the atoms in the monomer are present in the polymer, but condensation polymerisation is characterised by the elimination of small molecules such as water, carbon dioxide, methanol, etc.

ADDITION POLYMERISATION

This type of polymerisation is encountered with alkenes, some important examples being:

$$n\text{CH}_2{=}\text{CH}_2 \qquad \rightarrow \; {+}\text{CH}_2{-}\text{CH}_2{+}_n \qquad \text{poly(ethene)}$$
(polythene)

$$n\text{C}_6\text{H}_5 \cdot \text{CH}{=}\text{CH}_2 \qquad \rightarrow \; {+}\underset{\underset{\text{C}_6\text{H}_5}{|}}{\text{CH}}{-}\text{CH}_2{+}_n \qquad \text{poly(phenylethene)}$$
(polystyrene)

$$n\text{CH}_2{=}\text{CH} \cdot \text{CN} \qquad \rightarrow \; {+}\text{CH}_2{-}\underset{\underset{\text{CN}}{|}}{\text{CH}}{+}_n \qquad \text{poly(propenenitrile)}$$
(Acrilan or Orlon)

$$n\text{CH}_2{=}\text{CHCl} \qquad \rightarrow \; {+}\text{CH}_2{-}\underset{\underset{\text{Cl}}{|}}{\text{CH}}{+}_n \qquad \text{poly(chloroethene)}$$
(P.V.C.)

$$n\text{CH}_2{=}\text{CH} \cdot \text{OOC} \cdot \text{CH}_3 \rightarrow \; {+}\text{CH}_2{-}\underset{\underset{\text{OOC} \cdot \text{CH}_3}{|}}{\text{CH}}{+}_n \qquad \text{poly(ethenyl ethanoate)}$$
(polyvinyl acetate)

$$n\text{CF}_2{=}\text{CF}_2 \qquad \rightarrow \; {+}\text{CF}_2{-}\text{CF}_2{+}_n \qquad \text{poly(tetrafluoroethene)}$$
(P.T.F.E.)

The physical properties of the polymer can be varied to some extent by altering the polymerisation conditions, i.e. the temperature, pressure, or polymerisation catalyst. In general, three types of catalyst are used: free radical catalysts such as peroxides and salts of peroxoacids, acid catalysts such as concentrated sulphuric(VI) acid or boron trifluoride (a Lewis acid, page 365), and Ziegler or co-ordination catalysts.

The free radical catalysts readily split up into free radicals, an example being di(benzenecarbonyl)peroxide (dibenzoyl peroxide):

$$\underset{}{\text{C}_6\text{H}_5{-}\overset{\overset{\text{O}}{\|}}{\text{C}}{-}\text{O}{-}\text{O}{-}\overset{\overset{\text{O}}{\|}}{\text{C}}{-}\text{C}_6\text{H}_5 \rightleftharpoons 2\,\text{C}_6\text{H}_5{-}\overset{\overset{\text{O}}{\|}}{\text{C}}{-}\text{O} \cdot}$$

The free radical will, for simplicty, be represented as $R\cdot$ and it sets up a free radical chain reaction as follows:

$$R\cdot + CH_2\!\!=\!\!CH_2 \rightarrow R-CH_2-\overset{\cdot}{C}H_2 \xrightarrow{\ CH_2\!=\!CH_2\ }$$

$$R-CH_2-CH_2-CH_2-\overset{\cdot}{C}H_2, \quad \text{etc.}$$

The process stops by radicals joining up:

$$-\overset{\cdot}{C}H_2 + \overset{\cdot}{C}H_2- \rightarrow -CH_2-CH_2-$$

and by disproportionation:

$$-\overset{\cdot}{C}H_2 + \overset{\cdot}{C}H_2\!\!-\!\!CH- \rightarrow -CH_3 + CH_2\!\!=\!\!CH-$$

Acid catalysis occurs via protonation of the alkene, e.g.

$$H^+ + CH_2\!\!=\!\!CH_2 \rightarrow CH_3-\overset{+}{C}H_2 \xrightarrow{\ CH_2\!=\!CH_2\ }$$

$$CH_3-CH_2-CH_2-\overset{+}{C}H_2, \quad \text{etc.}$$

Boron trifluoride behaves similarly because it also is electron deficient (the boron has only six electrons in its outer shell).

Ziegler catalysts consist of titanium(IV) chloride and an organoaluminium compound, e.g. triethylaluminium, but their mechanism is not fully understood.

With alkenes such as phenylethene (styrene), $C_6H_5\cdot CH\!\!=\!\!CH_2$, there are three possible ways in which polymerisation could occur.

(a)   Head to tail, where the CH_2 is the head and CH the tail, i.e.

$$\begin{array}{ccc} CH_2\!\!=\!\!CH & + & CH_2\!\!=\!\!CH \\ | & & | \\ C_6H_5 & & C_6H_5 \end{array} \rightarrow \begin{array}{cc} -CH_2-CH-CH_2-CH- \\ \quad\quad | \quad\quad\quad\quad | \\ \quad\quad C_6H_5 \quad\quad\ \ C_6H_5 \end{array}$$

(b)   Head to head and tail to tail, i.e.

$$\begin{array}{l} CH\!\!=\!\!CH_2 + CH_2\!\!=\!\!CH + CH\!\!=\!\!CH_2 \rightarrow \\ | \qquad\qquad\quad | \quad\ | \qquad\qquad | \\ C_6H_5 \qquad\qquad C_6H_5\ C_6H_5 \qquad\ C_6H_5 \end{array}$$

$$-CH-CH_2-CH_2-CH-CH-CH_2-$$

(c)   A random arrangement of (a) and (b).

In fact, head to tail polymerisation occurs and types (b) and (c) are not encountered.

Stereochemistry of addition polymers

Alkenes of the type $CH_2\!\!=\!\!CHY$ can give polymers in which all the Y groups are on the same side of the chain, or on alternate sides, or randomly situated, and these are known as *isotactic*, *syndiotactic*, and *atactic* polymers respectively. These types of polymer are illustrated in Figure 19.1. Stereospecific polymers, i.e. isotactic or syndiotactic polymers, can be obtained by the use of various Ziegler catalysts.

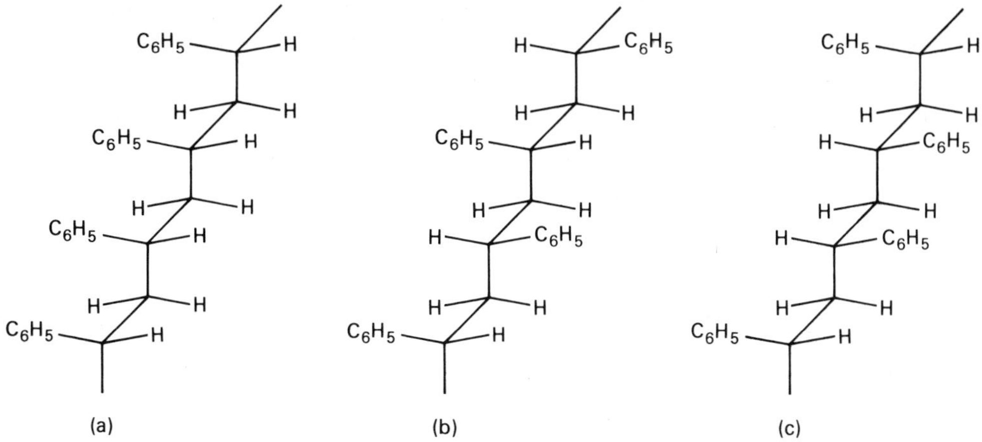

Figure 19.1   Illustration of (a) isotactic, (b) syndiotactic, and (c) atactic polymers.

Uses of addition polymers

Some important uses of addition polymers are given below.

Poly(ethene) (Polythene) — insulating, pipes, and food packaging.

Poly(phenylethene) (Polystyrene) — insulating, moulded objects, and decorations.

Poly(propenenitrile) (Acrilan, Courtelle, or Orlon) — synthetic fibre.

Poly(chloroethene) (P.V.C.) — insulator and protective clothing.

Poly(ethenyl ethanoate) (Polyvinyl acetate) — containers, water-based emulsion paints, and adhesives.

Poly(tetrafluoroethene)   (P.T.F.E.) — non-stick   kitchen   utensils,   non-lubricated bearings, and insulating.

Intermediate properties can be obtained by forming co-polymers, i.e. polymers containing two different monomers.

CONDENSATION POLYMERISATION

There are two main types of condensation polymer: polyesters, e.g. terylene, and polyamides, e.g. nylons.

Terylene is made by heating ethane-1,2-diol (ethylene glycol) and dimethyl benzene-1,4-dicarboxylate (dimethyl terephthalate):

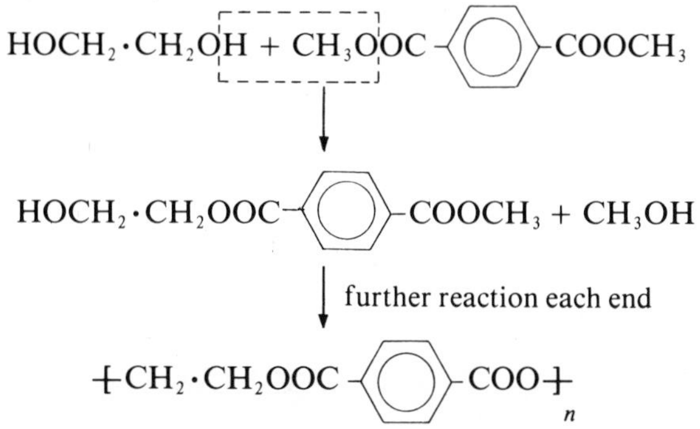

$$HOCH_2 \cdot CH_2OH + CH_3OOC \text{—} \bigcirc \text{—} COOCH_3$$

$$HOCH_2 \cdot CH_2OOC \text{—} \bigcirc \text{—} COOCH_3 + CH_3OH$$

further reaction each end

$$\text{—}(CH_2 \cdot CH_2OOC \text{—} \bigcirc \text{—} COO)_n$$

The dimethyl ester of benzene-1,4-dicarboxylic acid (terephthalic acid) is used instead of the free acid since this improves the efficiency of the polymerisation and it is more readily purified. The ester is manufactured by catalytic oxidation of 1,4-dimethylbenzene (*p*-xylene) followed by esterification of the resultant acid:

$$\text{CH}_3\text{-C}_6\text{H}_4\text{-CH}_3 \xrightarrow{\text{O}_2/\text{Co}^{3+}/140\,°C} \text{COOH-C}_6\text{H}_4\text{-COOH} \xrightarrow{\text{CH}_3\text{OH}/\text{H}^+} \text{COOCH}_3\text{-C}_6\text{H}_4\text{-COOCH}_3$$

Ethane-1,2-diol (ethylene glycol) is manufactured by catalytically oxidising ethene (ethylene) to epoxyethane (ethylene oxide) which is then treated with dilute acid:

$$\text{CH}_2{=}\text{CH}_2 \xrightarrow{\text{O}_2/\text{Ag}/300\,°C} \underset{\text{O}}{\text{CH}_2{-}\text{CH}_2} \xrightarrow{\text{H}^+/\text{H}_2\text{O}} \text{HOCH}_2{\cdot}\text{CH}_2\text{OH}$$

Various nylons can be made by using monomers containing different numbers of carbon atoms. Nylon 66, so called because each monomer contains six carbon atoms, is made by heating hexane-1,6-dioic acid (adipic acid) with hexane-1,6-diamine (hexamethylene diamine):

$$\text{HOOC}{\cdot}(\text{CH}_2)_4{\cdot}\text{CO}\boxed{\text{OH}} + \boxed{\text{NH}_2}{\cdot}(\text{CH}_2)_6{\cdot}\text{NH}_2$$

$$\downarrow \text{heat}$$

$$\text{HOOC}{\cdot}(\text{CH}_2)_4{\cdot}\text{CO}{\cdot}\text{NH}{\cdot}(\text{CH}_2)_6{\cdot}\text{NH}_2 + \text{H}_2\text{O}$$

$$\downarrow \text{further reaction each end}$$

$$\left[\!\!-\text{CO}{\cdot}(\text{CH}_2)_4{\cdot}\text{CO}{\cdot}\text{NH}{\cdot}(\text{CH}_2)_6{\cdot}\text{NH}-\!\!\right]_n$$

The hexane-1,6-dioic acid (adipic acid) is made from phenol:

$$\text{C}_6\text{H}_5\text{OH} \xrightarrow{\text{H}_2/\text{Ni}/300\,°C} \text{Cyclohexanol} \xrightarrow{[\text{O}]} \text{HOOC}{\cdot}(\text{CH}_2)_4{\cdot}\text{COOH}$$

Cyclohexanol

and some of this is used to prepare the diamine:

$$\text{HOOC}{\cdot}(\text{CH}_2)_4{\cdot}\text{COOH} \xrightarrow[\text{(ii) Heat/catalyst}]{\text{(i) NH}_3} \text{CN}{\cdot}(\text{CH}_2)_4{\cdot}\text{CN}$$

$$\xrightarrow{\text{H}_2/\text{Ni}/300\,°C} \text{NH}_2{\cdot}(\text{CH}_2)_6{\cdot}\text{NH}_2$$

Nylon 6 is made from phenol via the following series of reactions:

Cyclohexanol   Cyclohexanone   Cyclohexanone oxime

ε-caprolactam

Effect of cross-linking on physical properties of polymers

Cross-linking has a big effect on the physical properties of a polymer because it increases the relative molecular mass and so affects the solubility etc. Further, it curtails the movement of the chains with respect to one another. Cross-linking is essential if a polymer is to be of use as a fibre.

Polyamide chains are linked together by hydrogen bonding, e.g.

Terylene chains on the other hand are linked by dipole–dipole attraction, since the carbonyl groups are polarised $\overset{\delta+}{C}{=}\overset{\delta-}{O}$.

In contrast, poly(ethene) (polythene) chains are merely linked by van der Waals forces and, since these are relatively weak, the cross-linkage is not very effective. Poly(ethene) (polythene) is, therefore, not used as a fibre. Van der Waals forces are obviously also operative in nylons and terylene but they are not as important as the more powerful hydrogen bonding or dipole–dipole attractions.

Questions

19.1   Explain what is meant by the terms 'addition polymerisation' and 'condensation polymerisation', illustrating your answer in detail by reference to one important example of each type.

        Polymerisation processes usually yield long chain molecules. For a polymer to be useful, it must possess adequate 'cross-linkage' between such chains so as to assume a 3-dimensional structure. Outline how in certain types of polymers cross-linkage is achieved by hydrogen bonding.

[University of London]

19.2   Outline one series of reactions by which each of the following macromolecules is synthesised on an industrial scale from petroleum or coal tar. (a) Poly(ethene) (polyethylene), (b) Nylon 66. Indicate reagents and the conditions used and name the substances formed at the different stages.

        What is the essential difference in the polymerisation process by which these two macromolecules are made from monomeric materials?

        Compare the properties of poly(ethene) and nylon 66 and indicate why one of these materials is more suitable than the other for making fibres.   [J.M.B.]

19.3 (a)   An organic acid X has a relative molecular mass of 138 and a percentage composition by weight: C = 60.87, H = 4.35, O = 34.78. Treatment of X with a solution of hydrogen chloride in ethanol under reflux gives a compound Y of molecular formula $C_9H_{10}O_3$ which is soluble in cold, dilute, aqueous sodium hydrogencarbonate solution. Suggest three possible structures for X and write a general equation to explain the formation of Y.

     (b)   Give the basic chemical constitution of each of the following:
(i)   poly(ethene) (polyethylene)
(ii) poly(phenylethene) (polystyrene)
(iii) polyesters
(iv) polyamides.

        Describe briefly the chemical processes used in the synthesis of either the polyester or the polyamide.   [J.M.B.]

20 Acidity and Basicity

The definition of acids and bases has been broadened considerably over the years. Initially, Arrhenius described acids as substances which yield H^+ ions and bases as compounds which give HO^- ions. This definition was extended by Brønsted and Lowry who considered acids to be proton donors and bases to be proton acceptors. Many more compounds are now classified as acids and bases as a result of the Lewis definition which describes acids as molecules or ions capable of co-ordinating with lone pairs of electrons, whilst bases have lone pairs available for co-ordination ($BF_3 + :NH_3 \longrightarrow H_3N: \rightarrow BF_3$ is therefore an acid-base reaction). However, the Arrhenius definition will suffice for a simple study of acids and bases.

ACIDITY

It was seen in Chapter 7 that, when a weak acid is dissolved in water, the following equilibrium is set up:

$$HA + H_2\ddot{O} \rightleftharpoons H_3O^+ + A^-$$

and that:

$$K_a = \frac{[H_3O^+][A^-]}{[HA]}$$

where K_a is the acidity constant and [ ] stands for concentration. Further, since the values of K_a are very small, it is more convenient to compare acid strengths by the use of pK_a values where $pK_a = -lgK_a$, e.g. K_a for ethanoic acid (acetic acid) is 1.7×10^{-5} but $pK_a = 4.8$. With stronger acids, $[A^-]$ and $[H_3O^+]$ will be higher and $[HA]$ will be lower and so K_a will be larger and pK_a will be smaller.

Factors affecting acidity

The acidity of a compound, HA, is determined by (a) the electronegativity of A and (b) the stability of A^-. This is illustrated by the pK_a values for methane, CH_3—H, and methanol, CH_3O—H, being approximately 50 and 16 respectively. Methanol is the stronger acid because as oxygen is more electronegative than carbon, the O—H bond breaks more readily than the C—H bond, and CH_3O^- is more stable than CH_3^-.

General order of acidity

The following general order of acidity is observed:

alcohols $<$ phenols $<$ carboxylic acids $<$ sulphonic acids

Phenol, C_6H_5OH, is more acidic than alcohols because the charge on the phenoxide ion can be delocalised round the benzene ring, but in alkoxide ions it is situated entirely on the oxygen. The delocalisation of the charge in the phenoxide ion can be represented as a resonance hybrid of several Kekulé type structures:

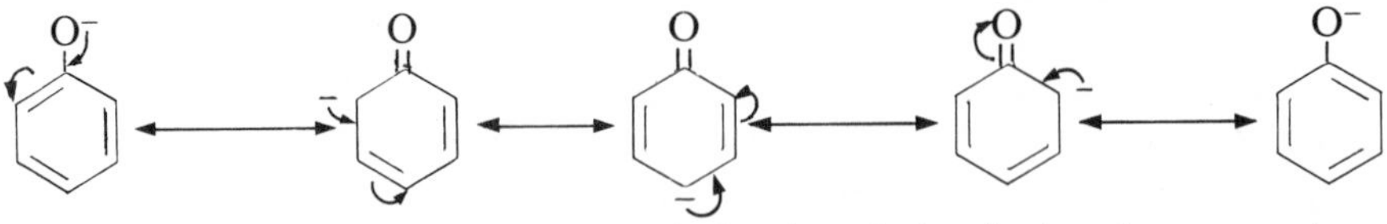

Alternatively, the picture from the orbital point of view is that the oxygen has an unhybridised $2p$ orbital in the same plane as those of the carbons, and so overlap occurs, and the charge is delocalised round the ring.

Carboxylic acids, $R-\overset{\overset{O}{\|}}{C}-OH$, are more acidic than phenol due, to some

extent, to $R-\overset{\overset{O^{\delta-}}{\|}}{C^{\delta+}}$ being more electron-withdrawing than C_6H_5. The main reason, however, is the greater stability of the carboxylate ion. Delocalisation of the charge in the phenoxide ion is not very efficient because, although the charge is shared with the ring, negative carbon is less stable than negative oxygen. In contrast, the carboxylate ion is effectively stabilised by the negative charge being equally shared between the two oxygen atoms:

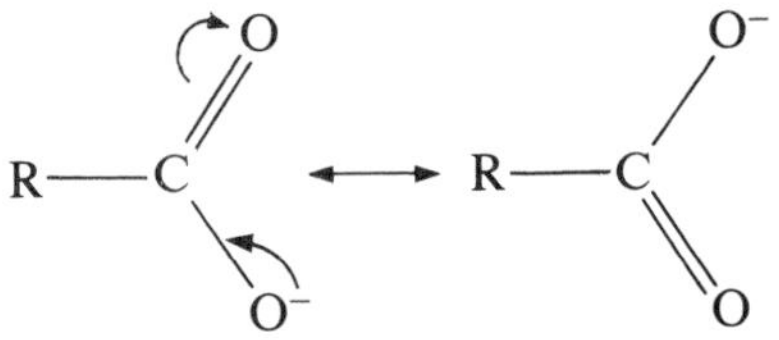

The stability of the carboxylate ion from the orbital stand point stems from delocalisation of the charge via a molecular orbital embracing the two oxygens and the carbon:

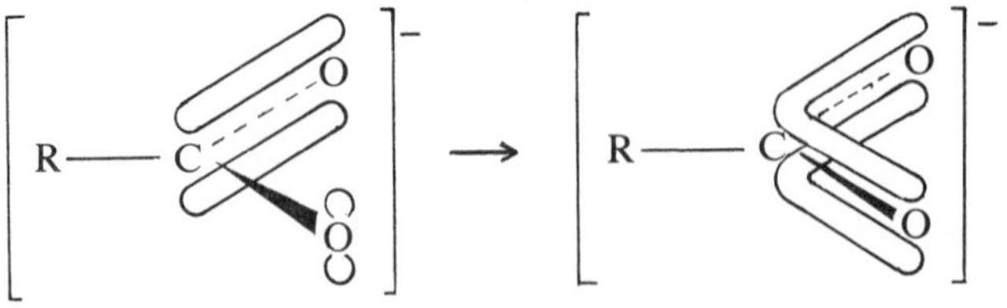

Sulphonic acids are similar in strength to sulphuric(VI) acid since the charge of the anion is equally shared by the three oxygens:

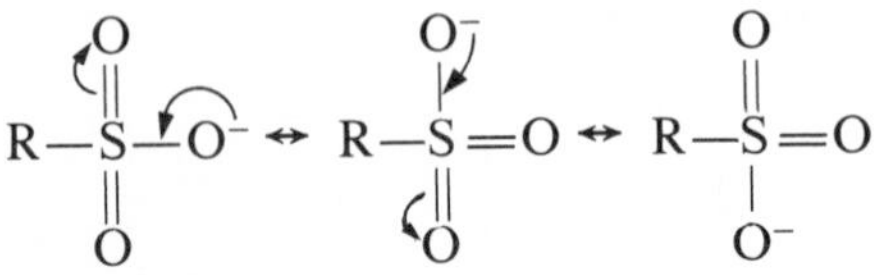

Effect of substituents

The ease of dissociation of an acidic compound, and the stability of the resultant anion, are both affected by substituent groups in the molecule. Thus, the introduction of an electron repelling group into a carboxylic acid molecule will result in a reduction of the acid strength for two reasons.

(a) It will push electrons towards the slightly electron-deficient carbonyl carbon atom

so reducing its charge. The hydroxyl oxygen will therefore have a better chance of attracting more than its fair share of electrons in the C—O bond, and so it will not pull so strongly on the O—H bond electrons, i.e. the O—H bond will not break so readily.

(b) When dissociation has occurred, the electron-repelling group will be pushing electrons towards the already electron-rich —COO⁻ group so reducing its stability.

Conversely, electron-withdrawing groups increase the acidity because they aid the formation of the anion by making the carbonyl carbon more positive and they increase the stability of the anion by helping to delocalise its charge, e.g.

Table 20.1 shows the variation of pK_a with various substituent groups.

Table 20.1 The pK_a values of some acids

Acid	pK_a
$CH_3 \cdot CH_2 \cdot COOH$	4.88
(cyclohexane with H and COOH)	4.87
$CH_3 \cdot COOH$	4.76
(benzene ring)—COOH	4.20
HCOOH	3.77
$CH_2Br \cdot COOH$	2.90
$CH_2Cl \cdot COOH$	2.86
$CH_2F \cdot COOH$	2.66
$CHCl_2 \cdot COOH$	1.29
$CCl_3 \cdot COOH$	0.65

It is apparent from the pK_a values of benzenecarboxylic (benzoic) and cyclohexanoic acids that the cyclohexyl group is electron-donating compared to the phenyl group. The reason for this is that the acid group is attached to an sp^3 hybridised carbon atom in cyclohexanoic acid and an sp^2 hybridised one in benzenecarboxylic acid (benzoic acid). The greater the p character of the hybridised orbital, the further are the electrons from the nucleus, and so the greater the electron-releasing properties.

The phenyl group is more electron-releasing than hydrogen and so benzenecarboxylic acid (benzoic acid) is a weaker acid than methanoic acid (formic acid). The acid group obtains electrons from the benzene ring as follows:

i.e. the π orbitals of the electron attracting carbonyl group are in the same plane as, and interact with, the π orbitals of the ring.

The pK_a values for the monohalogenated acids reflect the decreasing electronegativity of the halogens as their size increases.

BASICITY

If a base, $R_3N:$, is dissolved in water, the equilibrium set up is:

$$R_3N: + H_2O \rightleftharpoons R_3\overset{+}{N}H + HO^-$$

and, as seen in Chapter 7,

$$K_b = \frac{[R_3\overset{+}{N}H]\,[HO^-]}{[R_3N:]}$$

where K_b is the basicity constant. Again, pK_b (i.e. $-\lg K_b$) is used in preference to K_b since a more convenient scale results. As the base strength increases, $[R_3\overset{+}{N}H]$ and $[HO^-]$ increase and $[R_3N:]$ decreases, i.e. K_b will be larger and so pK_b will be smaller.

Since $pK_a + pK_b = 14$ (page 127), the strength of bases may be expressed in terms of pK_a, thus providing a continuous scale for acids and bases. K_a for a base is a measure of the ease with which $R_3\overset{+}{N}H$ will lose a proton:

$$R_3\overset{+}{N}H + H_2\overset{..}{O} \rightleftharpoons R_3N: + H_3O^+$$

$$K_a = \frac{[R_3N]\,[H_3O^+]}{[R_3\overset{+}{N}H]}$$

It may have been expected that the order of basicity of the alkylamines would be

$$\text{primary} < \text{secondary} < \text{tertiary}$$

since the alkyl groups push electrons towards the nitrogen and make it electron richer. However, this is not the observed order as the following pK_b values show:

NH_3	$CH_3 \cdot NH_2$	$(CH_3)_2NH$	$(CH_3)_3N$
4.76	3.36	3.23	4.20

The reason for the apparent anomaly is that basic strength depends not only on electron availability at the nitrogen atom, but also on the stability of the resultant cation. The alkyl groups not only increase electron availability at the nitrogen, but also decrease the stability of the cation by leaving less hydrogen atoms available for hydrogen bonding. Hence

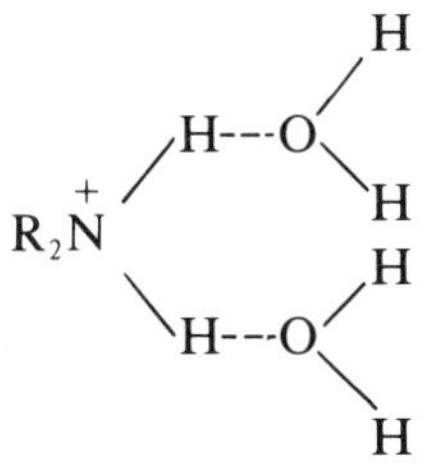

is more stable than

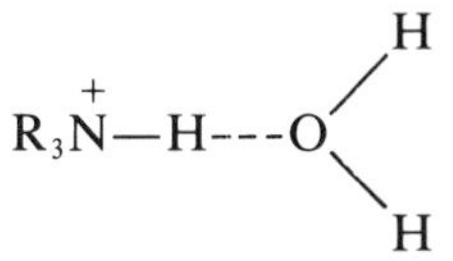

because its charge density is reduced by its greater solvation.

All quarternary ammonium hydroxydes, e.g. $(CH_3)_4\overset{+}{N}\overset{-}{O}H$, are strong bases since they are of necessity fully ionised.

Arylamines are less basic than alkylamines because the lone pair of electrons on the nitrogen interacts with the π orbitals of the ring, e.g.

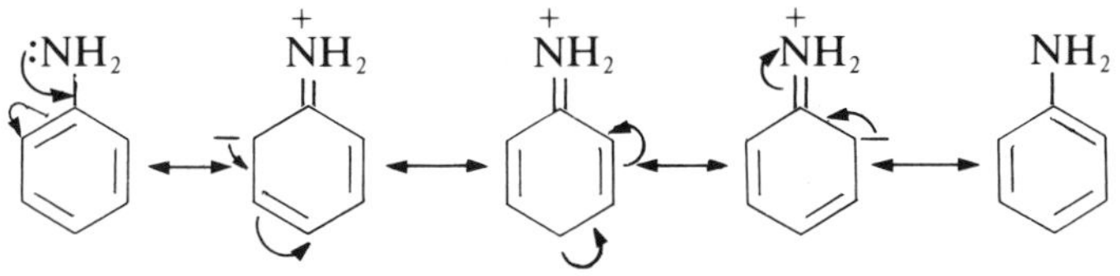

The reduced availability of the lone pair for co-ordination is reflected by pK_b for phenylamine (aniline) being 9.38 compared to 3.36 for methylamine. Secondary and tertiary aromatic amines, e.g. $C_6H_5 \cdot NH \cdot CH_3$ and $C_6H_5 \cdot N(CH_3)_2$, are slightly weaker bases than the primary amines owing to greater interaction of the lone pair with the ring and the decreasing ability of the cations to stabilise themselves by hydrogen bonding with water molecules.

The introduction of an electron-withdrawing group adjacent to the amino group greatly reduces the basicity of the compound owing to the decreased electron availability at the nitrogen. Amides are very weak bases owing to

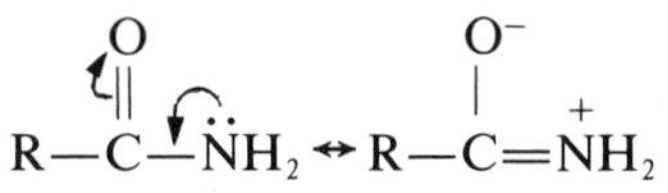

and they form salts only with strong acids, e.g. hydrochloric acid.

Questions

20.1   The pK_a for the ionisation of ethanoic (acetic) acid in aqueous solution at 25 °C is 4.8. Write an equation for the ionisation of ethanoic (acetic) acid in aqueous solution and hence derive a mathematical expression relating the pK_a to the concentration of the components of this system. Why is it necessary to state the solvent and temperature when giving a value of pK_a?

From the expression derived show that when the pH of a solution of ethanoic (acetic) acid is 4.8 the concentration of undissociated acid is equal to the concentration of anions.

Describe, giving brief experimental details, how the pK_a for ethanoic (acetic) acid may be determined experimentally.

Given that the pK_a for methanoic (formic) and ethanoic (acetic) acids are 3.8 and 4.8 respectively, explain quantitatively the pK_a values of the following organic acids:

(a)   chloroethanoic (chloroacetic) acid 2.9
(b)   benzenecarboxylic (benzoic) acid 4.2
(c)   phenol 10.0.   [J.M.B.]

20.2   To separate 10 cm³ samples of hydrochloric acid of varying concentrations were added 2 drops of thymol blue, or congo red, or bromophenol red. The colour of the indicator was observed and the following results were obtained:

Molarity of hydrochloric acid	10^{-1}	10^{-2}	10^{-3}	10^{-4}	10^{-5}	10^{-6}	10^{-7}
Thymol blue	red	orange	yellow	yellow	yellow	yellow	yellow
Congo red	blue	blue	blue	blue-red	red	red	red
Bromophenol red	yellow	yellow	yellow	yellow	yellow	orange	red

Deduce the effective pH range of each indicator.

0.1 M solutions of methanoic (formic) acid, ethanoic (acetic) acid, chloroethanoic (chloroacetic) acid, trichloroethanoic (trichloroacetic) acid and 4-chlorobutanoic acid were prepared.

To 10 cm³ of each solution were added 5 cm³ of 0.1 M sodium hydroxide and 2 drops of indicator. The colours observed were as follows:

	Thymol blue	Congo red	Bromophenol red
Methanoic (formic) acid	yellow	blue-red	yellow
Ethanoic (acetic) acid	yellow	red	yellow
Chloroethanoic (chloroacetic) acid	yellow	blue	yellow
Trichloroethanoic (trichloroacetic) acid	red	blue	yellow
4-chlorobutanoic acid	yellow	red	yellow

Deduce the approximate pK_a values of these acids. Suggest how these values can be correlated with the nature of the group R in the carboxylic acid $R \cdot COOH$.   [J.M.B.]

20.3   The following are the pK_a values, at a given temperature of some organic substances.

HCOOH	3.77
CH_3COOH	4.76
$ClCH_2COOH$	2.86
CH_3CH_2COOH	4.88
C_6H_5OH	9.95

(a)   Which is the strongest acid listed?

(b)   Which is the strongest base, $C_6H_5O^-$ or $CH_3CH_2COO^-$? Explain your answer briefly.

(c)   Explain qualitatively the difference in the pK_a values of $ClCH_2COOH$ and CH_3COOH.

(d)   Explain qualitatively the difference in the pK_a values of HCOOH and CH_3COOH.

(e)   Phenol forms a salt with aqueous sodium hydroxide but does not react with aqueous sodium hydrogencarbonate. Which is the strongest acid of the following: phenol (C_6H_5OH), carbonic acid (H_2CO_3), or water?   [J.M.B.]

Answers to Numerical Questions

Chapter 1

1	55.89
6	$\nu_1 = 2.467 \times 10^{15}$ Hz
	$\lambda_1 = 12.16 \times 10^{-8}$ m
	$\nu_2 = 2.924 \times 10^{15}$ Hz
	$\lambda_2 = 10.25 \times 10^{-8}$ m
	$\nu_3 = 3.084 \times 10^{15}$ Hz
	$\lambda_3 = 9.721 \times 10^{-8}$ m
	$\nu_4 = 3.158 \times 10^{15}$ Hz
	$\lambda_4 = 9.494 \times 10^{-8}$ m
7	1307 kJ mol^{-1}

Chapter 2

1	18.14 litres
3	12.15 g. N_2 14.28%, H_2 42.86%, NH_3 42.86%
4	134.9
5	90.4
6	116
10 (a)	25
(b)	0.84
(c)	8.6%
18	2.87×10^{-10} m

Chapter 4

1	-239.0 kJ mol^{-1}
2	-104.1 kJ mol^{-1}
3	-126.4 kJ mol^{-1}
4	$-84.9, +226.7,$ and -311.6 kJ mol^{-1}
5	-416.6 kJ
6	-278 kJ mol^{-1}
7	-253.8 kJ
8	$-586, -169,$ and -248 kJ mol^{-1}
9 (a)	333.7 kJ
(b)	$+31$ kJ
10	390.8 kJ mol^{-1}

Chapter 5

2 (b)	41.76 kJ mol^{-1}
(d)	70 °C
3	1.948×10^{-2} mol litre^{-1}

4	$C_4H_8O_4$
5	14.22 mm
6	364
8	S_8
10	Dimer in benzene
11	478 cm^3
12	0.2445 atm
13 (a)	0.04465 M and 0.05571 M
(b)	1.092 atm and 1.362 atm
14	1.871 g litre^{-1}
15 (a)	M
(b)	2M
16	80
(i)	0.002 mol litre^{-1}
(ii)	0.100 mol litre^{-1}
18	c. 158.4

Chapter 6

3	$N_2 = 0.0291$ vol, $NO = 0.0291$ vol
4 (b)	3.6×10^{-7} mol s^{-1}
(c)	9×10^{-4}
(d)	1.5×10^2
5	4.18×10^{-5} atm^{-2}

Chapter 7

1 (i)	1.698×10^{-10} mol^2 litre^{-2}
(ii)	2.437×10^{-5} g
2 (a)	AgCl 1.96×10^{-10} mol^2 litre^{-2}, Ag_2CrO_4 3.607×10^{-12} mol^3 litre^{-3}
(b)	$>1.96 \times 10^{-9}$ mol litre^{-1}
(c)	$>2.685 \times 10^{-5}$ mol litre^{-1}
3 (a)	1.683×10^{-2} g litre^{-1}
(b)	3.12×10^{-5} g litre^{-1}
4 (a)	1.529×10^{-4} mol^3 litre^{-3}
(b)	1.859×10^{-2} g
6 (a)	6.27×10^{-5} mol litre^{-1}
(b)	1.12×10^{-3} mol litre^{-1}
(c)	7.9%
7 (i)	4.3%
(ii)	2.43. 3.8
8 (a)	11.13
(b)	3.70

11 (a) $K_a = 3.19 \times 10^{-1}$ mol litre^{-1},
              $pK_a = 3.50$
14   0.0214 M

Chapter 8

1 (b) (i)   33.8 min
       (ii)   66.5 min
2   (i)   1st order
       (ii)   0.05 h^{-1}
       (iii) 2.625×10^{-3} mol litre^{-1} h^{-1}
3   103.3 kJ mol^{-1}

Chapter 12

12 (e)   1.595 g
     (f) $CuSO_4 \cdot 3H_2O$

Chapter 14

1 CH_4
2 C_3H_8O
3 C_2H_5NO
4 C_2H_5Br
5 CH_2Cl, $C_2H_4Cl_2$
6 (a) C_3H_6O
     (b) C_3H_6O
     (c) $C_3H_6O_2$
7 $C_2H_4Br_2$

Chapter 16

1   50% ethene (ethylene)
25 (a) C_4H_9Br
     (d)   33.3%

Units, Conversion Factors, and Physical Constants

SI (Système Internationale) units have been introduced to provide an unambiguous set of standard units for universal use. The system is based on seven basic units: metre — unit of length, kilogramme — unit of mass, second — unit of time, ampere — unit of electric current, kelvin — unit of thermodynamic temperature, mole — unit of amount of substance, and the candella — unit of luminous intensity. There are also two subsidiary units: radian — unit of angle, and the steradian — unit of solid angle. SI has the advantage over previous systems of measurement of being fully decimal and coherent. Thus all derived units are obtained by simple multiplication or division of the basic units without introducing numerical factors. If measurements in the basic units of SI are substituted in an equation, the answer will automatically be in the appropriate SI unit.

Some SI units and their symbols are given below:

Quantity	SI unit	Symbol
Length	metre	m
Mass	kilogramme	kg
Time	second	s
Electric current	ampere	A
Temperature (absolute)	kelvin	K
Frequency	hertz	Hz
Energy	joule	J
Force	newton	N
Pressure	pascal	Pa
Electrical charge	coulomb	C
Electromotive force	volt	V
Resistance	ohm	Ω

Preferred fractions for use with SI units are:

Fraction	Prefix	Symbol
10^{-1}	deci	d
10^{-2}	centi	c
10^{-3}	milli	m
10^{-6}	micro	μ
10^{-9}	nano	n
10^{-12}	pico	p

Some non-SI units are still acceptable or in common use; the conversion factors are:

Unit	Symbol	SI equivalent
Atmosphere	atm	1.013×10^5 N m^{-2}
Electron volt	eV	1.602×10^{-19} J
Litre	l	1.000×10^{-3} m^3
Calorie	cal	4.184 J
Ångstrom unit	Å	1.000×10^{-10} m
Inch	in	2.54×10^{-2} m

Physical constants

Quantity	Symbol	Value
Speed of light in vacuum	c	2.998×10^8 m s^{-1}
Avogadro constant	N_A	6.022×10^{23} mol^{-1}
Boltzmann constant	k	1.380×10^{-23} J K^{-1}
Rydberg constant	$R\infty$	1.097×10^7 m^{-1}
Planck constant	h	6.626×10^{-34} J s

RELATIVE ATOMIC MASSES (ATOMIC WEIGHTS)

Element	Symbol	Atomic number	Relative atomic mass	Element	Symbol	Atomic number	Relative atomic mass
Actinium	Ac	89	227.0	Mercury	Hg	80	200.6
Aluminium	Al	13	26.9	Molybdenum	Mo	42	95.9
Americium	Am	95	243.0	Neodymium	Nd	60	144.2
Antimony	Sb	51	121.8	Neon	Ne	10	20.2
Argon	Ar	18	39.9	Neptunium	Np	93	237.0
Astatine	At	85	210.0	Nickel	Ni	28	58.7
Arsenic	As	33	74.9	Niobium	Nb	41	92.9
Barium	Ba	56	137.3	Nitrogen	N	7	14.0
Berkelium	Bk	97	249.0	Osmium	Os	76	190.2
Beryllium	Be	4	9.0	Oxygen	O	8	16.0
Bismuth	Bi	83	209.0	Palladium	Pd	46	106.4
Boron	B	5	10.8	Phosphorus	P	15	31.0
Bromine	Br	35	79.9	Platinum	Pt	78	195.1
Cadmium	Cd	48	112.4	Plutonium	Pu	94	242.0
Caesium	Cs	55	132.9	Polonium	Po	84	210.0
Calcium	Ca	20	40.1	Potassium	K	19	39.1
Californium	Cf	98	251.0	Praseodymium	Pr	59	140.9
Carbon	C	6	12.0	Promethium	Pm	61	145.0
Cerium	Ce	58	140.1	Protactinium	Pa	91	231.0
Chlorine	Cl	17	35.5	Radium	Ra	88	226.1
Chromium	Cr	24	52.0	Radon	Rn	86	222.0
Cobalt	Co	27	58.9	Rhenium	Re	75	186.2
Copper	Cu	29	63.5	Rhodium	Rh	45	102.9
Curium	Cm	96	247.0	Rubidium	Rb	37	85.5
Dysprosium	Dy	66	162.5	Ruthenium	Ru	44	101.1
Einsteinium	Es	99	254.0	Samarium	Sm	62	150.4
Erbium	Er	68	167.3	Scandium	Sc	21	45.0
Europium	Eu	63	152.0	Selenium	Se	34	79.0
Fermium	Fm	100	253.0	Silicon	Si	14	28.1
Fluorine	F	9	19.0	Silver	Ag	47	107.9
Francium	Fr	87	223.0	Sodium	Na	11	23.0
Gadolinium	Gd	64	157.3	Strontium	Sr	38	87.6
Gallium	Ga	31	69.7	Sulphur	S	16	32.1
Germanium	Ge	32	72.6	Tantalum	Ta	73	180.9
Gold	Au	79	197.0	Technetium	Tc	43	99.0
Hafnium	Hf	72	178.5	Tellurium	Te	52	127.6
Helium	He	2	4.0	Terbium	Tb	65	158.9
Holmium	Ho	67	164.9	Thallium	Tl	81	204.4
Hydrogen	H	1	1.0	Thorium	Th	90	232.0
Indium	In	49	114.8	Thulium	Tm	69	168.9
Iodine	I	53	126.9	Tin	Sn	50	118.7
Iridium	Ir	77	192.2	Titanium	Ti	22	47.9
Iron	Fe	26	55.8	Tungsten	W	74	183.9
Krypton	Kr	36	83.8	Uranium	U	92	238.0
Lanthanum	La	57	138.9	Vanadium	V	23	50.9
Lead	Pb	82	207.2	Xenon	Xe	54	131.3
Lithium	Li	3	6.9	Ytterbium	Yb	70	173.0
Lutetium	Lu	71	175.0	Yttrium	Y	39	88.9
Magnesium	Mg	12	24.3	Zinc	Zn	30	65.4
Manganese	Mn	25	54.9	Zirconium	Zr	40	91.2

Logarithms

	0	1	2	3	4	5	6	7	8	9	1	2	3	4	5	6	7	8	9
10	0000	0043	0086	0128	0170						4	8	13	17	21	25	30	34	38
						0212	0253	0294	0334	0374	4	8	12	16	20	24	28	32	36
11	0414	0453	0492	0531	0569						4	8	12	15	19	23	27	31	35
						0607	0645	0682	0719	0755	4	7	11	15	18	22	26	30	33
12	0792	0828	0864	0899	0934						4	7	11	14	18	21	25	28	32
						0969	1004	1038	1072	1106	3	7	10	14	17	20	24	27	31
13	1139	1173	1206	1239	1271						3	7	10	13	16	20	23	26	30
						1303	1335	1367	1399	1430	3	6	9	13	16	19	22	25	28
14	1461	1492	1523	1553	1584						3	6	9	12	15	18	21	24	27
						1614	1644	1673	1703	1732	3	6	9	12	15	18	21	24	27
15	1761	1790	1818	1847	1875						3	6	9	11	14	17	20	23	26
						1903	1931	1959	1987	2014	3	6	8	11	14	17	19	22	25
16	2041	2068	2095	2122	2148						3	5	8	11	13	16	19	21	24
						2175	2201	2227	2253	2279	3	5	8	10	13	16	18	21	23
17	2304	2330	2355	2380	2405						3	5	8	10	13	15	18	20	23
						2430	2455	2480	2504	2529	2	5	7	10	12	15	17	20	22
18	2553	2577	2601	2625	2648						2	5	7	10	12	14	17	19	21
						2672	2695	2718	2742	2765	2	5	7	9	12	14	16	19	21
19	2788	2810	2833	2856	2878						2	5	7	9	11	14	16	18	20
						2900	2923	2945	2967	2989	2	4	7	9	11	13	15	18	20
20	3010	3032	3054	3075	3096	3118	3139	3160	3181	3201	2	4	6	8	11	13	15	17	19
21	3222	3243	3263	3284	3304	3324	3345	3365	3385	3404	2	4	6	8	10	12	14	16	18
22	3424	3444	3464	3483	3502	3522	3541	3560	3579	3598	2	4	6	8	10	12	14	15	17
23	3617	3636	3655	3674	3692	3711	3729	3747	3766	3784	2	4	6	7	9	11	13	15	17
24	3802	3820	3838	3856	3874	3892	3909	3927	3945	3962	2	4	5	7	9	11	12	14	16
25	3979	3997	4014	4031	4048	4065	4082	4099	4116	4133	2	3	5	7	9	10	12	14	15
26	4150	4166	4183	4200	4216	4232	4249	4265	4281	4298	2	3	5	7	8	10	11	13	15
27	4314	4330	4346	4362	4378	4393	4409	4425	4440	4456	2	3	5	6	8	9	11	13	14
28	4472	4487	4502	4518	4533	4548	4564	4579	4594	4609	2	3	5	6	8	9	11	12	14
29	4624	4639	4654	4669	4683	4698	4713	4728	4742	4757	1	3	4	6	7	9	10	12	13
30	4771	4786	4800	4814	4829	4843	4857	4871	4886	4900	1	3	4	6	7	9	10	11	13
31	4914	4928	4942	4955	4969	4983	4997	5011	5024	5038	1	3	4	6	7	8	10	11	12
32	5051	5065	5079	5092	5105	5119	5132	5145	5159	5172	1	3	4	5	7	8	9	11	12
33	5185	5198	5211	5224	5237	5250	5263	5276	5289	5302	1	3	4	5	6	8	9	10	12
34	5315	5328	5340	5353	5366	5378	5391	5403	5416	5428	1	3	4	5	6	8	9	10	11
35	5441	5453	5465	5478	5490	5502	5514	5527	5539	5551	1	2	4	5	6	7	9	10	11
36	5563	5575	5587	5599	5611	5623	5635	5647	5658	5670	1	2	4	5	6	7	8	10	11
37	5682	5694	5705	5717	5729	5740	5752	5763	5775	5786	1	2	3	5	6	7	8	9	10
38	5798	5809	5821	5832	5843	5855	5866	5877	5888	5899	1	2	3	5	6	7	8	9	10
39	5911	5922	5933	5944	5955	5966	5977	5988	5999	6010	1	2	3	4	5	7	8	9	10
40	6021	6031	6042	6053	6064	6075	6085	6096	6107	6117	1	2	3	4	5	6	8	9	10
41	6128	6138	6149	6160	6170	6180	6191	6201	6212	6222	1	2	3	4	5	6	7	8	9
42	6232	6243	6253	6263	6274	6284	6294	6304	6314	6325	1	2	3	4	5	6	7	8	9
43	6335	6345	6355	6365	6375	6385	6395	6405	6415	6425	1	2	3	4	5	6	7	8	9
44	6435	6444	6454	6464	6474	6484	6493	6503	6513	6522	1	2	3	4	5	6	7	8	9
45	6532	6542	6551	6561	6571	6580	6590	6599	6609	6618	1	2	3	4	5	6	7	8	9
46	6628	6637	6646	6656	6665	6675	6684	6693	6702	6712	1	2	3	4	5	6	7	7	8
47	6721	6730	6739	6749	6758	6767	6776	6785	6794	6803	1	2	3	4	5	5	6	7	8
48	6812	6821	6830	6839	6848	6857	6866	6875	6884	6893	1	2	3	4	4	5	6	7	8
49	6902	6911	6920	6928	6937	6946	6955	6964	6972	6981	1	2	3	4	4	5	6	7	8

Logarithms

	0	1	2	3	4	5	6	7	8	9	1	2	3	4	5	6	7	8	9
50	6990	6998	7007	7016	7024	7033	7042	7050	7059	7067	1	2	3	3	4	5	6	7	8
51	7076	7084	7093	7101	7110	7118	7126	7135	7143	7152	1	2	3	3	4	5	6	7	8
52	7160	7168	7177	7185	7193	7202	7210	7218	7226	7235	1	2	2	3	4	5	6	7	7
53	7243	7251	7259	7267	7275	7284	7292	7300	7308	7316	1	2	2	3	4	5	6	6	7
54	7324	7332	7340	7348	7356	7364	7372	7380	7388	7396	1	2	2	3	4	5	6	6	7
55	7404	7412	7419	7427	7435	7443	7451	7459	7466	7474	1	2	2	3	4	5	5	6	7
56	7482	7490	7497	7505	7513	7520	7528	7536	7543	7551	1	2	2	3	4	5	5	6	7
57	7559	7566	7574	7582	7589	7597	7604	7612	7619	7627	1	2	2	3	4	5	5	6	7
58	7634	7642	7649	7657	7664	7672	7679	7686	7694	7701	1	1	2	3	4	4	5	6	7
59	7709	7716	7723	7731	7738	7745	7752	7760	7767	7774	1	1	2	3	4	4	5	6	7
60	7782	7789	7796	7803	7810	7818	7825	7832	7839	7846	1	1	2	3	4	4	5	6	6
61	7853	7860	7868	7875	7882	7889	7896	7903	7910	7917	1	1	2	3	4	4	5	6	6
62	7924	7931	7938	7945	7952	7959	7966	7973	7980	7987	1	1	2	3	3	4	5	6	6
63	7993	8000	8007	8014	8021	8028	8035	8041	8048	8055	1	1	2	3	3	4	5	5	6
64	8062	8069	8075	8082	8089	8096	8102	8109	8116	8122	1	1	2	3	3	4	5	5	6
65	8129	8136	8142	8149	8156	8162	8169	8176	8182	8189	1	1	2	3	3	4	5	5	6
66	8195	8202	8209	8215	8222	8228	8235	8241	8248	8254	1	1	2	3	3	4	5	5	6
67	8261	8267	8274	8280	8287	8293	8299	8306	8312	8319	1	1	2	3	3	4	5	5	6
68	8325	8331	8338	8344	8351	8357	8363	8370	8376	8382	1	1	2	3	3	4	4	5	6
69	8388	8395	8401	8407	8414	8420	8426	8432	8439	8445	1	1	2	2	3	4	4	5	6
70	8451	8457	8463	8470	8476	8482	8488	8494	8500	8506	1	1	2	2	3	4	4	5	6
71	8513	8519	8525	8531	8537	8543	8549	8555	8561	8567	1	1	2	2	3	4	4	5	5
72	8573	8579	8585	8591	8597	8603	8609	8615	8621	8627	1	1	2	2	3	4	4	5	5
73	8633	8639	8645	8651	8657	8663	8669	8675	8681	8686	1	1	2	2	3	4	4	5	5
74	8692	8698	8704	8710	8716	8722	8727	8733	8739	8745	1	1	2	2	3	3	4	5	5
75	8751	8756	8762	8768	8774	8779	8785	8791	8797	8802	1	1	2	2	3	3	4	5	5
76	8808	8814	8820	8825	8831	8837	8842	8848	8854	8859	1	1	2	2	3	3	4	5	5
77	8865	8871	8876	8882	8887	8893	8899	8904	8910	8915	1	1	2	2	3	3	4	4	5
78	8921	8927	8932	8938	8943	8949	8954	8960	8965	8971	1	1	2	2	3	3	4	4	5
79	8976	8982	8987	8993	8998	9004	9009	9015	9020	9025	1	1	2	2	3	3	4	4	5
80	9031	9036	9042	9047	9053	9058	9063	9069	9074	9079	1	1	2	2	3	3	4	4	5
81	9085	9090	9096	9101	9106	9112	9117	9122	9128	9133	1	1	2	2	3	3	4	4	5
82	9138	9143	9149	9154	9159	9165	9170	9175	9180	9186	1	1	2	2	3	3	4	4	5
83	9191	9196	9201	9206	9212	9217	9222	9227	9232	9238	1	1	2	2	3	3	4	4	5
84	9243	9248	9253	9258	9263	9269	9274	9279	9284	9289	1	1	2	2	3	3	4	4	5
85	9294	9299	9304	9309	9315	9320	9325	9330	9335	9340	1	1	2	2	3	3	4	4	5
86	9345	9350	9355	9360	9365	9370	9375	9380	9385	9390	1	1	2	2	3	3	4	4	5
87	9395	9400	9405	9410	9415	9420	9425	9430	9435	9440	0	1	1	2	2	3	3	4	4
88	9445	9450	9455	9460	9465	9469	9474	9479	9484	9489	0	1	1	2	2	3	3	4	4
89	9494	9499	9504	9509	9513	9518	9523	9528	9533	9538	0	1	1	2	2	3	3	4	4
90	9542	9547	9552	9557	9562	9566	9571	9576	9581	9586	0	1	1	2	2	3	3	4	4
91	9590	9595	9600	9605	9609	9614	9619	9624	9628	9633	0	1	1	2	2	3	3	4	4
92	9638	9643	9647	9652	9657	9661	9666	9671	9675	9680	0	1	1	2	2	3	3	4	4
93	9685	9689	9694	9699	9703	9708	9713	9717	9722	9727	0	1	1	2	2	3	3	4	4
94	9731	9736	9741	9745	9750	9754	9759	9763	9768	9773	0	1	1	2	2	3	3	4	4
95	9777	9782	9786	9791	9795	9800	9805	9809	9814	9818	0	1	1	2	2	3	3	4	4
96	9823	9827	9832	9836	9841	9845	9850	9854	9859	9863	0	1	1	2	2	3	3	4	4
97	9868	9872	9877	9881	9886	9890	9894	9899	9903	9908	0	1	1	2	2	3	3	4	4
98	9912	9917	9921	9926	9930	9934	9939	9943	9948	9952	0	1	1	2	2	3	3	4	4
99	9956	9961	9965	9969	9974	9978	9983	9987	9991	9996	0	1	1	2	2	3	3	3	4

Antilogarithms

	0	1	2	3	4	5	6	7	8	9	1	2	3	4	5	6	7	8	9
0.00	1000	1002	1005	1007	1009	1012	1014	1016	1019	1021	0	0	1	1	1	1	2	2	2
0.01	1023	1026	1028	1030	1033	1035	1038	1040	1042	1045	0	0	1	1	1	1	2	2	2
0.02	1047	1050	1052	1054	1057	1059	1062	1064	1067	1069	0	0	1	1	1	1	2	2	2
0.03	1072	1074	1076	1079	1081	1084	1086	1089	1091	1094	0	0	1	1	1	1	2	2	2
0.04	1096	1099	1102	1104	1107	1109	1112	1114	1117	1119	0	1	1	1	1	2	2	2	2
0.05	1122	1125	1127	1130	1132	1135	1138	1140	1143	1146	0	1	1	1	1	2	2	2	2
0.06	1148	1151	1153	1156	1159	1161	1164	1167	1169	1172	0	1	1	1	1	2	2	2	2
0.07	1175	1178	1180	1183	1186	1189	1191	1194	1197	1199	0	1	1	1	1	2	2	2	2
0.08	1202	1205	1208	1211	1213	1216	1219	1222	1225	1227	0	1	1	1	1	2	2	2	3
0.09	1230	1233	1236	1239	1242	1245	1247	1250	1253	1256	0	1	1	1	1	2	2	2	3
0.10	1259	1262	1265	1268	1271	1274	1276	1279	1282	1285	0	1	1	1	1	2	2	2	3
0.11	1288	1291	1294	1297	1300	1303	1306	1309	1312	1315	0	1	1	1	2	2	2	2	3
0.12	1318	1321	1324	1327	1330	1334	1337	1340	1343	1346	0	1	1	1	2	2	2	2	3
0.13	1349	1352	1355	1358	1361	1365	1368	1371	1374	1377	0	1	1	1	2	2	2	3	3
0.14	1380	1384	1387	1390	1393	1396	1400	1403	1406	1409	0	1	1	1	2	2	2	3	3
0.15	1413	1416	1419	1422	1426	1429	1432	1435	1439	1442	0	1	1	1	2	2	2	3	3
0.16	1445	1449	1452	1455	1459	1462	1466	1469	1472	1476	0	1	1	1	2	2	2	3	3
0.17	1479	1483	1486	1489	1493	1496	1500	1503	1507	1510	0	1	1	1	2	2	2	3	3
0.18	1514	1517	1521	1524	1528	1531	1535	1538	1542	1545	0	1	1	1	2	2	2	3	3
0.19	1549	1552	1556	1560	1563	1567	1570	1574	1578	1581	0	1	1	1	2	2	3	3	3
0.20	1585	1589	1592	1596	1600	1603	1607	1611	1614	1618	0	1	1	1	2	2	3	3	3
0.21	1622	1626	1629	1633	1637	1641	1644	1648	1652	1656	0	1	1	2	2	2	3	3	3
0.22	1660	1663	1667	1671	1675	1679	1683	1687	1690	1694	0	1	1	2	2	2	3	3	3
0.23	1698	1702	1706	1710	1714	1718	1722	1726	1730	1734	0	1	1	2	2	2	3	3	4
0.24	1738	1742	1746	1750	1754	1758	1762	1766	1770	1774	0	1	1	2	2	2	3	3	4
0.25	1778	1782	1786	1791	1795	1799	1803	1807	1811	1816	0	1	1	2	2	2	3	3	4
0.26	1820	1824	1828	1832	1837	1841	1845	1849	1854	1858	0	1	1	2	2	3	3	3	4
0.27	1862	1866	1871	1875	1879	1884	1888	1892	1897	1901	0	1	1	2	2	3	3	3	4
0.28	1905	1910	1914	1919	1923	1928	1932	1936	1941	1945	0	1	1	2	2	3	3	4	4
0.29	1950	1954	1959	1963	1968	1972	1977	1982	1986	1991	0	1	1	2	2	3	3	4	4
0.30	1995	2000	2004	2009	2014	2018	2023	2028	2032	2037	0	1	1	2	2	3	3	4	4
0.31	2042	2046	2051	2056	2061	2065	2070	2075	2080	2084	0	1	1	2	2	3	3	4	4
0.32	2089	2094	2099	2104	2109	2113	2118	2123	2128	2133	0	1	1	2	2	3	3	4	4
0.33	2138	2143	2148	2153	2158	2163	2168	2173	2178	2183	0	1	1	2	2	3	3	4	4
0.34	2188	2193	2198	2203	2208	2213	2218	2223	2228	2234	1	1	2	2	3	3	4	4	5
0.35	2239	2244	2249	2254	2259	2265	2270	2275	2280	2286	1	1	2	2	3	3	4	4	5
0.36	2291	2296	2301	2307	2312	2317	2323	2328	2333	2339	1	1	2	2	3	3	4	4	5
0.37	2344	2350	2355	2360	2366	2371	2377	2382	2388	2393	1	1	2	2	3	3	4	4	5
0.38	2399	2404	2410	2415	2421	2427	2432	2438	2443	2449	1	1	2	2	3	3	4	4	5
0.39	2455	2460	2466	2472	2477	2483	2489	2495	2500	2506	1	1	2	2	3	3	4	5	5
0.40	2512	2518	2523	2529	2535	2541	2547	2553	2559	2564	1	1	2	2	3	4	4	5	5
0.41	2570	2576	2582	2588	2594	2600	2606	2612	2618	2624	1	1	2	3	3	4	4	5	5
0.42	2630	2636	2642	2649	2655	2661	2667	2673	2679	2685	1	1	2	2	3	4	4	5	6
0.43	2692	2698	2704	2710	2716	2723	2729	2735	2742	2748	1	1	2	3	3	4	4	5	6
0.44	2754	2761	2767	2773	2780	2786	2793	2799	2805	2812	1	1	2	3	3	4	4	5	6
0.45	2818	2825	2831	2838	2844	2851	2858	2864	2871	2877	1	1	2	3	3	4	5	5	6
0.46	2884	2891	2897	2904	2911	2917	2924	2931	2938	2944	1	1	2	3	3	4	5	5	6
0.47	2951	2958	2965	2972	2979	2985	2992	2999	3006	3013	1	1	2	3	3	4	5	5	6
0.48	3020	3027	3034	3041	3048	3055	3062	3069	3076	3083	1	1	2	3	4	4	5	6	6
0.49	3090	3097	3105	3112	3119	3126	3133	3141	3148	3155	1	1	2	3	4	4	5	6	6

Antilogarithms

	0	1	2	3	4	5	6	7	8	9	1	2	3	4	5	6	7	8	9
0.50	3162	3170	3177	3184	3192	3199	3206	3214	3221	3228	1	1	2	3	4	4	5	6	7
0.51	3236	3243	3251	3258	3266	3273	3281	3289	3296	3304	1	2	2	3	4	5	5	6	7
0.52	3311	3319	3327	3334	3342	3350	3357	3365	3373	3381	1	2	2	3	4	5	5	6	7
0.53	3388	3396	3404	3412	3420	3428	3436	3443	3451	3459	1	2	2	3	4	5	6	6	7
0.54	3467	3475	3483	3491	3499	3508	3516	3524	3532	3540	1	2	2	3	4	5	6	6	7
0.55	3548	3556	3565	3573	3581	3589	3597	3606	3614	3622	1	2	2	3	4	5	6	7	7
0.56	3631	3639	3648	3656	3664	3673	3681	3690	3698	3707	1	2	3	3	4	5	6	7	8
0.57	3715	3724	3733	3741	3750	3758	3767	3776	3784	3793	1	2	3	3	4	5	6	7	8
0.58	3802	3811	3819	3828	3837	3846	3855	3864	3873	3882	1	2	3	4	4	5	6	7	8
0.59	3890	3899	3908	3917	3926	3936	3945	3954	3963	3972	1	2	3	4	5	5	6	7	8
0.60	3981	3990	3999	4009	4018	4027	4036	4046	4055	4064	1	2	3	4	5	6	6	7	8
0.61	4074	4083	4093	4102	4111	4121	4130	4140	4150	4159	1	2	3	4	5	6	7	8	9
0.62	4169	4178	4188	4198	4207	4217	4227	4236	4246	4256	1	2	3	4	5	6	7	8	9
0.63	4266	4276	4285	4295	4305	4315	4325	4335	4345	4355	1	2	3	4	5	6	7	8	9
0.64	4365	4375	4385	4395	4406	4416	4426	4436	4446	4457	1	2	3	4	5	6	7	8	9
0.65	4467	4477	4487	4498	4508	4519	4529	4539	4550	4560	1	2	3	4	5	6	7	8	9
0.66	4571	4581	4592	4603	4613	4624	4634	4645	4656	4667	1	2	3	4	5	6	7	9	10
0.67	4677	4688	4699	4710	4721	4732	4742	4753	4764	4775	1	2	3	4	5	7	8	9	10
0.68	4786	4797	4808	4819	4831	4842	4853	4864	4875	4887	1	2	3	4	6	7	8	9	10
0.69	4898	4909	4920	4932	4943	4955	4966	4977	4989	5000	1	2	3	5	6	7	8	9	10
0.70	5012	5023	5035	5047	5058	5070	5082	5093	5105	5117	1	2	4	5	6	7	8	9	11
0.71	5129	5140	5152	5164	5176	5188	5200	5212	5224	5236	1	2	4	5	6	7	8	10	11
0.72	5248	5260	5272	5284	5297	5309	5321	5333	5336	5358	1	2	4	5	6	7	9	10	11
0.73	5370	5383	5395	5408	5420	5433	5445	5458	5470	5483	1	3	4	5	6	8	9	10	11
0.74	5495	5508	5521	5534	5546	5559	5572	5585	5598	5610	1	3	4	5	6	8	9	10	12
0.75	5623	5636	5649	5662	5675	5689	5702	5715	5728	5741	1	3	4	5	7	8	9	10	12
0.76	5754	5768	5781	5794	5808	5821	5834	5848	5861	5875	1	3	4	5	7	8	9	11	12
0.77	5888	5902	5916	5929	5943	5957	5970	5984	5998	6012	1	3	4	5	7	8	10	11	12
0.78	6026	6039	6053	6067	6081	6095	6109	6124	6138	6152	1	3	4	6	7	8	10	11	13
0.79	6166	6180	6194	6209	6223	6237	6252	6266	6281	6295	1	3	4	6	7	9	10	11	13
0.80	6310	6324	6339	6353	6368	6383	6397	6412	6427	6442	1	3	4	6	7	9	10	12	13
0.81	6457	6471	6486	6501	6516	6531	6546	6561	6577	6592	2	3	5	6	8	9	11	12	14
0.82	6607	6622	6637	6653	6668	6683	6699	6714	6730	6745	2	3	5	6	8	9	11	12	14
0.83	6761	6776	6792	6808	6823	6839	6855	6871	6887	6902	2	3	5	6	8	9	11	13	14
0.84	6918	6934	6950	6966	6982	6998	7015	7031	7047	7063	2	3	5	6	8	10	11	13	15
0.85	7079	7096	7112	7129	7145	7161	7178	7194	7211	7228	2	3	5	7	8	10	12	13	15
0.86	7244	7261	7278	7295	7311	7328	7345	7362	7379	7396	2	3	5	7	8	10	12	13	15
0.87	7413	7430	7447	7464	7482	7499	7516	7534	7551	7568	2	3	5	7	9	10	12	14	16
0.88	7586	7603	7621	7638	7656	7674	7691	7709	7727	7745	2	4	5	7	9	11	12	14	16
0.89	7762	7780	7798	7816	7834	7852	7870	7889	7907	7925	2	4	5	7	9	11	13	14	16
0.90	7943	7962	7980	7998	8017	8035	8054	8072	8091	8110	2	4	6	7	9	11	13	15	17
0.91	8128	8147	8166	8185	8204	8222	8241	8260	8279	8299	2	4	6	8	9	11	13	15	17
0.92	8318	8337	8356	8375	8395	8414	8433	8453	8472	8492	2	4	6	8	10	12	14	15	17
0.93	8511	8531	8551	8570	8590	8610	8630	8650	8670	8690	2	4	6	8	10	12	14	16	18
0.94	8710	8730	8750	8770	8790	8810	8831	8851	8872	8892	2	4	6	8	10	12	14	16	18
0.95	8913	8933	8954	8974	8995	9016	9036	9057	9078	9099	2	4	6	8	10	12	15	17	19
0.96	9120	9141	9162	9183	9204	9226	9247	9268	9290	9311	2	4	6	8	11	13	15	17	19
0.97	9333	9354	9376	9397	9419	9441	9462	9484	9506	9528	2	4	7	9	11	13	15	17	20
0.98	9550	9572	9594	9616	9638	9661	9683	9705	9727	9750	2	4	7	9	11	13	16	18	20
0.99	9772	9795	9817	9840	9863	9886	9908	9931	9954	9977	2	5	7	9	11	14	16	18	20

Index

Absorption coefficient, 86
  spectrum, 10–13
Acetaldehyde, 322
Acetals, 327
Acetamide, 335, 336, 339
Acetanilide, 343
Acetic acid, 97, 130, 330
  anhydride, 338–9
Acetone, 322
Acetophenone, 311, 322
Acetyl chloride, 311, 322, 333, 335–6
Acetylation, 335
Acetylene, 304
  structure of, 289
Acid anhydrides, 338–9
  chlorides, 334–6, see also acyl chlorides
Acidity, 365–8
  constant, 365
Acids, 121–2
  carboxylic, 330–4
  conjugate, 121
  strength of, 365–8
  strong, 122, 131, 132, 134–5
  weak, 122–3, 126–9, 135, 365–8
Activated complex, 157
Activation energy, 156, 157–8
Active mass, 111
Activity, 71
Acyl chlorides, 311, 334–6, 338
Acylation, 311, 335, 342–3
Adiabatic changes, 71
Addition polymerisation, 359–61
Adipic acid, 362
Alanine, 294, 356
Alcohols, 315–20
  hydrogen bonding in, 291, 316
Alcoholysis, 335
Aldehydes, 322–30
Aldol condensation, 329
Aldoses, 351
Alkali metals, 189
Alkaline earths, 189
Alkanes, 296–9
Alkene oxides, 302
Alkenes, 299–303
  structure of, 288
Alkyl groups, 296
  halides, 312–5, 342
  ease of substitution of, 313
Alkoxides, 316
Alkylamines, 341, 368–9
Alkylations, 311
Alkynes, 304, 315
  structure of, 289
Allotropy,
  of carbon, 49–50, 195
  of oxygen, 210
  of phosphorus, 208
  of sulphur, 233
  of tin, 195

Alpha particles, 6
  rays, 7
Alumina, 274, 317
Aluminium, 190, 227–8
  chloride, 178, 192, 309, 311
  compounds, 179, 190–3, 262, 263, 264
  halides, 191–2
  hydride, 178
  hydroxide, 190
  oxide, 177
  sulphide, 212
Amides, 336–7, 369
Amines, 340–3, 345
  basic strength of 368–9
Aminoacids, 356–8
p-Aminoazobenzene,
Aminoethanoic acid, 356
2-Aminopropanoic acid, 294, 356
Ammines, 207
Ammonia, 130, 188, 209, 314
  manufacture, 117, 237–8
  preparation and properties, 206–8
  shape of, 68
  solution, 130, 207, 257, 335, 339, 356
Ammonium acetate, 132
  chloride, 130, 132
  ethanoate, 132
  hydrogensulphide, 341
  hydroxide, see ammonia, solution
  nitrate (V), 202
  thiocyanate, 255
  vanadate (V), 250
Ammonolysis, 335
Amphiprotic compounds, 122
Andrews, 38
Anhydrides, 338–9
Anhydrite, 48–9
Aniline, 336, 341–3, 345
  hydrochloride, 342
Anils, 326
Anisole, 321
Anodising, 228
Arenes, 305–12
Argon, 220, 222–3
Aromatic hydrocarbons, 305–12
Arrhenius, 365
  equation, 158
  factor, 158
Arylamines, 341, 369
Association (see also hydrogen bonding), 97, 336, 342
Astatine, 214
Asymmetric carbon atom, 294
Atactic polymers, 360
Atoms, 1
Atomic number, 1
  radius, 167–9
  spectrum of hydrogen, 10–13
  structure, 1–19
  volume, 171–2

Atomic weights, see relative atomic masses
Atomicity, 1
Aufbau principle, 17
Autocatalysis, 159
Avogadro constant, 31, 48
Avogadro's law, 24, 31
Azeotropic mixtures, 105

Balmer series, 13
Barium and compounds, 184–9, 267
  chloride, 267
  sulphate (VI), 278
Bartlett, 222
Bases, 121–2, 365, 368–9
  conjugate, 121
  strong, 123, 131, 134–5
  weak, 122, 127, 130, 132, 133, 135, 368–9
Basicity, 368–9
  constant, 368
Bauxite, 227
Becker, 6
Beckmann method, 92–3
  thermometer, 90, 91
Becquerel, 6
Benzal chloride, 310
Benzaldehyde, 312
Bezamide, 336
Benzanilide, 342
Benzene, 305–12
  structure of, 305–7
Benzenecarbaldehyde, 312
Benzenecarbonitrile, 337
Benzenecarbonyl chloride, 334
Benzenecarbonylation, 335
Benzenecarboxamide, 336
Benzenecarboxylic acid, 312, 330, 368
Benzenediazonium chloride, 343
Benzene-1, 4-diol, 222
Benzenesulphonic acid, 309
Benzoic acid, 312, 330, 368
Benzontrile, 337
Benzophenone, 322
Benzotrichloride, 310
Benzoyl chloride, 334
Benzoylation, 335
Benzyl chloride, 310
Benzylamine, 341
Benzylidene chloride, 310
Berkeley and Hartley, 96
Beryllium, 179
  compounds, 179, 184–7
Bessemer process, 230
Beta rays, 7, 8
Bidentate groups, 245
Binding energy, 8
Birkeland-Eyde process, 117
Blast furnace process, physico-chemical principles of, 231–2
Body centred cubic packing, 44
Boiling point, 89, 171, 182–3

effect of size and shape on, 290
Boiling point constant, 90
Bond enthalpies (energies), 61, 74–6, 305–6
  length, 39–41
  order, 58
Boric (III) acid, 190
Born-Haber cycles, 76–9, 182, 186
Boron, 190
  and compounds, 190–2
  trifluoride, 58, 191
Bothe, 6
Boyle's law, 21, 22, 30
Brackett series, 13
Bragg equation, 40
Bromides, 266–9
Bromination, 309–10
Bromine, 214–6, 300, 320
Bromoethane, 161, 314–5
Bromoethene, 304
2-Bromo-2-methylpropane, 161
Brønsted, 121, 265
Brown ring test, 204–5
Brownian movement, 39
Buchner flask and funnel, 274
Buffer solutions, 129–31
Butanedione dioxime, 245
Butan-2-ol, 294
Butenedioic acids, 292–3
Butenedioic anhydride, 293, 311
6-Butyl alcohol, 315, 319
t-Butyl bromide, 161

Caesium and compounds, 184–9
  chloride lattice, 45–6
Calcium and compounds, 184–9
  carbide, 304
  carbonate, 115
  chloride, 207, 271
  dicarbide, 304
  fluoride, 218, 219
    lattice, 47
  formate, 331
  methanoate, 331
  phosphate (V), 236
  sulphate (VI), 271
Carbanions, 284
Carbocation, 284, 301, 317
Carbohydrates, 351–5
Carbon, 193–6, 287–9
  as a reducing agent, 138, 195, 231
  determination of, in organic compounds, 276–7
Carbon-12, 2
Carbon compounds, 193–201
  dioxide, 197–8
  disulphide, 200
  monoxide, 199
    as a reducing agent, 138
  tetrachloride, 200, 297 see also tetrachlormethane
Carbonates, 135–6, 187, 266–9
Carbonium ions, 284, 301, 317
Carbonyl chloride, 199
  compounds, 322–30

sulphide, 199
Carboxylic acids, 330–4, 366–8
Carius method, 278
Cast iron, 230
Catalysts, 159–60
Catenation, 194
Cell diagram, 142
Cells, 140
Cellulose, 354
Cerium (IV) salts, 234
Chadwick, 6
Charge transfer spectrum, 250
Charles' law, 21–2, 30
Chemical bonding, 56–68
  garden, 97
  kinetics, 149–62
Chlorate (I) ion, 219–20, 242–3
Chloric (I) acid, 215
Chlorides, 178, 266–9
Chlorination, 297, 309–10
Chlorine, 209, 214–16, 297, 300, 334
  as an oxidising agent, 255
Chloroacetic acid, 334
Chloroethanoic acid, 334
Chloroethene, 304
Chloroform, 297 see also trichloromethane
Chloromethane, 297
(Chloromethyl) benzene, 310
Chloro-2-and chloro-4-methyl-benzene, 310
Chlorostannous acid, 202
o- and p-Chlorotoluene, 310
Chromatography, column, 274
  gas-liquid or vapour phase, 275
  paper, 275
  thin layer, 275
Chrome alum, 252
Chromium, 241–3
  compounds, 245, 250–2, 263, 264
Chromium (III) hydroxide, 251
  potassium sulphate (VI)-12-water, 252
Chromium (VI) dichloride dioxide, 252
  oxide, 251
Chromyl chloride, 252
Cis-isomers, 292
Clathrate compounds, 222
Close packing, 41–4
Cobalt, 241–3
  compounds, 243, 244, 247, 262, 263, 264
Colligative properties, 88
Collision distances, 169
  geometry, 156
  theory, 156–7, 158
Colorimeter, 155
Common ion effect, 120
Complex ions, 243–7
Concentration, effect on reaction rate, 111, 148–53
Condensation polymerisation, 359, 361–2
  reactions, 325
Conductance, 127
Conductometric titrations, 128

Constant boiling point mixtures, 105
Constitutional isomerism, 283
Contact process, 117, 235
Convergence limit, 12, 163
Co-ordinate bonds, 60
Co-ordination number, in crystals, 44–6
  in complex ions, 244–5
Copper, 236, 241–3
  compounds, 244, 245, 246, 255–7, 262, 263, 264
Copper (I) chloride, 256
  cyanide, 257
  iodide, 256
  oxide, 256
  sulphate (VI), 257
  sulphide, 145
Copper (II) chloride, 256
  hexacyanoferrate (II), 94
  hydroxide, 257
  oxide, 256
  sulphate (VI), 256, 357
Covalent bonds, 57–61
  radius, 168
Cracking, 299
Critical pressure, 38
  temperature, 38
  volume, 38
Cryolite, 146, 227
Cryoscopic constant, 92
Crystal structure, 41–51
Crystallisation, 273–4
Cubic close packing, 42–3
Curie, 6
Cyanohydrins, 324
Cyclisation, 299
Cyclohexane, shape of, 289–90
Cyclohexanol, 362
Cyclohexanone, 362
Cyclohexene, 305
Cyclopentane, shape of, 290

d-Block elements, 241
Dalton's law of partial pressures, 23, 30
Daniell cell, 140–1, 142
Dative covalent bonds, see co-ordinate bonds
de Broglie, 14
Decarboxylation, 331
Dehydration, of alcohols, 317
  of amides, 337
Dehydrogenation, 311, 319
Dehydrohalogenation, 315
Delocalisation, 65–6, 366
  energy, 307
Depression of freezing point, 92–3
Detergents, 298
Devarda's alloy, 237
Dewar structures, 306
Diagonal relationships, 179
Diamagnetism, 59
Diamond, 49–50
Diaquatetraammine copper (II) hydroxide, 257
Diastereoisomers, 294
Diazonium salts, 154, 343–5
Dibenzoyl peroxide, 359
1, 1-Dibromoethane, 304

Dicarboxylic acids, 339
Dichlorine oxide, 177
Dichloromethane, 297
(Dichloromethyl) benzene, 310
Diethyl ether, 271, 318
Dihydric alcohols, 320
Dilead (II) lead (IV) oxide, 197, 199–200
Dimethyl benzene-1,4-dicarboxylate, 361, 362
Dimethyl ether, 321
   terephthalate, 361, 362
Dimethylamine, 336, 341
N,N-Dimethylaniline, 345
2,2-Dimethylpropane, effect of structure on physical properties, 290
1, 3-Dinitrobenzene, 308
Dinitrogen oxide, 202
2,4-Dinitrophenylhydrazine, 326
Diphenylmethanone, 322
Dipole, 64
   moment, 65
Dipole–dipole attractions, 36
Dipole–ion attractions, 207
Disodium oxide, 177
Disodium titanium (IV) oxide, 248
Disproportionation, 216, 234, 253, 256, 257
Dissociation constant, 122
Distillation, 272
Distribution coefficient, 99
   law, 99
Drying agents, 271
Dumes relative molecular mass determination, 27–8

Ebullioscopic constant, 90
Effective nuclear charge, 164
Einstein's equation, 8, 14
Electrochemical cells, 140–4
   series, 143
Electrode, 141–2
   calomel, 141
   glass, 144
   hydrogen, 141, 144
   silver, 144
Electrolysis, 140
Electrolytes, relative molecular masses of, 98
   weak,
Electrolytic conductivity, 127
Electromagnetic spectrum, 11
Electromotive force (e.m.f.), 141
Electron affinity, 167
   charge, 3
   diffraction, 41
   impact method, 163–4
Electronegativity, 63
Electronic theory, 56–68
Electrons, charge on, 3
   and magnetic properties, 59
   see also diamagnetism and paramagnetism
   and periodicity, 163–71
   and van der Waals forces, 36
   delocalised, 49, 65–6, 288, 306

diffraction of, 41
discovery, 2–3
energy levels, 14–19
in covalent bonds, 57–60
in intermediate bonds, 61–4
in ion formation, 57
in metals, 65
in oxidation and reduction, 136–44
in radioactivity, 8–9
localised, 287
lone pairs of, 66
spin, 17
unpaired and colour, 243
wave/particle nature of, 14
Electrophiles, 285
Electrophilic substitution reactions, 308–11
Electropositivity, 144, 176, 179
Elevation of boiling point, 89–91
Elimination, 315
Ellingham diagram, 231
e/m ratio, 2–3
Emission spectrum, 10–13
Empirical formula, 276–9
Enantiomers, 294
Enantiomorphs, 294
Enantiotropy, 195
Endothermic, 69
Energetics, 69–84
Energy-temperature relationship, 31
Enthalpy change, 69, 71–80
Enthalpy of atomisation, 72
   of combustion, 71–3
   of formation, 71–9, 305
   of fusion, 51, 86, 171, 184
   of hydrogenation, 305
   of reaction, 73
   of solution, 80
   of sublimation, 86
   of vaporisation (evaporation), 86, 171, 184
Entropy, 81–2
Enzyme, 159, 258, 353
Epoxyethane, 304
Equilibria, chemical, 111–19
   ionic, 120–48
   phase, 85–110
Equilibrium constant, 112, 115–16
   law, 111–12
Esterification, 318, 332
Esters, 318
Ethanal, 322
Ethanamide, 335, 336, 339
Ethane, 296
Ethanedioates, 139
Ethanedioic acid, 159, 339
Ethane-1, 2-diol, 304, 320, 361, 362
Ethanoic acid, 330
   anhydride, 338–9
Ethanol, 104, 304, 318, 319
Ethanoyl chloride, 311, 322, 333, 335–6
Ethanoylation, 335
Ethene, 299–303
   structure of, 288
Ethers, 321–2

Ethoxybenzene, 321
Ethoxyethane, 271, 318
Ethyl acetate, 332
   alcohol, 304, see also ethanol
   bromide, 161
   ethenoate, 332
Ethylbenzene, 311
Ethylene, 299–303
   dichloride, 304
   glycol, 304, 320, 361, 362
   oxide, 304
   structure of, 288
Ethylidene bromide, 304
Ethyne, 304
   structure of, 289
Exothermic, 69
Extensive properties, 81
Extraction of metals, 144–6

Face centred cube, 43
Fajan's rules, 63
Fehling's solution, 328
Flame tests, 188–9
Fluorides, 266–9
Fluorine, 214–16, 222
   monoxide, 215, 216
Fluorite, 47
Fluoroboric acid, 191
Fluorospar, 227
Formaldehyde, 322
Formamide, 336
Formic acid, 330, 331
Fractionating column, 103
Free energy, 231–2
Free radicals, 160, 284, 298
Freezing point, 92
Freezing point constant, 92
Friedel–Crafts reactions, 311
Fructose, 352–3
Fumaric acid, 292–3
Functional groups, 280

Gamma rays, 7
Gay–Lussac, 21–2
Geiger, 5
Geometrical isomerism, 292–3
Germane, 197
Germanium, 193–6
   and compounds, 193–202
Germanium (II) halides, 201
   oxide, 199
Germanium (IV) halides, 200–1
   oxide, 197–8
Glucose, 351–2
Glycine, 356
Graham's law, 30, 32
Graphite, 49–50
Ground state, 12
Group, 19
   I and II, 184–9
   III, 190–3
   IV, 193–202
   V, 202, 204–10
   VI, 210–14
   VII, 214–20
   O, 220, 222–3
Guldberg and Waage, 111
Gypsum lattice, 48–9

Haber process, 117, 237
Haematite, 229
Halates(I), 219

Half-cells, 140
  life, 9–10, 152
  reactions, 137–40
Halic(I) acids, 219–20
Halide ions, 214–15
Halides, 218–19
Haloform reaction, 319–20
Halogen carrier, 309
Halogenoalkanes, 312–15, 342
  ease of substitution of, 313
Halogenobenzenes,   309–10,
  312, 313, 314
Halogens, 214–16
  detection in organic com-
  pounds, 277, 278
Head elements, differences of,
  189
Heat, 69–71
Heat of atomisation, 72
  combustion, 71–3
  formation, 71–9, 305
  fusion, 51, 86, 171, 184
  hydrogenation, 305
  reaction, 73
  solution, 80
  sublimation, 86
  vaporisation (evaporation),
  86, 171, 184
Heisenberg uncertainty
  principle, 14
Helium, 220
Hemiacetals, 327
Henry's law, 88
Hess's law, 72
Heterolytic fission, 284
Hexachlorostannic acid, 201
Hexafluorosilicic acid, 201
Hexagonal close packing, 42–5
Hexamethylene diamine, 362
Hexane-1, 6-diamine, 362
Hexane-1, 6-dioic acid, 362
Hofmann degradation, 337
Homologous series, 283
Homolytic fission, 284
Hund's rule, 17
Hybridisation, 287
Hydration, 80, 303
  energy, 80, 184–5
Hydrazine, 326
Hydrazones, 326
Hydrides, 181–4
  of Group IV elements, 196–7
  of oxygen and sulphur,
  211–3
  of period 3 elements, 178
Hydriodic acid, 217
Hydrobromic acid, 217
Hydrochloric acid, 217, 220,
  263–4
  oxidation of, 200, 253
Hydrofluoric acid, 217
Hydrogen, 138, 181
  atomic spectra, 10–13
  detection in organic compounds, 276–7
Hydrogen   bonds,   37, 97,
  182–4, 291 *see   also*
  association
  in acids, 292
  in alcohols, 291, 316
  in amines, 342

in crystal lattices, 48–9, 50–1
in hydrogen fluoride, 183
in polymers, 363
in water, 37, 211
Hydrogen bromide, 217–18
  chloride, 217–18
  cyanide, 324
  dichlorocuprate (I), 256
  difluorides, 218
  fluoride, 183, 217–8
  hexachloroplumbate (IV),
  200
  hexachlorostannate (IV),
  201
  hexachlorotitanate (IV), 248
  hexafluorisilicate (IV), 201,
  218
  halides, 217–18, 301, 318
  iodide, 217–18
  ion concentration, 124–6
  molecule ion, 58–9, 61
  peroxide, 138, 140, 249, 250
  sulphide, 138, 211–13, 234,
  252
  tetrachlorocuprate (II), 256
  tetrachlorostannate (II), 202
  tetrafluoroborate (III), 191
Hydrogenation, catalytic, 160,
  301
Hydrogencarbonates, 187, 188
Hydrolysis, 131
  of acid chlorides, 335
  of amides, 336–7
  of anhydrides, 339
  of carbonates and
  hydrogencarbonates, 187
  of esters, 333
  of halides, 178, 191, 192,
  200–1, 209–10, 218, 219
  of halogenoalkanes, 313–14
  of halogens, 215–16
  of hydrides, 181, 197
  of ionic oxides, 210
  of nitriles, 340
  of salts, 131–2
  of sugars, 351
  of sulphides, 212
Hydrophilic group, 298
Hydroquinone, 222
Hydroxide ion, 124
Hydroxy compounds, 315–21
Hydroxylamine, 326
Hydroxynitriles, 324
2-Hydroxypropanoic acid,
  294, 324
Hypochlorite ion, 219–220,
  242–3
Hypochlorous acid, 220
Hypohalites, 219
Hypohalous acids, 219–20

Ice lattice, 50–1
Ideal gas equation, 22
Ideal gases, 29
Indicators, acid-alkali, 132–5
Inductive effect, 284
Inert gases, 220, 222–3
  pair effect, 194
Infrared spectroscopy, 281
Inhibitors, 159
Intensive properties, 81

Internal energy change, 69,
  70–1
Iodides, 266–9
Iodine, 214–16, 309, 310
  as an oxidising agent, 234
  in the haloform (iodoform)
  reaction, 320
  lattice, 50
Iodoform, 320, 329
Ionic bonding, 57, 60–5
  crystals, 45–9, 51
  packing, 45–9
  product of water, 123–4
  radius, 169
Ionisation, 164
  energy (potential) 163–7
Ions, 1, 57
Iron, 229–32, 241–3
  cast, 230
  compounds, 254–5, 262,
  263, 264
  conversion into steel, 230
  corrosion of, 232
  extraction, 229
  physico-chemical
  principles of, 231–2
  pyrites, 229
  wrought, 230
Iron(II) diiron(III) oxide, 232
Iron(II) hydroxide, 255
  oxide, 231–2
  sulphate(VI), 205
  sulphide, 211
Iron(III) halides, 309
  hydroxide, 255
  oxide, 228
Isobars, 2
Isobutene, 319
Isoelectronic, 204
  series, 169
Isoprophyl alcohol, 315
Isotactic polymers, 360
Isothermal, 38
Isotonic, 94
Isotopes, 2

Joule–Thomson effect, 35

Kekulé, 305
Ketones, 322–30
Ketoses, 351
Kieselguhr, 198
Kinetic energy, 29, 31
Kinetic theory of gases, 29–31
  of liquids, 38–9
Kinetics, 149–62
  and mechanism, 161
Kjeldahl method, 277
Krypton, 220, 222–3

Lactic acid, 294, 324
Landsberger method, 90–1
Lassaigne sodium fusion,
  277
Latent heats, *see* heats
Lattice energy, 61, 62, 76–9
Law of mass action, 111
Le Chatelier's principle,
  116–17
Lead, 193–6
  compounds, 193–202, 262,
  263, 264

Lead(II) nitrate(V), 199, 205
  oxide, 199
Lead(IV) chloride, 200, 201
  oxide, 197–8
Lewis acids, 192, 365
  bases, 192, 365
Ligands, 219, 243
Light and rate of reaction,
  160
Limonite, 229
Liquefaction of gases, 38
Lithium, 168
  and compounds, 179,
    184–90
  aluminium hydride, 184,
    196, 325, 332, 337, 340
  carbonate, 187
  tetrahydridoaluminate(III),
    184, 196, 325, 332, 337,
    340
Litmus, 133
Loadstone, 229
Lowry, 121, 365
Lyman series, 13
Lyophilic group, 298

Macromolecular structures,
  49–50, 52
Magnesium, 146, 237
  and compounds, 179, 184–8
  carbonate, 188
  chloride, 178
  hydride, 178
  oxide, 177
  sulphate(VI), 271
Magnetite, 229
Maloic acid, 292–3
  anhydride, 293, 311
Maltose, 353–4
Manganate(VI), 253
Manganate(VII), 253
Manganese, 241–3
  and compounds, 252–4
Manganese(III) oxide, 148
Manganese(IV), oxide, 253,
  254, 312
Markownikoff's rule, 301
Marsden, 5
Mass defect, 8
  number, 2
  spectrograph, 4
  spectrometer, 4–5, 281–2
Melting point, 51, 52, 92, 171
  effect of size and shape on,
    290
  mixed, 276
Mesomeric effect, 284
Meta-isomers, 307
Metallic bonds, 65
  crystals, 41–4, 51
  radius, 168
Metamerism, 321
Metaphosphoric acid, 177
Methanal, 322
Methanamide, 336
Methane, 196–7, 296, 297–8
  structure of, 287
Methanoic acid, 198, 330,
  331
Methanol, 315
Methoxybenzene, 321

Methoxymethane, 321
Methyl alcohol, 315
  orange, 133, 136
  red, 133
Methylamine, 341, 342
p-N-Methylaminoazobenzene
N-Methylaniline, 341, 345
Methylbenzene, 305, 308–12
2- and 4-Methylbenzene-
  sulphonic acids, 309
Methyl-2-nitrobenzene, 308
Methyl-4-nitrobenzene, 308
N-Methylphenylamine, 341,
  345
2-Methylpropan-2-ol, 315, 319
2-Methylpropene, 319
Millikan oil drop
  experiment, 2–3
Molecular crystals, 50–1, 52
  depression constant, 92
  elevation constant, 90
  formula, 276, 280
  geometry, 287–295
  velocities, 33–5
  weights, *see* relative
    molecular masses
Molecularity, 151
Molecules, 1
Monodentate groups, 245
Monomer, 359
Monotropy, 195

Naphthalene lattice, 50
Naphthalen-2-ol, 344
β-Naphthol, 344
Natural gas, 298
Neon, 220
Neutrons, 5–6
Nickel, 241–3, 302
  carbonyl, 199
  compounds, 243–5, 262,
    263, 264
Nitrate(V) ions, 66, 266–9
Nitrates(III), 266–9
Nitration, 298, 308
Nitric(III) acid, 140, 343, 357
Nitric(V) acid,
  as an oxidising agent, 138,
    195, 196, 236–7
  in nitrations, 298, 308, 320
  structure of, 65–6
Nitric oxide, 204
Nitriles, 340
Nitrites, 266–9
Nitroalkanes, 298
Nitrobenzene, 308
Nitrogen, 202, 208
  detection in organic
    compounds, 277
  and compounds, 202,
    204–10, 236–8
  dioxide, 205–6
  oxide, 204
  trichloride, 209
Nitronium ion, 308
2- and 4- Nitrophenol, 320
Nitrosonium ion, 204
Nitrosyl cation, 204
  halides, 204
Nitryl cation, 308
o- and p-Nitrotoluene, 308

Nitrous acid, 140, 343, 357
  oxide, 202, 204
Noble gases, 220, 222–3
Nuclear magnetic resonance
  spectroscopy, 283
Nucleophiles, 284
Nucleophilic substitution, 313
Nylon 6, 362
Nylon 66, 362

Oleum, 309
Optical isomerism, 293–5
Orbitals, 14–17
  anti-bonding, 58–60
  bonding, 58–60
Order of reaction, 151
Ortho-isomers, 307
Orthophosphoric acid, 210
Osmosis, 23–7
Osmotic pressure, 94
  determination of, 96
Ostwald process, 238
Ostwald's dilution law, 123
  method, 86
Oxalates, 139
Oxalic acid, 159, 339
Oxidation, 136–40
  of alcohols, 318–19
  of aldehydes and ketones,
    327–8
  of alkanes, 298
  of alkenes, 302
  of benzene, 311
  number or state, 170–1
Oxides, of period 3 elements,
  177
Oxidising agents, 137–40
Oximes, 326
Oxonium ion, 210, 211
Oxygen, 138, 210–1, 222
  difluoride, 215, 216
Ozone, 210, 302, 308
Ozonides, 302–3
Ozonolysis, 303

p- Block elements 190
Palladium, 301
Para-isomers 307
Paramagnetism, 59, 243
Partial pressures, 23
Partition coefficient, 99
  law, 99
Paschen series, 13
Pasteur, 294
Pauli exclusion principle, 17
Pauling, 63
Pentane, effect of structure on
  physical properties, 290
Perdisulphates, 235
Perdisulphuric acid, 235
Period, 19
Periodic table, 17–19
Periodicity, 163–80
Permanganate, 253
Peroxodisulphates(VI), 235
Peroxodisulphuric(VI) acid,
  235
Peroxotrifluoroethanoic acid,
  302
Persulphates, 235
Petrifluoroethanoic acid, 302

Petroleum, crude, 298–9
Pfeffer, 94
Pfund series, 13
pH, 124–6
  measurement of, 136, 144
  meter, 144
Phase, 85
  equilibria, 85–110
Phenetole, 321
Phenol, 316–17, 320–1, 335, 343, 362, 366
Phenolphthalein, 133
Phenyl benzenecarboxylate, 332, 335
Phenyl benzoate, 332, 335
Phenylamine, 336, 341–3, 345
Phenylammonium chloride, 342
N-Phenylbenzenecarboxamide, 343
N-Phenylethanamide, 343
Phenylethanone, 311, 322
Phenylethene, 311
Phenylhydrazine, 326
(Phenylmethyl) amine, 341
Phosgene, 199
Phosphine, 209
Phosphonic acid, 177, 210
Phosphonium iodide, 209
Phosphoric(V) acid, 210, 218, 317
Phosphorous acid, 177, 210
Phosphorus, 202, 208, 334
  compounds, 208–10
  halides, 333–4
  oxychloride, 210
  pentachloride, 178, 209–10, 333
  pentoxide, 177
  red, 218
  trichloride, 178, 209–10, 334
  trichloride oxide, 210
  trihalides, 218
  trioxide, 177
Phosphorus(III) oxide, 177
Phosphorus(V) oxide, 177
Photochemical reactions, 160
Photoelectric effect, 10
Photon, 10
Physical properties, effect of functional groups on, 291
Pi bonds, 288
Pig iron, 229
pK_a 126–9
pK_b 127
Planck's constant, 10
  equation, 14
Platinum, 301
Plutonium, 10
Polarisation of covalent bonds, 63–5
  of ions, 61–3
Polarised light, 293
Polyamides, 361
Poly(chloroethene), 359, 361
Polyesters, 361
Poly(ethene), 359, 361
Poly(ethenyl ethanoate), 359, 361
Polyhalides, 219
Polymerisation, 328, 359–62

Polymers, 359
Polymorphism, 46
Polypeptides, 357
Poly(phenylethene), 359, 361
Polyphosphoric(V) acid, 177
Poly(propenenitrile), 359, 361
Polysaccharides, 351
Polystyrene, 359, 361
Poly(tetrafluoroethene), 359, 361
Polythene, 359, 361
Polyvinyl acetate, 359, 361
Polyvinyl chloride, 359, 361
Positorns, 9
Potash, 238
Potassium and compounds, 184–9
  chlorate(I), 216
  chlorate(V), 216, 253
  cyanide, 219
  dichromate(VI), 137, 252
  ferricyanide, 205, 255
  ferrocyanide, 247, 255
  hexacyanoferrate(II), 247, 255
  hexacyanoferrate(III), 205, 255
  hydroxide, 188, 216, 271, 315
  hypochlorite, 216
  iodide, 138
  manganate(VII), 137–8, 159, 254, 302, 318, 319
  nitrate(V), 253
  permanganate, 137–8, 159, 254, 302, 318, 319
  thiocyanate, 255
Potentiometric titrations, 144
Promoters, 159
Propan-2-ol, 315
Propanone, 322
Propene, 299–303
Propylene, 299–303
Proteins, 357
Protons, 4–5, 181
P.T.F.E., 359, 361
Purification of organic compounds, 271–6
P.V.C., 359, 361
Pyknometer, 86

Quantum, 10
  numbers, 14
Quaternary ammonium salts, 342, 369

R_f value, 275
Racemic mixture, 294
Radial probability diagram, 15
Radioactivity, 6–10
Radium, 5, 7, 220
Radius ratios, 47–8
Radon, 220, 222
Raney nickel, 301
Raoult's law, 88
  negative deviation, 104
  positive deviation, 104
Rate constant, 148
  determining step, 151
  of reaction, 111, 149–55
Real gases, 35–7
Red lead, 199

Redox reactions, 137, 139–40
  potentials, 142
  series, 143
Reducing agents, 138–40
Reduction, 136–40
  of acids, 332
  of aldehydes and ketones, 325, 328
  of alkenes, 301
  of amides, 337
  of nitriles, 340
  of nitro compounds, 341
Relative atomic masses, 2
  table of, 375
Relative density, 24–5
Relative molecular masses, 25
  determination of, 24–9, 88–98
Resistivity, 127
Resolution, 294
Resonance hybrids, 66, 191, 198, 199, 204, 205, 206, 306, 230, 343
Rosenmund reduction, 322
Rotational motion, 31
Rubidium and compounds, 184–9
Rust, 232
Rutherford, 5–6, 8
Rydberg, 12

s- Block elements, 184
Salt bridge, 140
Salting out, 271
Saponification, 333
Saturated compounds, 297
Scandium, 241–3
Schiff's bases, 326
  reagent, 329
Schmidt, 6
Schrödinger, 14
Semicarbazide, 326
Semi-permeable membranes, 93
Shapes of molecules, 66–8
SI units, 373
Siderite, 229
Sigma bonds, 287
Silane, 176, 196
Silica (silicon dioxide), 218
  gel, 198
Silicon, 193–6
  compounds, 193–202
  tetrachloride, 178, 200, 219
  tetrafluoride, 201
Silicon(II) oxide, 199
Silicon(IV) oxide, 177, 197–8, 218
Silicones, 194
Silver acetylide, 304
  bromide, 266
  chloride, 266
  compounds, 262, 266–7
  dicarbide, 304
  fluoride, 219
  halides, 266
  iodide, 266
  nitrate(V), 266, 304, 313
  oxide, 262
Smelting, 229
Soda-lime, 297, 331

Soddy, 8
Sodium, 145, 271
  and compounds, 184–9
  and ethanol, 337, 340
  acetate, 297
  alkylsulphonates, 298
  aluminate, 227
  amalgam, 328
  borohydride, 184, 325
  carbonate, 263, 331
  chlorate(I), 253
  chloride, 218
    lattice, 45
  ethanoate, 297
  formate, 199
  germanate(II), 199
  germanite, 199
  hexafluoroaluminate(III),
    146
  hydride, 182
  hydrogensulphate(IV), 324
  hydrogensulphite, 324
  hydroxide, 188, 227, 262,
    315
  hypochlorite, 253
  hypophosphite, 209
  methanoate, 199
  methoxide, 314, 316, 321
  monoxide, 177
  nitrate(III), 343
  nitrite, 343
  nitrosyl, 205
  orthovanadate, 250
  pentacyanonitrosylferrate
    (II), 205
  peroxide, 177, 251, 253
  phenoxide, 317, 321
  phosphinate, 209
  plumbate(II), 199
  plumbite, 199
  silicate(IV), 97
  stannate(II), 199
  stannite, 199
  sulphate(VI), 271
  tetrahydridoborate(III), 184,
    325
  thiosulphate(VI), 139
  titanate, 248
  vanadate(V), 250
Solubility of gases, 86–8
  of salts, 117, 185
  product, 120–1
Solutions of gases in liquids,
    86–8
  of liquids in liquids, 102–5
  of solids in liquids, 88–102
Solvation, 79–80
Solvent extraction, 271
Spontaneous reactions, 81–2
Stability constants, 246
Standard electrode potential,
    141–2, 167, 185, 193
Stannane, 197
Starch, 354
States of matter, 21–52
  gaseous state, 21–38
  liquid state, 38–9
  solid state, 39–51
Stationary phase, 275
Steam distillation, 105–7,
    272–3

Steel, 230
Stereoisomerism, 292
Steric hindrance, 325
Strontium and compounds,
    184–9
Structural formula, 276, 280–3
  isomerism, 283
Styrene, 311
Sublimation, 86
Sucrose, 353
Sugars, 351
Sulphate(IV) ions, 234, 266–9
Sulphate(VI) ions, 66, 266–9
Sulphites, 234, 266–9
Sulphonation, 298, 309
Sulphonic acids, 309, 366
Sulphur, 138, 210, 233–4
  detection in organic
    compounds, 277, 278
  compounds, 211–4, 234–6
  dichloride dioxide, 298
  dichloride oxide, 334
  dioxide, 177, 252
  hexafluoride, 67, 234
  trioxide, *see* sulphur(VI)
    oxide
Sulphur(VI) oxide, 177, 309
Sulphuric(IV) acid, 139
Sulphuric(VI) acid, 213–4,
    218, 225–6, 267–8
  as a catalyst, 332
  as a dehydrating agent, 213,
    317, 352
  as an oxidising agent, 138,
    195, 213, 218
  in sulphonations, 309
  manufacture, 235
  uses, 236
Sulphurous acid, 139
Sulphuryl chloride, 298
Superphosphate, 236
Syndiotactic polymers, 360
Synthetic macromolecules,
    359–64

Temperature, effect on
    reaction rate, 156
  and kinetic energy, 31
Terylene, 361
Tetracarbonylnickel(O), 199
Tetrachloromethane, 200, 297
Tetrathionates, 234
Theoretical plates, 104
Thermite process, 228
Thermodynamics, first law,
    69–80
  second law, 81–2
Thionyl chloride, 334
Thiosulphate(VI), 234
Thomson, 2, 6
Tin, 193–6
  compounds, 193–202
Tin(II) chloride, 139, 201, 202
  ethanedioate, 199
  oxalate, 199
  oxide, 199
  sulphate(VI), 196
Tin(IV) chloride, 201
  oxide, 197–8
Titanium, 145, 241–3, 247–9
  compounds, 247–9

Titanium(II) oxide
  sulphate(VI), 248
Titanium(III) chloride, 248,
    249
Titanium(IV) chloride, 248
  oxide, 248
Titanyl sulphate (VI), 248
Titrations, acid-base, 134–5
  conductometric, 127
  potentiometric, 144
Tollen's reagent, 327–8
Toluene, 305, 308–12
o- and *p*- Toluenesulphonic
  acids, 309
Trans-isomers, 292
Transition elements, 18,
    241–61
  state, 157
  theory, 157–8
Translational motion, 31–2
2, 4, 6- Tribromophenol, 320
Trichloromethane, 297
(Trichloromethyl) benzene,
    310
Triethylamine, 341
Triiodomethane, 320, 329
Trioxygen, 210, 302, 308
Triple point, 86

Ultraviolet light, 10, 160
Unit cell, 42
Universal gas constant, 22
  indicator, 136
Unsaturated compounds, 300

Van der Waals' equation, 37
  forces, 36, 49, 50, 184
  radius, 108–9
Vanadium, 241–3, 249–50
  compounds, 249–50
Vanadium(IV) chloride, 249
  fluoride, 250
  oxide sulphate(VI), 250
Vanadium(V) fluoride, 250
  oxide, 250, 311
Van't Hoff, 94, 98
  factor, 98
Vapour density, 24–5
  pressure, 85
Velocity constant, 111, 148
Vibrational motion, 31–2
Victor Meyer relative
    molecular mass deter-
    mination, 25–7
Vinyl bromide, 304
  chloride, 304
Viscosity of alcohols, 291
Voltaic cells, 140

Water, 211
  amphiprotic behaviour, 122
  and hydrogen bonds, 37,
    211
  as a solvent, 79–80
  co-ordinated, 211, 243
  ionic product of, 123–4
  properties, 211
  shape, 68
Wilson cloud chamber, 7
Work, 69–72
Wrought iron, 230
Wurtzite lattice, 46

Xenon, 220, 222–3
   compounds, 222
   tetrafluoride, 222
Xenon(VI) oxide, 222
X-ray diffraction, 38–41

Zartmann experiment, 34–5
Ziegler catalysts, 359, 360
Zinc, 241–3, 250
   amalgam, 328
   blende. 46
   compounds, 243, 244
   sulphide, 5, 46
Zwitterions, 356